Springer Series in Experimental Entomology

Thomas A. Miller, Editor

Springer Series in Experimental Entomology
Editor: T.A. Miller

Insect Neurophysiological Techniques
By T.A. Miller

Neuroanatomical Techniques
Edited by N.J. Strausfeld and T.A. Miller

Sampling Methods in Soybean Entomology
Edited by M. Kogan and D. Herzog

Neurohormonal Techniques in Insects
Edited by T.A. Miller

Cuticle Techniques in Arthropods
Edited by T.A. Miller

Functional Neuroanatomy
Edited by N.J. Strausfeld

Techniques in Pheromone Research
Edited by H.E. Hummel and T.A. Miller

Measurement of Ion Transport and Metabolic Rate in Insects
Edited by T.J. Bradley and T.A. Miller

Neurochemical Techniques in Insect Research
Edited by H. Breer and T.A. Miller

Methods for the Study of Pest *Diabrotica*
Edited by J.L. Krysan and T.A. Miller

Insect-Plant Interactions
Edited by J.R. Miller and T.A. Miller

Immunological Techniques in Insect Biology
Edited by L.I. Gilbert and T.A. Miller

Lawrence I. Gilbert
Thomas A. Miller
Editors

Immunological Techniques in Insect Biology

With Contributions by

Joanne Ballarino   Ad M.T. Beenakkers   Lawrence I. Gilbert
Walter G. Goodman   Noelle A. Granger   Bruce W. Hermann
Rainer Keller   Joseph G. Kunkel   Michael Ma
Rob C.H.M. Oudejans   Hugo Schooneveld
Thomas K.F. Schulz   Klaus-Peter Sieber   J.A. Veenstra
Hans Voshol   James T. Warren   Shuenn-Jue Wu

With 71 Figures

Springer-Verlag
New York Berlin Heidelberg
London Paris Tokyo

Lawrence I. Gilbert
Department of Biology
CB 3280, Coker Hall
University of North Carolina
Chapel Hill, North Carolina
27599-3280
U.S.A

Thomas A. Miller
Department of Entomology
University of California
Riverside, California 92521
U.S.A

Library of Congress Cataloging-in-Publication Data
Immunological techniques in insect biology / Lawrence I. Gilbert,
Thomas A. Miller, editors.
p. cm.—(Springer series in experimental entomology)
Bibliography: p.
Includes index.
1. Insects—Physiology. 2. Insects—Immunology. 3. Immunology—Technique.
I. Gilbert, Lawrence I. (Lawrence Irwin), 1929–
II. Miller, Thomas A. III. Series.
QL495.I46 1988
595'.20429—dc19 87-32057
CIP

Softcover reprint of the hardcover 1st edition 1988

Typeset by Asco Trade Typesetting Ltd., Hong Kong.

9 8 7 6 5 4 3 2 1

ISBN-13: 978-1-4612-8356-0 e-ISBN-13: 978-1-4612-3798-3
DOI: 10.1007/978-1-4612-3798-3

Series Preface

Insects as a group occupy a middle ground in the biosphere between bacteria and viruses at one extreme, amphibians and mammals at the other. The size and general nature of insects present special problems to the study of entomology. For example, many commercially available instruments are geared to measure in grams, while the forces commonly encountered in studying insects are in the milligram range. Therefore, techniques developed in the study of insects or in those fields concerned with the control of insect pests are often unique.

Methods for measuring things are common to all sciences. Advances sometimes depend more on how something was done than on what was measured; indeed a given field often progresses from one technique to another as new methods are discovered, developed, and modified. Just as often, some of these techniques find their way into the classroom when the problems involved have been sufficiently ironed out to permit students to master the manipulations in a few laboratory periods.

Many specialized techniques are confined to one specific research laboratory. Although methods may be considered commonplace where they are used, in another context even the simplest procedures may save considerable time. It is the purpose of this series (1) to report new developments in methodology, (2) to reveal sources of groups who have dealt with and solved particular entomological problems, and (3) to describe experiments which may be applicable for use in biology laboratory courses.

THOMAS A. MILLER
Series Editor

Preface

The utilization of immunochemical analysis for a wide variety of anatomical, physiological, biochemical, and molecular biological research projects has increased logarithmically in recent years, and ELISA readers have become almost as commonplace in the laboratory as spectrophotometers. In the case of insect biology, radioimmunoassays for insect hormones have revolutionized the field of insect endocrinology, and immunocytochemistry has accomplished the same for insect morphology and physiology. Although a number of reference series exist detailing these emerging technologies, for the most part the techniques are not available in a single text, and in no case are they explained for the entomologist or biologist using insects as model systems. This is a "how-to" book, and we believe that it will be of great help to our colleagues who have not used immunological analyses and to those who have but may have encountered problems specific to the use of insect material.

The editors attempted to ensure that each chapter adhere to a uniform organization without modifying the writing style of individual contributors, but this was difficult because of the diversity of approaches. For example, some contributors use the research data of their own laboratories to exemplify a technique, whereas others utilize an extensive literature base. Although common abbreviations are used throughout this volume, there is also significant overlap between contributions—e.g., immunization techniques. This was done purposefully, because we wished each chapter to stand by itself so that the reader interested in a particular topic can obtain complete information by reading a single chapter rather than scanning the entire volume.

In general, the editors are pleased with the volume and feel that it will be of great value to insect biologists, from graduate students to senior investigators, who wish to utilize immunoassays in studies ranging from systematics to molecular biology.

Chapel Hill, North Carolina LAWRENCE I. GILBERT

Contents

Contributors

Joanne Ballarino
Department of Entomology, University of Maryland, College Park, MD 20742, U.S.A.

Ad M.T. Beenakkers
Department of Experimental Zoology, University of Utrecht, 3508 TB Utrecht, The Netherlands

Lawrence I. Gilbert
Department of Biology, CB 3280, Coker Hall, University of North Carolina, Chapel Hill, NC 27599-3280, U.S.A.

Walter G. Goodman
Department of Entomology, University of Wisconsin, Madison, WI 53706, U.S.A.

Noelle A. Granger
Department of Cell Biology and Anatomy, CB 7090 Swing, University of North Carolina, Chapel Hill, NC 27599-7090, U.S.A.

Bruce W. Hermann
Research Analytical Division, R.J. Reynolds Tobacco Company, Winston-Salem, NC 27102, U.S.A.

RAINER KELLER
Institut für Zoophysiologie, Rheinischen Friedrich-Wilhelms-Universität, D-5300 Bonn, Federal Republic of Germany

JOSEPH G. KUNKEL
Department of Zoology, University of Massachusetts, Amherst, MA 01003, U.S.A.

MICHAEL MA
Department of Entomology, University of Maryland, College Park, MD 20742, U.S.A.

ROB C.H.M. OUDEJANS
Department of Experimental Zoology, University of Utrecht, 3508 TB Utrecht, The Netherlands

HUGO SCHOONEVELD
Department of Entomology, Agricultural University, 6700 EH, Wageningen, The Netherlands

THOMAS K.F. SCHULZ
Department of Experimental Zoology, University of Utrecht, 3508 TB Utrecht, The Netherlands

KLAUS-PETER SIEBER
Department of Entomology, University of Maryland, College Park, MD 20742, U.S.A.

J.A. VEENSTRA
Department of Biological Organic Chemistry, C.S.I.C., 08034 Barcelona, Spain

HANS VOSHOL
Department of Experimental Zoology, University of Utrecht, 3508 TB Utrecht, The Netherlands

JAMES T. WARREN
Department of Biology, University of North Carolina, Chapel Hill, NC 27599-3280, U.S.A.

SHUENN-JUE WU
Department of Entomology, University of Maryland, College Park, MD 20742, U.S.A.

Chapter 1

Analytic Immunologic Techniques

Joseph G. Kunkel

I. Introduction

Analytic immunologic procedures have become important parts of the arsenal of techniques for describing and elucidating physiologic and developmental changes in naturally occurring antigens (Ags) from insects. Antisera, which are solutions of the natural defense secretions produced by vertebrate immune systems in response to foreign antigens, can be used as analytic reagents to make quantitative as well as qualitative measurements of these insect Ags. An antiserum contains antibodies (ABs) which are used to identify the components that are present in a complex mixture and measure their relative or absolute titers at the same time. Immunology also provides objective methods for deciding on the homology or degree of homology between similar macromolecules from different species or different stages or tissues from a single species.

This chapter will address the past and continuing use of polyclonal antisera of interest to insect biologists. No evaluation of the emerging monoclonal antibody (mAB) technology will be attempted, since its full implications for insect investigations are not yet known. Cautionary notes will be made, however, where a current use of polyclonal AB may or may not be replaced by mABs.

Before investigators embark on using a particular immunologic technique, they should carefully define their needs and resources. A lucid general description of antigenicity and immunity (Berzofsky, 1985; Darnell et al., 1986) is essential reading prior to attempting to understand the details of technique. The investigator should then become aware of the range of techniques available and the powers and limitations of each technique. Reading a handbook of immunology can be confusing in the variety of available methods (Ouchterlony, 1968; Axelsen, 1983); however, there are several techniques that have been particularly useful (or popular) among insect physiologists and biochemists.

Reading a selection of recent applications of immunologic techniques to insect problems, such as the ones included in this book and the accompanying bibliographies, is a good way of delineating the range of approaches that have been useful in the past. Some familiarity with past successful immunologic approaches is recommended before attempting a novel method. Diagrams are provided here describing idealized results, but interested investigators are referred to original published results in the accompanying references. Reading a recent handbook of immunology and taking a training course in practical aspects of the subject are highly recommended for those who want to maximize the information that can be obtained with the available techniques.

Immunologic techniques can be enlightening but can be expensive in time, animals, equipment, and materials. A great deal of time and energy can be saved by seeking expert advice and training in the early stages of a study. The levels of analysis are described beginning with the least complicated and least expensive. In initial exploratory investigations, where not a great deal is known about the complexity of the system, the simplest applicable approach should obviously be tried first. The same careful and graded approach is useful as a didactic tool in a class or laboratory setting in which one wants to introduce students to the concepts of immunology in the simplest and most economical manner.

II. General Comments on the Scope of an Immunologic Approach

Several properties of polyclonal antisera do not apply to monoclonal antisera: First, the mixture of ABs found in a polyclonal antiserum recognize a variety of structural entities (determinants) on the macromolecule that is acting as an immunizing Ag (Fig. 1.1). These entities include native determinants on the Ag surface and buried "cryptic" determinants in the Ag's interior. A mAB is by definition reactive with primarily one determinant of an immunizing Ag, although its affinity to specific determinants on other Ags may be higher or lower and must be assessed by experience.

Second, the avidity (apparent affinity) of the AB population for an Ag can vary over a wide range (Fig. 1.1). Avidity is an average affinity exhibited by a complex antiserum that contains hundreds of individual antibodies of different concentrations, each with its own affinity to the Ag. A mAB has a specific affinity for an Ag which can be described precisely by an equilibrium constant of the binding reaction.

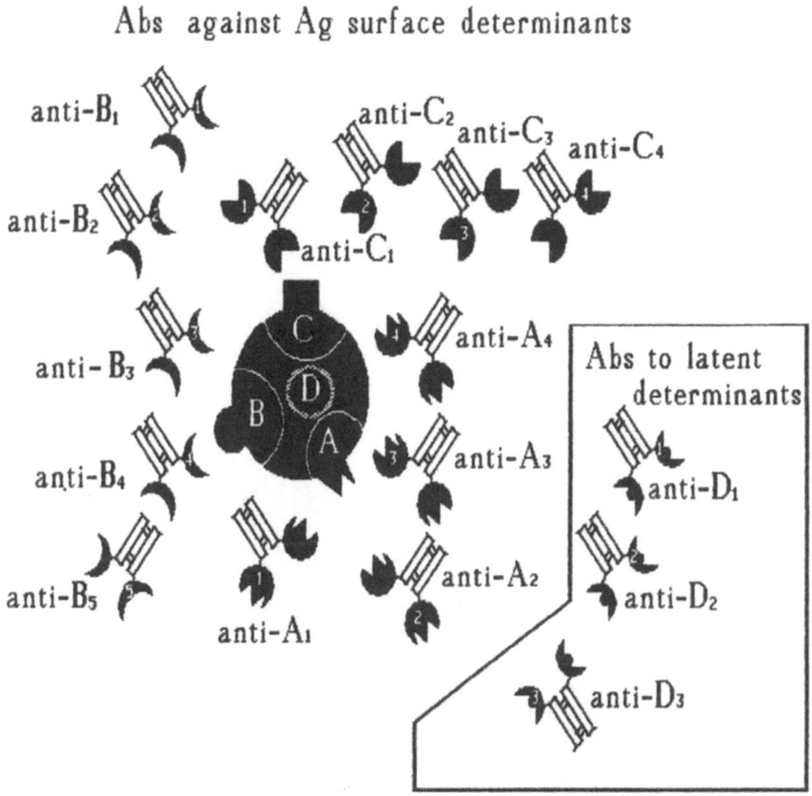

Figure 1.1. Antigenic determinants and polyclonal antisera. A polyclonal antiserum is derived from multiple clones of AB-producing cells and may have multiple distinct bivalent ABs (A_1, A_2, A_3, A_4, B_1, ..., D_3) reacting to each determinant (A, B, C, D) on a single native Ag, depicted in the center. Determinants A, B, and C are depicted on the surface of a polyvalent globular native protein. Cryptic internal determinant, D, may be buried within an internal core of the native protein, and antibodies reactive with this latent determinant may react with it only when the Ag is in a denatured state. Each animal immunized with the Ag may make antibodies to some or all of these determinants. Avidity of a particular polyclonal antiserum is determined by a frequency-weighted average of individual affinities of the population of secreted ABs.

Third, in the one-to-one stoichiometry of the precipitation reaction (Fig. 1.2), polyclonal antisera have no problem linking large Ags into extended networks that tend to precipitate. There is, however, a problem with monoclonal AB's becoming involved in the precipitate unless the Ag is immobilized in some other way, either to another biological structure such as another subunit, a membrane, or to an inert support such as a plastic surface or

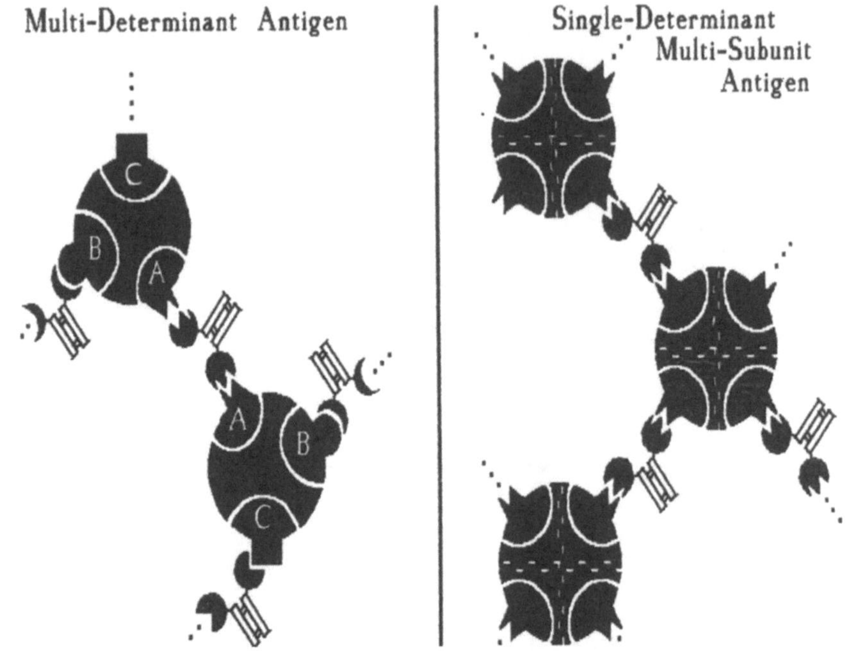

Figure 1.2. The basis of Ag precipitation by ABs is cross-linking and multivalency. An extended network of cross-linking can result in a precipitate. Every Ag is linked to another Ag through an AB, and every AB is linked to another AB through an Ag. The stoichiometry of Ag:AB to achieve complete Ag precipitation is approximately 1:1. Minimal requirements for precipitation include either particles with multiple subunits bearing a single determinant (right) or multiple determinants on a single particle (left).

nitrocellulose membrane. Specificity allows an AB to recognize a determinant on an Ag even in the presence of excess unrelated Ag. A highly avid (high average affinity) reaction allows many semipermanent (noncovalent) bonds between Ag and AB to be formed which are not reversed frequently enough to allow the Ag to dissolve.

One of the dangers of using monoclonal antibodies is that the AB affinity and specificity may not be high enough to form semipermanent bonds that survive washing procedures, and since there is not a substantial network of other ABs involved in an extended precipitate, the monoclonal staining of structures may be temporary. For polyclonal antisera, the stoichiometry of the precipitation reaction is usually equimolar, and thus one is able to measure specific Ag concentrations relative to the AB solution titer by identifying the equivalence concentration at which maximum precipitate is achieved, even in the presence of vast excesses of other Ags. Most polyclonal immunologic techniques take these preceding properties into account.

1. Measuring the Immunologic Complexity of a Problem

Antibodies can be used to measure and delineate the antigenic complexity of a situation. In a sense, one asks a vertebrate, such as a rabbit, how its immune system perceives a mixture of insect Ags. The rabbit's ability to recognize foreign macromolecules is another way of partitioning the complexity of an unknown system. Since insect macromolecules are almost all totally foreign to a vertebrate, the rabbit immune system usually provides a good source of antisera against individual or mixtures of insect proteins. This approach can be compared with other analytic techniques such as electrophoresis, chromatography, or enzymology. In each of these latter techniques, the concentrations of the molecules or activities to be separated are important. None of these techniques on its own is capable of a definitive identification of a particular mixture of proteins. In some cases a judicious combination of the techniques allows a strong inference concerning the identities of a mixture of molecules.

Immunologic techniques, on the other hand, are potentially capable of identifying and measuring individual Ags or mixtures of Ags all in one procedure. However, some immunologic techniques are aimed at, or limited to, measuring a single Ag; some are designed for, or adaptable to, looking at mixtures. A thorough understanding of the chemistry of the Ags under investigation and ABs in general is advisable. Be sure to read extensively about the general properties and mechanism of the particular immunologic technique one is using, its objectives, and limitations.

I prefer to start off a study with a polyclonal antiserum to the complex mixture in which the Ag of interest is normally found. For example, if one is interested in the major Ags in a system, such as insect hemolymph or insect yolk, an exploratory study of the major Ags is often useful in indicating the most practical stage for purification of the focal Ag. Such a study will also introduce one to the major contaminants to be contended with at later stages of study.

The literature is full of examples of overstatements based on too narrow an initial focus. It is easy once one has an antiserum specific to a single protein to concentrate on that one protein and forget the remainder. For instance, a focus on the variation in titer of a particular larval-specific serum protein during the life cycle of a cockroach (Kunkel and Lawler, 1973; Kunkel, 1975) revealed a major fluctuation associated with the molting cycle; a few years later, a more comprehensive study found that all the other major serum proteins of the cockroach were fluctuating in concert (Duhamel and Kunkel, 1978). If a general antiserum to all the serum proteins had been used initially, this emphasis on the cyclical fluctuation of the single serum protein would not have occurred.

Another caution related to immunologic complexity is worth emphasizing. Certain techniques such as microcomplement fixation, dot blots, and ELISA inherently require monospecific antisera in order for their high sensitivity in measurement to be reliable. Experience with the antigenic and AB complexity of the system under study is necessary to avoid artifacts. The specificity of monospecific antisera should be certified by one or several techniques such as immunoelectrophoresis (IEP), quantitative IEP (QIEP), Ouchterlony (spotting ABs and Ags separately on agar, migration forms a precipitation line in between), or Western blot (proteins identified by placing them on nitrocellulose and probed with antibodies).

2. Quantitative Measurement of Individual Ag Titers

A. ABs as Analytic Reagents: Physical Properties, Stability, Storage

a. Relative concentration of Ag and AB. A ratio of one AB molecule to one Ag molecule optimizes most immune precipitation assays. This is the basis of using antisera to measure the concentration of an Ag. However, this is also the basis for possible artifacts. Since an AB molecule has two combining sites, a precipitate is usually formed only if the antiserum is polyclonal and the Ag has multiple determinants on its surface or if the Ag has multiple subunits with the same determinant (Fig. 1.2). In general, monoclonal AB is not used for precipitate forming reactions, because Ags cannot be counted on to have multiple combining sites.

Another necessary requisite for precipitation is that sufficient Ag be cross-linked by AB into an insoluble lattice. Outside the range of approximate equivalence, on the side of Ag excess, the probability increases of finding an AB molecule bound to two Ag molecules with no additional AB molecules available to extend a lattice. This Ag–AB–Ag aggregate may be soluble (Fig. 1.3). On the side of AB excess, there tends to be less of a problem, particularly with polyclonal antisera against large Ags. The multiple determinants of a large Ag can adsorb a large number of AB molecules and still have determinants left to cross-link with another Ag molecule. After formation of a precipitate and when all the Ag is involved in a precipitate, additional AB can continue to add to the precipitate at other determinants without endangering the possibility of dissolving the precipitate by completely coating the Ag with AB on all its determinants. On a small Ag however, AB excess can lead to a soluble AB–Ag–AB complex, since if an Ag has only a few determinants, it may be completely coated with AB, be still soluble, and be blocked from cross-linking with other AB–Ag–AB complexes, since all their determinants are also covered with AB.

b. Monitoring Ag synthesis and processing. Antisera can be used to follow Ag synthesis or postribosomal modification by combining selective precipitation of an Ag and pulse or continous labeling with radioactive precursors. Several ways of using this approach can lead to substantial insight into the

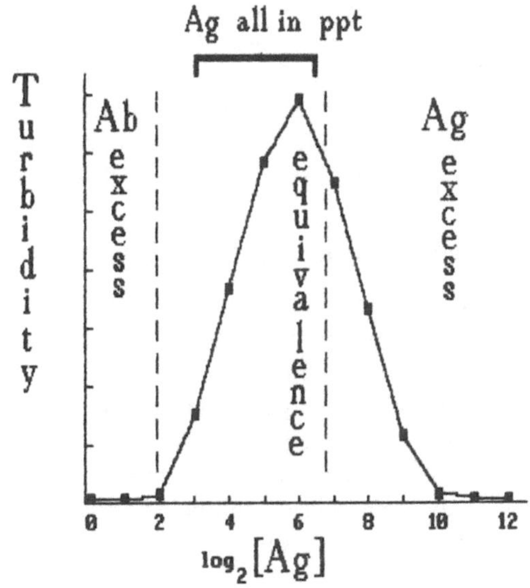

Figure 1.3. Classical Ag/AB equivalence in liquid. A fixed amount of AB is placed in a series of test tubes. A serial increasing concentration of Ag is added to successive tubes. At the low Ag concentration end of the series, no precipitate forms, because the excess of AB completely coats the determinants of the Ag, and there are no free determinants for an AB free end to cross-link with. In a region from 3 to 6 log units of Ag, all the Ag in the tubes is precipitated with available AB, generally at a higher than 1 : 1 ratio. This situation is possible because many Ags have far more than two AB combining sites. As the Ag concentration is increased beyond equivalence, the frequency of soluble Ag aggregates linked by one or a few AB molecules increases. Finally, every AB molecule is linked in a soluble complex with two Ag molecules (Ag–AB–Ag), but there are no free AB molecules to cause further cross-linking.

synthesis as well as postribosomal processing events in a protein's function (Hagedorn et al., 1978; Levenbook and Bauer, 1984; Wojchowski et al., 1986).

B. Lab and Field Ag Samples

A polyclonal immunological test for an Ag is robust in many ways, because it does not depend entirely on the Ag of interest retaining 100% of its native configuration. Some immunological tests are forgiving of many degradative phenomena such as the activity of an enzyme. In fact, immunological tests can be run on denatured material adsorbed to inert supports such as nitrocellulose. This property considerably extends the variety of situations from which samples can be obtained, including field studies and refrigerator- or freezer-stored material. Indeed, useful antisera have been made against fossil Siberian mammoth material which were used to show a closer affinity of mammoth to Indian elephant than to African elephant.

Ecological and taxonomic studies would be aided by the ability to take samples in the field and be assured that they will not degrade before they are analyzed in the laboratory. In this regard an organic compound, phenoxyethanol, is available that can be added to samples in the field and will sufficiently preserve protein samples at ambient temperature without deterioration of immune responses (Nakanishi et al., 1969).

C. Range of Sensitivity and Capacity of Immunological Procedures

Immunological techniques range in sensitivity over several orders of magnitude from milligrams to nanograms of an Ag. In the least sensitive method, precipitation in liquid, the upper limit of Ag to be precipitated depends on how much AB one can afford to use per precipitation. This technique is often used where large amounts of precipitated Ag are available, such as when synthesis of a protein is being monitored. In such situations there may already be substantial previously synthesized, unlabeled protein which must also be precipitated in order to ensure that all the newly synthesized, labeled protein is found in the precipitate.

Thus, maximum sensitivity is not always the objective of an immunological technique. When maximum capacity is the objective, it is sometimes possible to extend an antiserum's capacity by using it numerous times. Binding of AB to matrices such as agarose or dextran beads is one approach to being able to reuse an AB conveniently. When maximum sensitivity of measurement is needed, several techniques are available. Very small amounts of Ag can be detected using equivalently small amounts of AB whose location and amount are then amplified and quantified by layers of secondary AB which are then visualized by enzymatic or fluorographic means. In many of the more sensitive techniques, the amount of Ag precipitated may be so small that the maximum radioactivity attainable by incorporation studies is not able to be detected above background. An investigator must have a reasonable idea about the sensitivity or capacity required to answer the questions of interest.

3. Quantification of Degree of Ag Difference

A. Intraspecific Variation

a. Processing during stages of synthesis and function. During the ribosomal synthesis and postribosomal processing of proteins, antibodies can interact with the nascent and developing structure in a preparative or analytic fashion. Antibodies to minor cellular proteins can be used to precipitate the polysomes on which the nascent chains of the protein are being synthesized, thus providing a way of purifying mRNAs of minor proteins (Shapiro et al., 1974). The processing of proteins sometimes involves destruction of determinants whose loss can be detected and followed immunologically (Storella et al., 1985; Wojchowski et al., 1986; Kunkel et al., 1987). Immunological techniques are sometimes useful in analyzing developmental steps by providing a way of purifying the protein before and after a change so that they can be analyzed by some other technique such as SDS-PAGE (sodium dodecyl sulfate-polyacrylamide gel electrophoresis) and autoradiography (Imboden and Law, 1983; Engelmann, 1986; Wojchowski et al., 1986).

b. Identification of homologous electromorphs. Seemingly related proteins or enzymes within a species can differ in electrophoretic mobility for several reasons. They may be totally unrelated, be alleles of the same gene, be developmental stages, be physiological forms, or be differential processing artifacts. Unrelated proteins would not cross-react immunologically. Monospecific antisera are able to identify the immunologically related or distinct electromorphs of an isozyme on Western blots of the electrophoretically separated proteins onto a blotting medium such as nitrocellulose or diazotized paper (Towbin et al., 1979; Hawkes et al., 1982; Smith and Fisher, 1984). In addition to using monospecific antisera, complex antisera that have antibodies to several peptides or proteins can be affinity-purified by reacting them to electrophoretically separated peptides blotted onto nitrocellulose filters. The peptide-specific antibodies can be eluted from the filters and reused to identify or test for cross-reaction with other peptides or proteins (Olmsted, 1981).

B. Interspecific Variation

a. Phylogenetic relationships. The use of immunology to measure phenotypic differences between species is an established method of carrying out taxonomic or phylogenetic research (Champion et al., 1978; O'Brien et al., 1985). Antibodies to serum proteins of insects cross-react most highly to closely related species, and that cross-reaction falls off with phylogenetic distance (Kunkel and Lawler, 1974). A quantitative approach to measuring the degree of cross-reaction of a drosophilid serum protein has been used to estimate evolutionary times of divergence among insects (Beverley and Wilson, 1984, 1985).

b. Homology of structure and function. Many insect Ags that have been studied have no known enzymatic activity, and therefore their homology with a protein from another species must be based on physical and chemical characteristics including their immunological cross-reactivity. When two species are sufficiently distant, a lack of cross-reacting Ag does not mean that they share no homologous Ag. The homology between Ags from different families or orders may be more sensitively measured when more conservative internal protein structures are probed with AB. Denaturation of the protein of interest on a nitrocellulose filter, for instance, has allowed the homology between arylphorins of a lepidopteran and a hymenopteran to be studied (Ryan et al., 1984.)

III. Specific Techniques

1. Obtaining Immunological Reagents

A. Choice of Animals for AB Production

Animals used for immunization with Ags include rabbit, goat, mouse, and chicken. Each has its advantages and disadvantages. The mouse is the easiest and cheapest to house but perhaps the most difficult from which to obtain large amounts of antiserum reliably. Some analytical techniques, however, need only small amounts of antiserum. With practice, large volumes of antiserum can be obtained from mice by induction of ascites fluid during the immunization procedure (Tung et al., 1976). One should be able to obtain 10–30 ml of ascites fluid per mouse and repeat that several times for some mice. In addition, mice are the basic animal for production of mABs. It is possible that mABs will replace many of the former functions of polyclonal ABs in the near future, particularly the roles involving measuring trace amounts of Ag.

The rabbit has been a mainstay of AB production for insect physiologists and biochemists; they will reliably produce 50 ml of blood weekly, indefinitely. It is yet to be determined whether considerations of animal care costs and changing technology will displace the rabbit with the mouse or another AB source. Goats are also useful when very large amounts of antiserum are needed; one can obtain 250 ml weekly from a goat without overly stressing it (Brookes, 1986). Chickens are only rarely used in insect biochemical contexts, but they provide certain conveniences; the antibodies are harvested from the eggs they lay. The chicken continues to provide neatly packaged aliquots of its immunoglobulins for titer testing, and when the AB titer is high, one pools a series of eggs for one's final harvest. Overall, the rabbit still stands out as the most often used AB source when judged by numbers of publications.

B. Choice of Immunization Schedule

a. A standard immunization protocol. Immunization of a rabbit will be described as an example. The most frequently used protocol used in my laboratory lasts for 3 months. An initial bleeding is useful to provide control serum as needed in some techniques and for cross-matching of serotypes when pooling of sera between rabbits is anticipated. The initial injection is made with 250 μg of Ag in 1 ml of PBS (phosphate-buffered saline) emulsified with an equal volume of Freund's complete adjuvant. The emulsion is made by forcing the mixture through a fine tubing connecting two syringes. If the emulsion does not disperse when dropped into a dish of water, it is ready to be injected. Allowing the mixture to freeze and thaw a few times during the emulsification will help along the procedure.

The mixture is injected subcutaneously along one side of the back of the rabbit in four or five locations. The second injection is done 4 weeks later in a similar manner along the other side of the back, however, using Freund's incomplete adjuvant. Four weeks later a series of five daily injections of 50 μg Ag in 1 ml PBS are given intravenously via the marginal ear vein.

Seven days later the rabbit is ready to be harvested for the first time. The animal is anesthetized and vasodilated with a mixture of droperidal/fentanyl injected subcutaneously (Tillman and Norman, 1983). In 20 min 50 ml of blood can be drained effortlessly by puncturing the central artery of the ear with a 20-gauge needle and collecting the stream in a 50 ml siliclad-coated centrifuge tube. The blood is incubated at 37°C for 2 h to clot and then placed in the refrigerator overnight to allow the clot to contract maximally. The clot can be removed and the sera centrifuged to remove blood cells. The sera should be precipitated with ammonium sulfate $\frac{1}{3}$ saturated and redissolved in PBS twice. This step is essential in preventing proteases, released by minor hemolysis, from progressively digesting the immunoglobulins. Further purification of the IgG fraction on DEAE cellulose is routine in some procedures but not essential for the majority of methods described in this chapter.

b. Antisera for discriminating qualitative differences. For taxonomic or homology studies, different strategies must be taken in the production of antisera. Immunization of a rabbit should be carried out for at least 10 weeks if antisera cross-reacting with the most distant of relatives are desired (Champion et al., 1978). On the other hand, if one is interested in differentiating between very closely related Ags, the earliest ABs produced after the first weeks of immunization tend to be the most discriminating.

C. Monospecific Antisera

Some immunological assays require monospecific antisera or are simpler to analyze if monospecific antisera are used. Monospecific antisera are polyclonal antisera that are reactive with one macromolecular entity, be it a multisubunit native protein or a single polypeptide derived from a larger entity. Monospecific antisera can often be obtained by taking care to purify the Ag used for immunization and using a minimum of that Ag in the immunization procedure. Often rabbits are perverse in that they make antibodies to minor impurities in a hard-won "purified" Ag. The resultant polyclonal, multispecific antiserum may be useless as is, but such antisera may be improved or be used to improve the situation. For example, the unwanted antibodies may be absorbed by a solution of the Ag impurities if such a solution is available (Tanaka, 1973; Kunkel and Lawler, 1974). In this procedure, one uses a soluble or insoluble form of the offending Ags to precipitate the unwanted ABs.

In a different approach, one can use an initial multispecific antiserum and the techniques of crossed QIEP (Section III.D.2) to provide a pure Ag/AB precipitin localized in a gel which can be discretely cut out of a gel and used for immunization of another rabbit (Harboe and Closs, 1983). Still another approach is to use nitrocellulose-immobilized Ag of the desired type to prepare the corresponding AB by affinity binding to the immobilized Ag followed by elution of the specifically bound AB and use in other reactions (Olmstead, 1981; Ryan et al., 1985a).

2. Analysis of Immunological Titer

A. Ag Titer vs. AB Titer

a. Ag/AB equivalence: importance for avoidance of false negatives. A false impression of the specificity and complexity of an antiserum may be obtained if attention is not paid to the multiple Ag/AB equivalence concentrations operating in a potentially complex antiserum. Every polyclonal antiserum must be first treated as if it were a complex antiserum before it is proven to be monospecific to the Ag of interest. Each Ag will have its own concentration in a particular Ag mixture, and each corresponding AB will have its own titer in the complex antiserum. A titration curve may demonstrate that there is no one Ag/antiserum ratio in which all Ag/AB reactions are at equivalence.

Some techniques such as liquid precipitin and the Ouchterlony test are unforgiving of this situation, and false-negative conclusions about complexity may be obtained. That is, one may conclude falsely that the antiserum is more specific than it actually is, or one may conclude that a given Ag solution is simpler antigenically than it actually is. Other techniques such as QIEP or crossed QIEP are better able to deal with major imbalances in Ag/AB equivalences. QIEP can routinely accommodate 64-fold differences in Ag titer among unknown solutions whereas, using Ouchterlony or IEP, this order of difference could cause one to miss or neglect the existence of particular Ags in a complex mixture.

Avoidance of the aforementioned false-negative conclusion about complexity is important, because techniques that depend on the specificity of an antiserum such as Western blotting or in situ localization (Tanaka and Ishizaki, 1974; Raikhel, 1984; Graf et al., 1986; Peferoen et al., 1986; Raikhel and Lea, 1986) are in some cases intrinsically unable to reject false-positive reactions that they may detect. For that reason all putative monospecific polyclonal antisera and mABs to be used for in situ localizations should be tested for specificity using techniques such as Western blotting of tissue extracts (Graf et al., 1986; Peferoen et al., 1986; Raikhel et al., 1986). It is clear, for instance, that the majority of mABs do not distinguish between the several *Drosophila* vitellogenin gene products (Wu and Ma, 1986).

b. Adjustment of technique and AB titer for appropriate sensitivity. The technique of choice should be appropriate for the problem at hand. If large amounts of Ag are available, then the major concern might be to conserve valuable antiserum. If neither Ag nor AB is limiting, then the choice of technique is one of convenience and accuracy. Extremely sensitive techniques require very low amounts of Ag to be tested. At very low Ag titers Ag may be lost from a diluted solution by absorption onto glass or plastic surfaces. In some cases a less sensitive assay is the more reliable approach. A practical rule of thumb is to not use a technique that requires excessive dilution. If high dilution is essential, then all buffer solutions can be supplemented with protein such as bovine serum albumin and/or the plastic, and glass surfaces in contact with the dilute solutions may be coated with a silicone layer to minmize adsorption of proteins.

B. Measuring Ag/AB Relative Concentration, Titering

a. Classical Ag/AB equivalence titer by precipitin in liquid. A complete precipitation analysis of equivalence in liquid requires a relatively large amount of antiserum as well as Ag. In order to be quantitatively precise with this approach, one must have a reasonably monospecific antiserum and do a complete curve of precipitation from AB excess to Ag excess. The peak of precipitate occurs close to molar equivalence between Ag and AB. This peak is localized by analyzing the amount of precipitate. The precipitate can be washed to remove contaminating molecules and then dissolved in alkali. The protein content can then be estimated by numerous techniques. In this way the titer of the Ag can be related to the titer of the AB by how much dilution was necessary to make the two solutions react maximally, i.e. at their equivalence point. If the antiserum and Ag solutions are complex, there is a high probability that the peaks of equivalence may overlap or at best form a bimodal precipitate curve, since the concentrations used are logarithmic series which compress any titer and concentration differences.

b. Ag/AB equivalence in agar. Equivalence of Ag and AB in agar is determined by a combination of the concentration and diffusion coefficient of the Ag. The appropriate equivalence relationship can be determined by a serial dilution of Ag in a linear series of parallel Ouchterlony test wells (Fig. 1.4). The wells with equivalent titers of Ag and AB produce a crisp fine precipitin line which does not move with time. On either side of this equivalence relationship the excess of Ag or AB creates an unstable precipitin line which thickens and blurs with time. The majority of poor immunological illustrations of Ouchterlony reactions in research papers are due to not running this equivalence test prior to a final, "for publication" test.

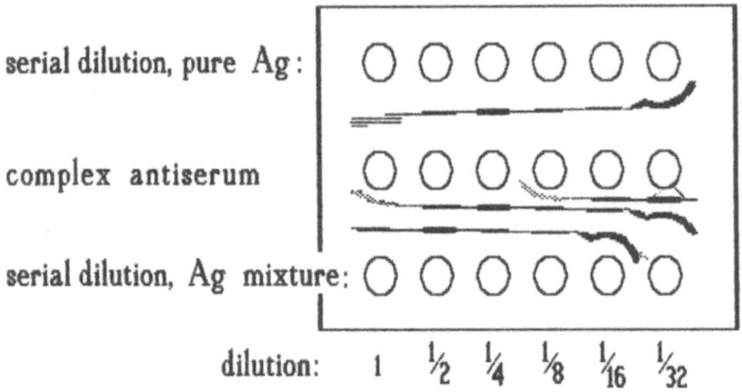

Figure 1.4. An Ag/AB equivalence test in agar is a necessary prerequisite of any successful Ouchterlony double-diffusion test. Parallel sets of wells are punched in the same agar medium in which a final "for publication" test will be done subsequently. Care is taken to space the wells identically to the future spacing of the final test pattern. A constant amount of antiserum is added to a central tier of wells. Serial dilutions of Ag are placed in parallel tiers of wells, and the Ags and ABs are allowed to diffuse together for 1 to a few days. The sharpest precipitin lines form between equivalence concentrations of Ag and AB; a pure Ag will have one or two best dilutions forming a crisp line which will intensify over the span of a few days. In a complex mixture, the equivalence dilution of each Ag may be different; initially several dilutions will show sharp lines, but with time, Ag or AB excess is indicated by a blurring of the precipitin lines. The blurring is more enhanced by staining.

3. Quantitative Estimation of Concentration

A. Precipitin Techniques (minimal equipment, instructional level)

a. Fluid precipitin. Analytical estimation of precipitates using liquid phase precipitation of Ag and AB is the simplest immunological reaction to interpret. It is usually carried out using a fixed dilution of antiserum and varying the Ag dilution from a concentration of Ag excess to a concentration of AB excess. The dilution of the Ag solution at the peak of the precipitin curve is a relative measure of the Ag titer. The higher the Ag titer in the original sample, the more it must be diluted to achieve the peak in a precipitin curve.

When the relative titer of the Ag solution is the principal information desired, this technique is not the recommended method; however, the analysis of fluid precipitins can be a valuable adjunct to the qualitative analysis of an Ag (Wojchowski et al., 1986). When one desires the approximate equivalence titer of an antiserum, it is possible to use techniques other than a large-scale liquid precipitin curve to estimate it. For instance, if the major information sought is the capacity of an antiserum to completely precipitate an Ag, one can do a microtitration from Ag excess to Ag equivalence in small, 250-μl microcentrifuge tubes (Fig. 1.5). The criterion for complete precipitation would be

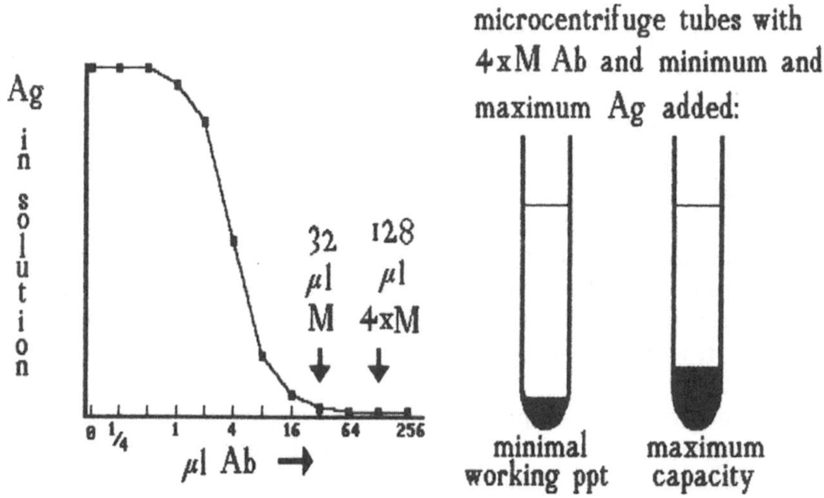

Figure 1.5. Microtitration to determine volume of antiserum and carrier Ag needed to achieve total precipitation of an Ag sample in a fluid precipitin reaction. This technique is carried out in microcentrifuge tubes. A fixed amount of carrier Ag (5–10 μg is sufficient) is added to each tube. An additional objective of this preliminary experiment is to find a convenient amount of carrier Ag providing a workable amount of precipitate that can be seen in the bottom of the tube after centrifugation, washed by resuspension in PBS, and repelleted several times without loss of precipitate. Increasing volumes of antiserum are added to the series of tubes; precipitate is given time to form and is centrifuged out of solution. The supernatant is then tested for any Ag remaining in solution using any available micromethod, such as QIEP. The volume of antiserum that completely precipitates the carrier amount of Ag allows one to calculate an AB equivalence amount. Using a multiple of the equivalence volume of antiserum ($4 \times 32 = 128$ μl) provides a liquid precipitin system that will always produce at least the minimal working precipitate with carrier Ag (tube on the left). This guards against AB excess. This system has the capacity to precipitate threefold more of "unknown" Ag without leaving any in solution (tube on right). If only the unknown Ag is labeled with radioactivity, the precipitate can be analyzed by diverse means to establish the nature and extent of labeling.

to leave no Ag in solution. The Ag remaining in solution can be measured by a more sensitive assay such as QIEP or ELISA. Although this microtechnique does not conform to a classical liquid precipitin curve, it answers the question: How much antiserum must I add in a particular situation to ensure complete precipitation of a specified amount of Ag?

Armed with the knowledge of how much AB is necessary to precipitate an Ag in solution, one can proceed to study the synthesis and postribosomal processing of a protein. The tissue or organ site of synthesis of an Ag can be explored this way. The fat body was identified as the site of vitellogenin synthesis in the seminal paper using this approach (Pan et al., 1969). Labeling with ^{3}H or ^{14}C amino acids (Pan, 1977; Hagedorn et al., 1978; deBianchi et al., 1983; Brookes, 1986; Borovsky and Whitney, 1987) or ^{35}S-methionine (Chen et al., 1978; Brennan et al., 1982; Isaac and Bownes, 1982; Kawooya et al., 1986; Peferoen and DeLoof, 1986; Wojchowski et al., 1986) for protein synthesis, ^{32}P for phosphorylation (Masuda and Oliveira, 1985; Della-Cioppa and Engelmann, 1987), and 2-(^{3}H)-mannose for oligosaccharide attachment (Kunkel et al., 1980; Wojchowski et al., 1986) has been combined with immunoprecipitation and scintillation counting or fluorographic analysis of the precipitates after SDS-PAGE. This method of immune precipitation of a labeled macromolecule has become a preferred approach for analyzing protein synthesis, structure, and function.

b. The ring test. The simplest test for the presence of an Ag (or an AB) which gives a quantal answer to whether Ag is absent or present in a solution (or whether an AB exists in an antiserum to an Ag solution) is the ring test. In this assay one overlays, in a narrow tube, a dense solution (adjusted to 5% sucrose) of Ag (or AB) with a light solution (no sucrose) of unknown to be tested. In a positive response, a precipitate accumulates at the interface of the two solutions within a half hour (Fig. 1.6). The interface is viewed on edge

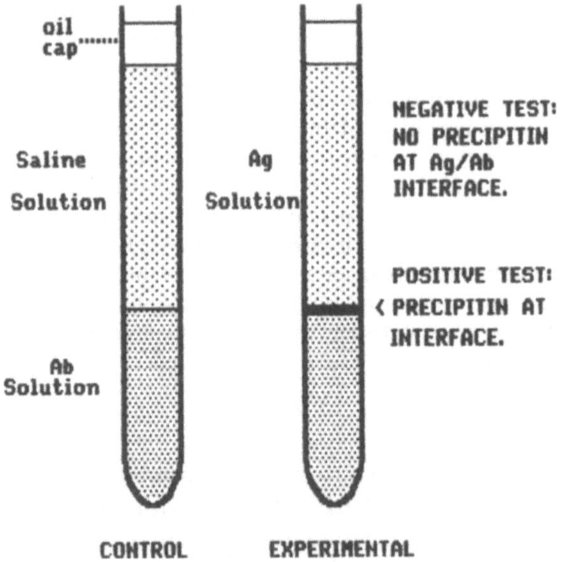

Figure 1.6. The ring test is a simple, relatively rapid test of immunological reactivity which is read by eye as positive or negative against a control tube with no reactive Ag in it. The bottom solution of AB is made more dense by addition of sucrose or glycerol. The top solution is overlaid carefully, and the position of the initially crisp interface is noted. Ag and AB diffuse together and precipitate at their liquid interface. In a "positive test" a precipitin is easily visible at the interface within a half hour.

against a light source and compared to a control tube which has the unknown Ag (or AB) solution replaced with PBS. A strongly positive reaction produces a thick plug of precipitate within an hour.

This quick assay may be useful for screening whether serum from an immunization of a particular animal with an Ag is a success or failure, before more careful evaluation of the antiserum is attempted. It has also been used by some authors to screen chromatography fractions for an Ag prior to pooling fractions (Borovsky & Whitney, 1987). For a yes/no answer, the ring test is expensive if the antiserum or Ag is valuable since the test relies on observing a precipitate which requires relatively large volumes of Ag and antiserum compared to other assays listed below. However, if time or immunological supplies are the more important factors, this assay is useful to have available.

B. Diffusion-Based Estimation of Ag Concentration

a. Oudin one-dimensional, single diffusion. The simplest quantitative measure of Ag titer using diffusion in agarose was devised by Oudin (Becker et al., 1951). This is a simple but elegant technique which requires very little specialized equipment. A researcher could carry out this assay on a low budget and achieve high sensitivity and accurate determinations. The theory underlying determination of the diffusion coefficient of Ags using this technique is also given by Becker and associates (1951). Students in a teaching laboratory can carry out this test with a high proportion of successes. One advantage of the technique is that it does not have to be read at any specific time after the initiation of the reaction and requires no special chemicals or dyes beyond purified agar or agarose, salt, and phosphate buffer. The only specialized equipment needed is the tubes in which the reaction is run. These tubes can be manufactured from $\frac{1}{4}''$ or $\frac{3}{16}''$ soft glass tubing stock, available from most scientific supply houses. The manufacture of the tubes is simple, involving closing one end of each tube with a Bunsen burner or more quickly with an acetylene torch.

In the Oudin technique AB is dispersed evenly in 0.5% agarose and placed in a narrow tube. The AB in agar is overlaid with an Ag solution which must be at least more concentrated than the equivalence titer of the AB in agar. The Ag diffuses into the gel and forms a zone of visible precipitate as the free Ag meets and precipitates with its corresponding AB (Fig. 1.7). The rate at which this precipitate front moves down the tube is proportional to both the concentration and the diffusion coefficient of the Ag. The rate of front migration changes predictably, being proportional to the square root of time. For this reason, all samples should be applied at approximately the same time, including the standards. The tubes are overlaid with mineral oil to prevent evaporation and incubated at a constant temperature in a horizontal position.

Heat shocks or prolonged temperature changes are to be avoided, as the diffusion rate is temperature dependent and the density of the precipitate will be different at different temperatures. Temperature artifacts can give the

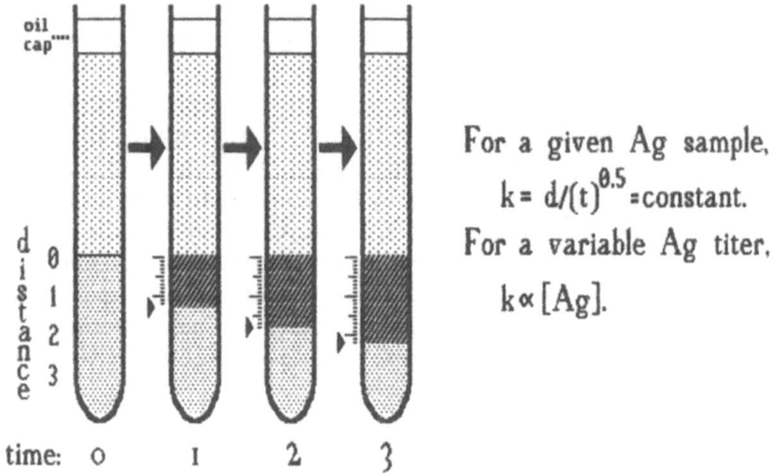

Figure 1.7. The Oudin one-dimensional, single-diffusion assay. The antiserum is placed in an ionagar or agarose gel layer and is considered to be relatively stationary. An Ag dilution, of higher concentration than the equivalence amount of AB, serves as the single dominant diffusing component. The Ag solution is added to the top of the agar at a recorded time zero, and the Ag starts diffusing into the AB-agar phase. A precipitin front is seen to advance with time. For each individual tube the distance, d, of front nigration divided by the square root of the elapsed time, t, gives a constant value, k. This k-value is proportional to the Ag concentration, and a dilution series provides a calibration curve of k versus [Ag] for determining relative or absolute titers of unknown Ag.

appearance of an artificial front which might be confusing in a complex antiserum. Luckily, the Ag/AB fronts move with time, but the temperature artifact "fronts" remain in place. The unknown Ag concentrations are usually determined by comparison to standard curves of known absolute or relative concentration that are run at the same time. Standard curves are highly reproducible, given constant technique and physical conditions.

If the antiserum being used is complex and the Ag solution contains several of the Ags that react with the antiserum, this technique is fraught with problems, since several fronts of precipitate will be moving along the tube and may not be able to be distinguished. The technique is limited to relatively high-titer Ag and AB solutions since the result is a visible precipitate that cannot be enhanced by protein stains.

Despite its relatively low sensitivity, the Oudin technique has been used in numerous insect studies for a variety of purposes. Prior to the advent of QIEP, the Oudin test was a preferred method of measuring the concentration of an Ag in a solution whether that solution was a serum sample (Kunkel and Lawler, 1974; Pan, 1977; Tojo et al., 1980, 1981; Telfer et al., 1983), a tissue extract (Ogawa and Tojo, 1981; Tojo et al., 1980, 1981), a sucrose gradient fraction (Kunkel and Lawler, 1974), or a chromatography column fraction (Duhamel and Kunkel, 1983, 1987). The technique is still useful, since it does not require samples to be diluted, desalted, or dialyzed prior to testing, as is often necessary with QIEP.

b. Radial immunodiffusion. This technique is a two-dimensional analogue of the Oudin technique (Inglid, 1983). A plate of AB in agar is poured, and Ag wells are cut in the plate. A circle of precipitate develops around the well as the Ag diffuses out and encounters AB; the circular front grows wider with time (Fig. 1.8). It works on the same theoretical basis but does not achieve the same quantitative elegance as the Oudin, since the reservoir of Ag is usually limited, and the system runs out of Ag. This results in departure from ideal theoretical behavior, which depends on an infinite Ag reservoir. However, the depletion of Ag provides an end point for the assay since the ring stops growing and visibly darkens at its rim signaling that it is finished expanding. The timing of this end point depends on the diffusion coefficient and concentration of the Ag and AB but usually occurs within a few days. This delay in the ability to read the assay reserves this technique for applications in which data analysis

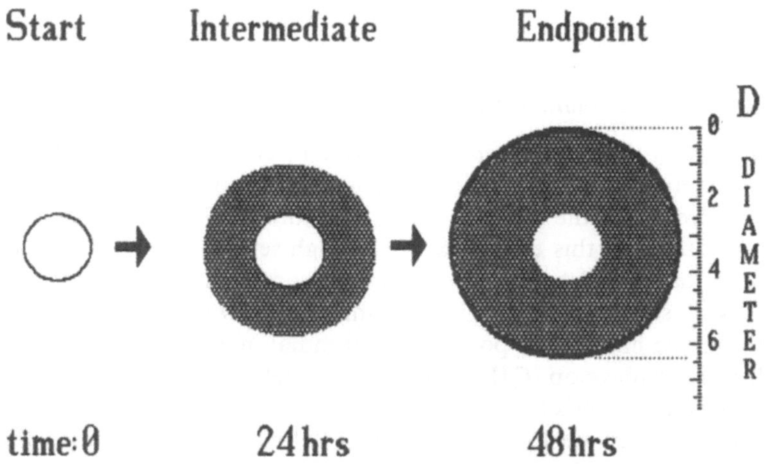

Figure 1.8. Radial immunodiffusion is a two-dimensional, single-diffusion assay similar in nature to the Oudin test. As in the Oudin test, the AB usually serves as a stationary phase, and the Ag is usually the mobile or dominant diffusing component. Antiserum is dispersed in agar and coats a slide or Petri dish. Samples of Ag are placed in wells cut in the agar surface. Ag diffuses out radially, forming a precipitate front as it advances. The system runs out of Ag, and a darker rim of precipitate develops at the edge of the final precipitate. The diameter of this disk of precipitate is determined after staining and is compared to a standard curve established for that antiserum by a previous or concurrent set of serial dilutions of known Ag titer.

is not a priority. Ag concentration is proportional to the diameter of the precipitate front. Ag concentration is usually calculated from a standard curve established separately or from standards run with the unknown samples. Concentrations in the microgram range are measurable. A single $3\frac{1}{4} \times 4''$ plate can accommodate up to 20 unknowns.

This technique requires less equipment than the Oudin test; it can be performed on a microscope slide or in disposable or reusable Petri dishes. It is about as costly in antisera as the Oudin test if it is evaluated at the unstained precipitin stage. However, it is possible to read radial immunodiffusion plates after staining. Thus a substantial increase in sensitivity and economy of antiserum can be achieved by dilution of the antiserum below the level of convenient visibility of unstained precipitate. Less Ag is intrinsically required to run the assay, since the Ag reservoir is limited to the volume that fits in the sample well. Complex antisera and Ag solutions present the same problem of confusion of the precipitin fronts as with the Oudin test, although if the Ag is an enzyme for which a suitable colorimetric staining reaction is available, this assay can be both sensitive and specific (Bog-Hansen and Brogren, 1983).

Several studies of insect serum proteins have found this technique economic, sensitive, and effective in quantifying changes in protein concentration during development (Irie and Yamashita 1983; Marinotti and deBianchi, 1983; deBianchi and Marinotti, 1985; Haunerland and Bowers, 1986; Trost and Goodman, 1986).

C. Electrophoretic Immunoprecipitin Techniques

a. Quantitative immunoelectrophoresis, QIEP, or "rocket" IEP. The problem of quantitative determinations of complex Ag solutions was to a certain degree solved with the advent of QIEP (Laurel, 1972; Verbruggen, 1975). The advantages of this technique are its high sensitivity (ng of Ag) and its ability to quantify a large series of samples at one time. Routinely 50 samples per $3\frac{1}{4} \times 4''$ slide can be run in a chamber that will hold multiple slides. The technique has an end point, as does radial immunodiffusion, and thus it will not overdevelop. QIEP adds to the method of radial diffusion the force of electrophoresis (Fig. 1.9). The equipment involved can be moderately expensive and may preclude the use of this technique for laboratories that are not suitably equipped. A standard electrophoresis power supply is required as well as a horizontal electrophoresis chamber. The electrophoresis is carried out at pH 8.3, which is close enough to the average pI of the immunoglobulins that they remain approximately stationary. As long as the Ag is mobile at that pH (and most are), this results in a rocket-shaped precipitin arc being laid down in the agarose. If the Ag is not sufficiently mobile at pH 8.3, the proteins may have to be carbamylated prior to the electrophoresis (Dillwith and Chippendale, 1984). One of the benefits of the QIEP technique is that all of the Ag in the sample well is forced to interact with AB relatively quickly, within

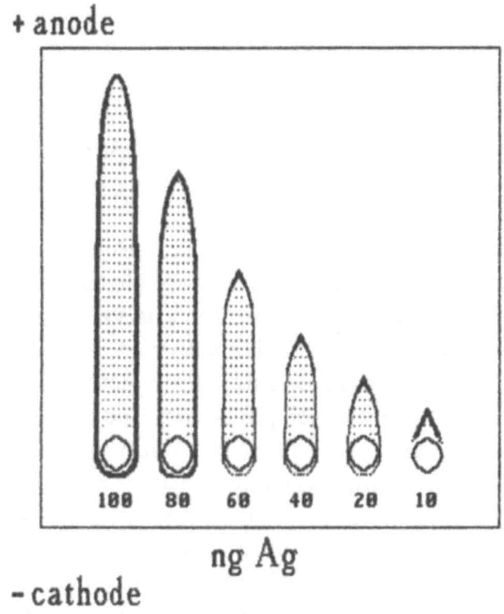

Figure 1.9. Quantitative immunoelectrophoresis (QIEP) or "rocket" IEP. The antibody is dispersed evenly in the thin layer of agarose on the slide. The quantity of Ag placed in each well can be estimated by the amount of AB area that is swept into a "rocket" of precipitin by the electrophoretically propelled Ag. This method, like radial diffusion, has a self-imposed end point which occurs when all of the Ag in the well has reacted with its equivalence amount of AB.

several hours for reasonably mobile Ag. The volume of agar that is cleared of AB by the moving Ag is equal to the equivalence volume for that amount of Ag. Although with higher concentrations of antisera the "rockets" of precipitin are immediately visible, the general routine procedure to obtain maximal sensitivity with this technique is to use a lower AB concentration in the agar and stain the rockets after drying down the agar onto the slide. This technique is about 100-fold more sensitive in measuring Ag than radial immunodiffusion, especially for large Ag molecules which are trapped by inwardly diffusing AB in the sample well at low concentrations in radial immunodiffusion. Complex antisera and Ag solutions are often resolved into different density rockets such that with a little experimentation and logic, the components of a complex system can be identified and measured at the same time.

QIEP has been used to analyze the titer of numerous insect proteins in physiological fluids or tissues. The titer of the protein vitellogenin has been followed in hemolymph and tissues of normal or experimentally treated animals for numerous insect species (Greenberg et al., 1978; Hagedorn et al., 1978; Kunkel, 1981; Buschor and Lanzrein, 1983; Koenig and Lanzrein, 1985;

Kunkel and Nordin, 1985; Benford and Bradley, 1986; Oliveira et al., 1986; Zhu et al., 1986; Imboden et al., 1987). Various other serum proteins including storage proteins (deBianchi and Marinotti, 1984; Levenbook and Bauer, 1984; Ryan et al., 1985b; deKort and Koopmanschap, 1987), diapause-associated protein (Dillwith et al., 1985), and lipophorin (Venkatesh and Chippendale, 1986) are able to be monitored using QIEP.

b. QIEP estimation of protein synthesis and processing. Using a monospecific antiserum to an Ag of interest, it is possible to do a radioactive tracer incorporation study into an Ag which is then analyzed by QIEP in which the labeled Ag is precipitated in a rocket (Hagedorn et al., 1978; Levenbook and Bauer, 1984). To eliminate false counts from electrophoretically immobile sample components, the sample well gel is kept free of AB as in fused rocket QIEP (Fig. 1.13, wells 1 and 2). The maximum amount of precipitate in each rocket can be increased by using higher antiserum concentration in the AB gel. As a rule of thumb, each doubling of antiserum concentration halves the area of rocket created by a given Ag amount. Rockets containing 2–10 μg of protein are not unreasonable. Care should always be taken in this technique to run the electrophoretic step long enough to remove all nonreactive substrates and proteins from the gel. The visible precipitin "rocket" is cut out of the agar gel slab on the slide after the electrophoresis step and dissolved in a scintillation solution for counting. This method compared favorably in accuracy with precipitin techniques in liquid (Hagedorn et al., 1978).

D. *Quantitative Immunotransfer Technique (maximal sensitivity)*

Immobilization of Ag or AB on a support such as a plastic well of a microtiter plate or a nitrocellulose filter and subsequent probing of that immobilized protein with AB or Ag, respectively, is a rapid and sensitive technique for quantifying Ags. When the support is a filter of some sort, the technique has been dubbed dot blotting (Hawkes et al., 1982; Jahn et al., 1984). The probing of the immobilized protein may be accomplished by a fluorescence-, enzyme-, or radionuclide-linked probe. The primary probe is most often an AB or a lectin (Haunerland et al., 1986; Marinotti and Bianchi, 1986). Often the AB probe cannot be conveniently directly visualized. A second all-purpose reagent is often used such as a second AB, which recognizes the species class of the primary AB, or protein A—a bacterial protein that binds to immunoglobulins. This secondary reagent has a covalently attached reporter group such as a fluorescent molecule or a radionuclide or a chromogenic enzyme that can be localized by convenient assays. When the support is a microtiter plate and the method of developing the probe is via an enzyme-linked AB, the developing industry surrounding this technology has dubbed the technique ELISA (enzyme-linked immunosorbent assay) (see Chapter 2). This technique works best with monospecific antisera and is particularly well suited to mABs, since an artificial immobilization for the Ag is provided, obviating the importance of cross-linking.

4. Analysis of Complexity

The analysis of complexity is a particularly strong feature of immunology. With the appropriate antiserum(a), an investigator may identify and measure the titer of the major and minor Ags in an extract. This is helpful in following the changes in particular Ag titers during a physiological or developmental cycle or during purification of particular Ags. Particular techniques are more or less useful in this respect.

A. The Ouchterlony Test, Double Diffusion

Double immunodiffusion, or the Ouchterlony test, is not often the technique of choice for demonstrating complexity, but it is a powerful technique when properly applied. Unfortunately, the technique is often carried out under conditions in which the ability to measure complexity is minimized. How to maximize the visualization of complexity with Ouchterlony is described by its namesake (Ouchterlony, 1968). This maximization involves making sure that the Ag and AB reservoirs are large enough to maintain infinite pools of reactants effectively. In that instance, the place in the gel at which equivalence of Ag and AB is achieved will tend to be stationary, since precipitate for a particular Ag/AB pair accumulates at that line in the gel for the longest time. This situation can rarely be achieved with Ouchterlony run on a microscope slide in which the wells tend to be small (3–10 μl) and the distance between wells is also small. In most instances a complex antiserum used in Ouchterlony will produce a pattern of precipitin lines difficult to interpret (Fig. 1.10 (1)). Most of the Ag/AB pairs will be unbalanced and produce precipitin lines curving toward either the Ag well or the AB well. It is rare that all antibodies of a complex antiserum are sufficiently balanced to produce crisp, straight precipitin lines as in (Fig. 1.10 (2)). For maximum resolution, the well pattern shape is suggested to be rectangular; the gel should be poured in a Petri dish, and the wells placed far enough apart to allow several Ag/AB pairs to form distinct precipitin lines. The greater distances between wells make higher AB and Ag titers in the sample wells necessary to achieve visible precipitates in the gel. The resultant precipitin lines can be enhanced by a number of modifications. Placing an osmotically active component such as high-molecular-weight dextran solution in reactant wells will prevent the well from drying out by losing liquid to the gel. This will maintain for longer the apparent infinite source of reactants in the wells. The unstained precipitin reaction is more intense at colder temperatures, and the density of the precipitate can be enhanced by inclusion of PEG6000 (Verbruggen, 1975) in the agar.

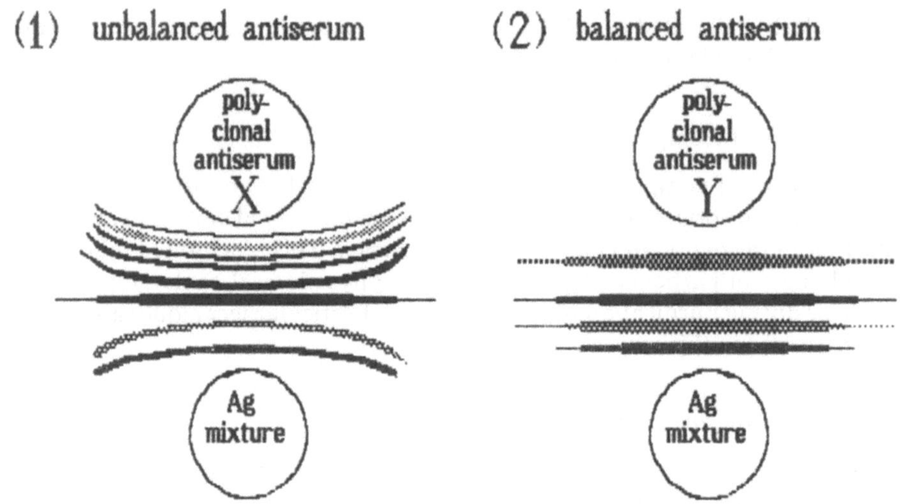

Figure 1.10. Ouchterlony double-diffusion assay. (1) Unbalanced serum. The multiple Ag/AB precipitin lines are not straight and curve away from the source of the excess component. (2) Balanced serum. The multiple Ag/AB precipitin lines are straight and crisp. This occurs when each Ag and AB are at approximate equivalence. Balanced complex antisera for Ouchterlony are difficult to achieve routinely.

A number of studies of insect Ags have used the Ouchterlony technique effectively for demonstrating antigenic complexity of particular Ag sources and testing membership of a particular Ag to a particular stage. For instance, using separate complex antisera made against female and male locust serum, it was shown that the major Ag of locust oocyte vitellin reacted only with anti-female serum (Gellissen et al., 1976). Another approach often used is to use a monospecific antiserum against an Ag such as vitellin purified from eggs and then show via Ouchterlony that it is present in various complex mixtures such as egg extract and female hemolymph but not male hemolymph (Mundal and Law, 1978).

B. Electrophoretic Analysis of Immunologic Complexity

a. Immunoelectrophoresis (IEP). By far the most effective way of improving the analysis of antigenic complexity is to combine or follow an electrophoretic step or steps with an immunologic step. Classical IEP separates Ags in agarose by electrophoresis on a slide prior to placing antibodies in a trough parallel to the lane of electrophoresis. Electrophoretic preseparation, followed by diffusion toward a reservoir of antiserum, provides separate foci of diffusion for electrophoretically separable Ags, and this improves reso-

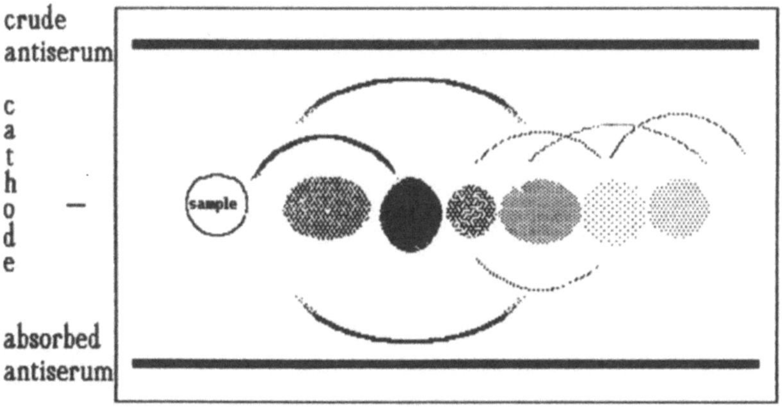

Figure 1.11. Immunoelectrophoresis (IEP). Prior electrophoresis of a mixture of antigens in agarose is combined with a subsequent step similar to Ouchterlony double diffusion. After the electrophoretic separation in the first dimension, a trough is cut in the agar parallel to the direction of electrophoresis, and antiserum is added to the trough and allowed to diffuse toward and precipitate with the electrophoretically separated Ags. A mixture of Ags is shown in this example reacting with a crude antiserum (upper trough) and an absorbed antiserum (lower trough).

lution by reducing the congestion of bands seen in Ouchterlony (Fig. 1.11, top). In some cases, a complex antiserum can provide valuable evidence for the involvement of one antigen of a mixture in a physiological process such as insect serum clotting (Bohn and Barwig, 1984). Absorption of complex antisera with less complex Ag solutions can result in more specific antisera with names such as female-specific (Telfer, 1954; Pan and Wyatt, 1971; Tanaka, 1973; Kunkel and Pan, 1976; Mundall and Law, 1979; Ogawa and Tojo, 1981), larval-specific (Kunkel and Lawler, 1974), or egg-specific (Yamashita, 1986) antiserum (Fig. 1.11, bottom). Absorption of an antiserum involves precipitation of the majority of ABs common to a heterologous Ag source. In the most favorable situation, the heterologous source is expected to differ from the homologous source in one or a few Ags such that the absorbed antiserum may retain AB to one or only a few distinct Ags.

b. Variations on QIEP. Immunological complexity can be dealt with in a number of ways using modifications to the basic QIEP technique according to Laurell (Laurell, 1972; Axelsen, 1983). As in IEP, one can combine the immunological detection step (QIEP) with a prior separation step such as simple slide or slab agarose electrophoresis in the first dimension to allow one to resolve Ags in the first dimension prior to the QIEP step applied at right angles (Fig. 1.12). Usually the first dimension is not done in acrylamide or

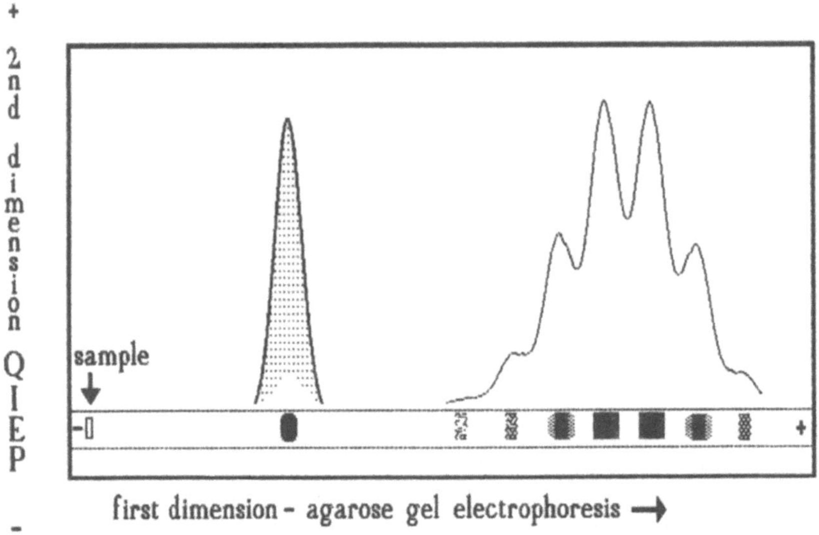

Figure 1.12. Crossed QIEP is the logical extension of IEP to QIEP. The first dimension can be run similar to IEP of Figure 1.11 above, or, for better resolution, the first dimension can be done in a vertical or horizontal agarose slab apparatus. The sliver of agarose from the first-dimension gel is mounted on a slide and supplemented with AB-containing agarose as in QIEP. The second-dimension electrophoresis of the Ags at a right angle creates AG/AB precipitin "rockets" with characteristic density of precipitate. Complex Ag solutions can be analyzed with complex antisera using this method. Fusion of adjacent rockets can be used to identify homologous electromorphs or isozymes.

media other than agarose, because there are incompatabilities of other media with agarose unless special care is taken. This technique, crossed QIEP, has been effectively used in following and evaluating the purification of proteins. If an antiserum to the crude starting mixture is available, one can apply purification steps and used crossed QIEP to check at each successive step how much the major contaminants have been reduced (Barwig, 1985). In extremely complicated antigenic solutions such as lepidopteran serum and yolk extracts, this method has been used effectively to identify and follow the different components during development (Telfer et al., 1981).

Another type of step prior to QIEP is simple diffusion of the samples in adjacent wells in plain agar for a few hours prior to the addition of AB agar to the remainder of the slide. This approach, called fused QIEP, results in fusion of precipitin rockets of adjacent identical proteins (Fig. 1.13, wells 3 and 4). This fusion is often enough to confirm the existance of multiple Ag/AB reactions in the adjacent samples (Storella et al., 1985; Wojchowski and

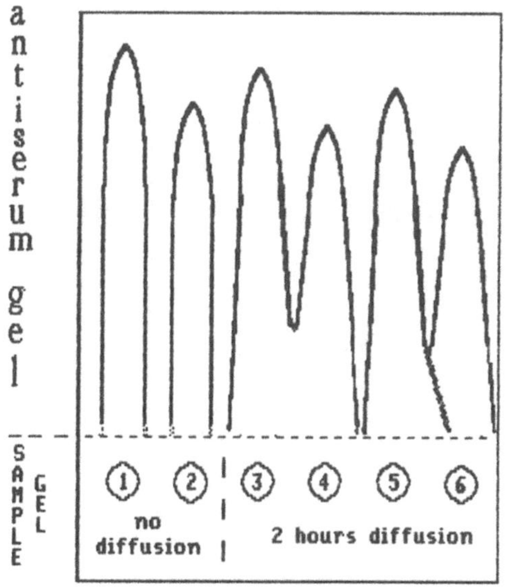

Figure 1.13. Fused QIEP is a logical extension of the Ouchterlony double-diffusion test to QIEP. The samples with potentially related Ags are placed in their individual sample wells (1–6) cut in a basal strip of agar containing electrophoresis buffer but no antiserum. The Ags are allowed to diffuse for a time determined by experiment to allow sufficient comingling of the Ags. An antiserum gel, similar to that in QIEP, is added to the remainder of the slide after 2 h of diffusion, and the samples are electrophoresed into the antiserum agar. Ags in wells 3 and 4 depict an identity reaction; they form a continuous precipitin line similar to a line of identity in the Ouchterlony test. Ags in wells 5 and 6 show partial identity by forming a "spur" of precipitate rather than a continuous line. Samples 1 and 2, placed in their respective wells just prior to electrophoresis, form distinct "rockets" of precipitin, untested for homology, but they leave any particulate matter contaminating the sample back in the gel surrounding the sample well.

Kunkel, 1987). One can also analyze one set of Ags with several layers of more or less specific antisera (Wojchowski et al., 1986).

c. PAGE electroblotting, Western blot. A powerful technique that has been added to the arsenal of techniques for analyzing complexity is the technique of electroblotting (Olmsted, 1981; Towbin et al., 1979). This combines the quantitative technique of dot-blotting with electrophoretic transfer from a one- or two-dimensional electrophoretically separated Ag sample (Fig. 1.14 (1)) onto a nitrocellulose filter. This filter can be probed directly with a specific AB solution or a lectin solution and the Ags recognized by linked-fluorescence, linked-enzyme reactions or, linked-radioactivity methods (Fig.

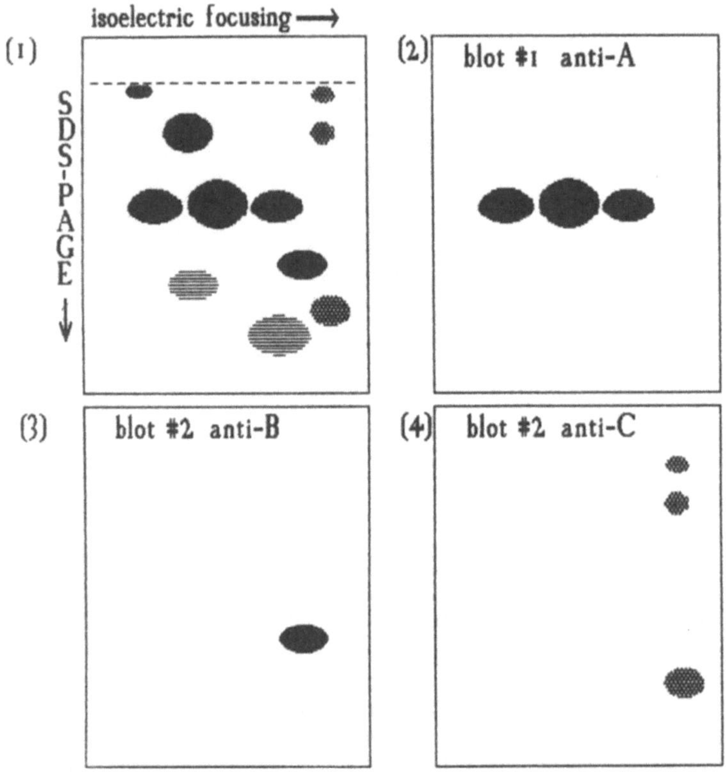

Figure 1.14. One example of PAGE electroblotting, the Western blot. (1) O'Farrell (1975) 2D electrophoresis of antigens by isoelectric focusing in the first dimension followed by SDS-PAGE in the second dimension. (2, 3) Separated Ags are blot or electrophoretically transferred to one or more nitrocellulose membranes and probed with specific antisera. (4) Removal of the probing AB from the nitrocellulose membrane is possible, and reprobing of the still attached Ag with another AB is carried out.

1.14 (2–4)). It has been used extensively in the invertebrate literature to identify specific peptides (Brock and Roberts, 1983; Sharrock, 1983; Ryan et al., 1984, 1985a,b, 1986; Smith and Fisher, 1984; Osir et al., 1986), to identify processing pathways (Wojchowski et al., 1986; Della-Cioppa and Engelmann, 1987) and to investigate complex Ag mixtures (Cox, 1987). In addition, the antibodies can be eluted from the nitrocellulose sheets and the same sheet probed with additional antisera (Fig. 1.14 (3, 4)) (Legocki and Verma, 1980). Alternatively, the antibodies recognizing a particular spot or band can be recovered, and these "affinity-purified" antibodies can be used to probe further Western blots (Olmsted, 1981; Smith and Fisher, 1984). This latter technique allows for unlimited ability to sort out the complexity of Ag mixtures.

5. Qualitative Differences and Similarities between Ags

A. Diffusion Techniques to Demonstrate Qualitative Homologies

a. Ouchterlony (minimum equipment, instructional level). The most frequently used technique for demonstrating identity or departures from identity in two Ags is the double-diffusion technique of Ouchterlony (1965). This application of the Ouchterlony technique works best with monospecific antisera, since complex antisera, as suggested in Section III.D.1.a, can be confusing to interpret. The Ag to which the antiserum was made is called the homologous Ag. An Ag from a different source, such as a different tissue or stage or species, is termed a heterologous Ag. The Ags are placed in peripheral wells and allowed to diffuse toward a well containing an antiserum (Fig. 1.15 (1-3)). Crossing of the precipitin lines of the two systems without any indications of interaction is an indication of immunological distinctness (Fig. 1.15 (1)). The complete fusion of the precipitin lines is an indication that the homologous and heterologous Ags in the two Ag wells are indentical, at least at the level of this technique's ability to discern identity (Fig. 1.15 (2)). Departures from identity are seen as "spurs" at the point of fusion of the two precipitin lines (Fig. 1.15 (3)). Such spurs are indications that a determinant on the homologous Ag is missing or dramatically changed on the heterologous Ag. Hierarchies of spurring can be developed into a similarity matrix, and such matrices have been used to build phylogenetic trees.

Homology between an Ag derived from different stages or tissue sources within a species can often be established with this technique (Pan, 1977; Chino and Yazawa, 1986; Venkatesh and Chippendale, 1986). This can take the form of absolute identity claims which are often overstated, since the Ouchterlony test sometimes does not detect differences of as much as 5% amino acid sequence difference. It is often stated, for instance, that the vitellogenin and vitellin of a particular species are immunologically identical based on a lack of spurring in the Ouchterlony double-diffusion assay. Such low-resolution lack of differences will survive until proved or disproved by a more discriminating test. The immunological method of choice for detecting minimal departures from identity is microcomplement fixation (MC/F).

Homology between Ags from different insect species has often been demonstrated with the Ouchterlony test. The simple existence of homologous Ag in other insects is relatively easily surveyed using the Ouchterlony technique (Kunkel and Lawler, 1974; deBianchi et al., 1983; Dillwith et al., 1985). Some proteins such as vitellogenins have changed rapidly during evolution (Kunkel and Nordin, 1985), and thus when cross-reaction is found, it is evidence for a relatively close phylogenetic relationship. Thus the observations of a strong identity reaction between vitellogenins of *Leucophaea* and *Nauphoeta* (Imboden et al., 1987) and between *Bombyx* and *Philosamia* (Izumi et al., 1980) are particularly important to physiological experiments concerning the species pairs. Homology between particular purified arylphorins of two well-

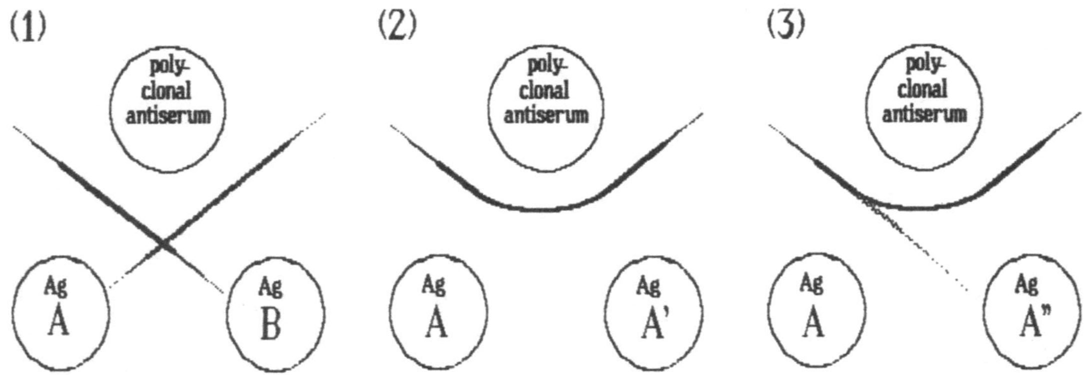

Figure 1.15. Ouchterlony double-diffusion comparison of Ags. A specific or complex polyclonal antiserum is placed in a central well in an agarose slab and allowed to interact with Ags diffusing toward it from two peripheral wells. (1) Ouchterlony demonstration of lack of homology between antigen A and antigen B. (2) Ouchterlony demonstration of complete identity of antigen A with A′. (3) Ouchterlony demonstration of partial identity between antigen A and A″.

studied "model" species, *Hyalophora cecropia* and *Manduca sexta*, was determined with antisera to each others' protein (Telfer et al., 1983). Two serum proteins of the little-studied lepidopteran *Papilio polyxenes* (Ryan et al., 1986) were shown to be homologous to two well-studied serum proteins of *Manduca sexta*. Arylphorins of the corn earworm and tobacco hornworm were shown to be homologous using the available *Manduca* AB (Haunerland and Bowers, 1986). In this way, once the homologies between proteins are established, someone working on a relatively unstudied species can tap into the experience of past researchers on properties and purification techniques developed using proteins of well-studied "model" species. Lack of an immunological cross-reaction with the Ouchterlony test should be viewed cautiously. Anticockroach LSP's lack of a reaction with cricket serum (Kunkel and Lawler, 1974) is a function of phylogenetic distance to be expected in such an ancient divergence. Similarly, *Hyalophora* and *Manduca* arylphorins do not cross-react with *Calliphora* arylphorin (Telfer et al., 1983), nor does *Drosophila melanogaster* LSP_2 cross-react with similar proteins as far removed as *Ceratitus* (Mintzas and Reboutsicas, 1984). Of course these negative results are not interpreted as proof that a homologously related and functioning protein does not exist in distant relatives unless immunological evidence is available that immunological cross-reaction does extend that far. In some instances, other immunological techniques can be used to establish more distant relationships (see Section III.E.2.c below) (Ryan et al., 1984).

b. Oudin demonstration of homology and physical property differences. A monospecific antiserum can be used to measure the degree of homology between two reactive Ags using the Oudin test (Hayden and Becker, 1960). If two Ags have identical physical characteristics and differ only in amino acid sequence, the degree of immunological cross-reaction can be estimated by comparing the density of the precipitin front formed by each Ag with the same AB in Oudin tubes. This is a simple procedure that can be carried out with any gel-scanning apparatus, and it results in quantitative comparisons that are parallel to the measurements made by MCF (Kunkel and Lawler, 1974; Kunkel and Nordin, 1985).

The Oudin single-diffusion assay itself is also able to detect the physical differences or changes in Ags that affect the Stoke's radius and the diffusion coefficient (Becker et al., 1951). Such changes must be relatively large to be detected, such as the difference between a monomer and a dimer of an Ag. When the Oudin test does detect substantial differences in the diffusion coefficient of two Ags, the use of Oudin tests to compare degree of cross-reaction is counterindicated. However, it should be noted that this counterindication applies to complications created with other immunological techniques as well. It is possible in some of these cases to force all of the Ag into a uniform monomeric state for the purpose of quantitative comparisons (Beverley and Wilson, 1982).

B. Electrophoretic Techniques for Demonstrating Qualitative Differences

a. Immunoelectrophoresis (IEP). IEP has been used effectively to demonstrate electrophoretic mobility differences of an Ag (Bohn and Saks, 1986). When performed in purified 1% agarose, mobility difference can be interpreted as differences in charge, since globular macromolecules up to 1 megadalton do not interact with the gel matrix. If the buffer pH for the electrophoretic step is adjusted to allow an estimation of the pH of reversal of antigen direction in the electrophoretic field, the isoelectric points of Ags can be determined (Kunkel and Pan, 1975).

b. Fused QIEP. Immunologic identity or similarity can be demonstrated in QIEP in a technique analogous to Ouchterlony spur formation. As in Ouchterlony, attention to the possibility of artifacts is essential. Using this procedure, the loss or change of a determinant on a protein during a physiologic, developmental, or evolutionary step may be detected. When the Ags are substantially changed, a decrease in the density of the precipitin line for the heterologous Ag may be noticeable in addition to the precipitin spur (Fig. 1.13, samples 5 and 6).

c. Crossed QIEP. Electrophoretic mobility of homologous Ags differs by substantial amounts owing to allelic differences or postribosomal processing. Isozymes or electromorphs may be tested for immunological cross-reaction by allowing them to separate sufficiently in one-dimensional electrophoresis such that they produce separate staining bands in their first dimension. By electrophoresing them into an antiserum dimension, one can test if they are immunologically distinct proteins or if they are different mobility forms of a related protein (Fig. 1.12). Such a technique demonstrated seven immunologically related, random oligomeric structures of the hexameric insect serum protein LSP (Duhamel and Kunkel, 1983). A similar study of *Manduca* demonstrated that three electromorphs of vitellogenin are immunologically identical (Imboden and Law, 1983). Of course such electromorphs could have up to 5% amino acid sequence difference and still not give rise to a spur difference.

d. Western blot assay, homologies. The combination of electrophoretic separation and immunological identification is combined in the Western blot technique (Towbin et al., 1979). This technique has been used in several cases to establish homology between insect proteins from different tissues and stages of development. The stages of vitellin peptide processing were established by Purcell and co-workers (1986) by making a monospecific antiserum to later-stage peptides and using those ABs to probe earlier-stage precursor peptides. The homologies between serum and tissue proteins from different species (Brock and Roberts, 1983; Ryan et al., 1985a; Kozma and Bownes, 1986) and even between different insect orders (Ryan et al., 1984; Robbs et al.,

1985) have been demonstrated using Western blotting. Homology is not a trivial problem, because insects have been shown to have many distinct serum proteins with different native electrophoretic mobilities and different but similar molecular weights (Wyatt and Pan, 1979). The reuse of the original protein blot onto nitrocellulose with different probing antibodies is a substantial help in sorting out the complexity of the gels (Legocki and Verma, 1980). Also, the salvage of AB from blots of specific Ag spots on 1D or 2D native or SDS PAGE gels is extremely useful in assigning homologies between Ags.

C. Quantification of Qualitative Differences

a. Quantitative precipitin reactions. The use of liquid precipitin reactions to compare Ags is quite sensitive, and its results have been shown to parallel the more demanding and more sensitive MCF technique, but the liquid precipitin technique is wasteful of Ag and AB solutions. The technique done properly requires that a complete precipitin curve from AB excess to Ag excess be done for each AB/Ag pair (Leone, 1947). The homologous Ag/AB pair will give the largest integral of area under the precipitin curve. This technique is relatively easy to do in a teaching lab using a simple spectrophotometer, such as a Bausch & Lomb Spectronic-20, to measure turbidity of the liquid precipitin reaction. The reactions develop relatively quickly, and a comparison between two Ags can be carried out easily in the space of one laboratory period. However, it is not recommended as a general research approach unless the Ags and ABs involved are available in large supply.

A great deal more detailed information about the degree of immunological relatedness between Ags can be obtained using Ouchterlony double-diffusion spur hierarchies (Ouchterlony, 1968), which are interpretable in terms of shared determinants and Ven diagrams (Moore and Goodman, 1968). Thus if two different techniques or processes destroy a protein determinant of a particular Ag, comparing the two modified forms with Ouchterlony can demonstrate whether or not they have been modified within the same determinant. Three related biological species with cross-reacting Ag can be ordered by how recently they shared a common Ag, evidence for their phylogenetic relationship. For instance, the Dictyoptera share an Ag, LSP (Kunkel and Lawler, 1974), which allows the relative phylogenetic affinities of the cockroaches, termites, and preying mantids to be investigated. It is clear from the strength of the spurs in those group comparisons that termite LSP is more closely related to cockroach LSP than might earlier have been suspected and that preying mantid LSP is more distantly related than that of termites.

There is a multidimensionality in the spur data from Ouchterlony that is not available through other comparison techniques. Besides estimating the proportion of shared determinants with a related Ag, the Ouchterlony test allows one to compare whether the shared determinants with two other related Ags are the same. The quantification via size of spurs has been shown to be highly correlated with MCF (Moore and Goodman, 1968). There is also the possibility of quantifying the precipitin spur reactions by densitometric scanning of precipitins (Butler and Leone, 1968). This technique has been used to demonstrate the affinity of lepidopteran vitellins within the speciose genus *Catocalla* (Kunkel et al., 1976), concluding that vitellin cross-reaction may be useful in differentiating closely related species.

b. Microcomplement fixation (MCF). The ability of MCF to detect single amino acid differences between two homologous proteins has established it as the immunological technique of choice in detection and quantification of qualitative differences between Ags. The technique uses dilute Ag solutions that need not be pure and monospecific antisera. The AB concentration used to achieve a peak of complement fixation with dilutions of the homologous Ag is used as a reference to calculate the necessary increase in AB concentration to obtain an equivalent fixation curve for a related Ag. The log of the required increase in concentration is termed the immunologic distance (ID) and is proportional to the percentage difference in amino acid sequence. Immunologic cross-reaction disappears in a series of well-studied Ags with departures in amino acid sequence approaching 30–40% (Champion et al., 1974).

Although single amino acid mutations in enzymes have been detected consistently with MCF (Champion et al., 1974), it must be added that various postribosomal modifications such as peptide cleavage or phosphorylation may also have large effects on MCF, and lack of identity is therefore not proof that a heterologous Ag recognized as different by MCF represents a distinct gene product.

MCF has not been applied extensively to insect problems, but when applied it has resulted in some interesting conclusions. A major serum protein of drosophilids has been studied with MCF suggesting far reaching conclusions for the mechanism of evolution of the Hawaiian Drosophilinae (Beverley and Wilson, 1985; Lewin, 1985). This same protein is found in all dipterans and could be a basis for a unified protein-clock time scale of dipteran evolution (Beverley and Wilson, 1984). Similarly, a major serum protein of cockroaches, LSP (Kunkel and Lawler, 1974), related to the arylphorins of holometabolous insects, was studied with MCF (Kunkel and Nordin, 1985) in a broad selection of cockroaches showing LSP to have a long and conservative history including a strong relationship to a termite serum protein. Thus immunological relatedness studied with this particular Ag argues that termites occupy a position within the order Dictyoptera equivalent to families of cockroaches. Although some taxonomists would argue that the evolution of sociality gives termites as a group the right to occupy their own order, a phylogenetic perspective concentrating on the time and branch points of divergences might consider termites to have originated as a social family of cockroaches. Using antisera against this and similar proteins, one might be able to unify the phylogenetic study of arthropods in general. The use of immunologic distances provides another quantitative measure with which to study organismal evolution.

IV. Conclusions

The techniques described in this chapter by no means exhaust the variety of possible immunological approaches to the study of insect Ags. They are rather a selected list of the frequently and personally applied approaches that have, in my view, had a useful impact on the field of insect biochemistry and physiology. Even among the techniques discussed, there are an abundance of variations that may be useful for answering a particular question. Such variations can be gleaned from the general literature on immunological techniques (see Axelsen, 1983) after one has a firm understanding of the basic principles. Each year new techniques and products become available, and the person familiar with the general procedures is more adept with new techniques or variants of old methods.

Acknowledgments. The literature search and experience with above-mentioned techniques were gained and reduced to practice under the auspices of grants from the National Science Foundation (DCB8517781) and the U.S. Department of Agriculture (86-CRCR-1-2153). I am thankful to R. Dompencial, E. Bowdan and G.I. Kunkel for readings of the manuscript.

References

Axelsen, N.H. (1983). *Handbook of Precipitation-in-Gel Techniques.* Oxford, U.K. Blackwell.

Barwig, B. (1985). Isolation and characterization of plasma coagulogen (PC) of the cockroach *Leucophaea maderae* (Blattaria). *J. Comp. Physiol. B* **155**:135–143.

Becker, E.L., Munoz, J., Lapresle, C., and LeBeau, L.J. (1951). Antigen-antibody reactions in agar, II. Elementary theory and determinations of diffusion coefficients of antigens. *J. Immunol.* **67**:501–511.

Benford, H.H., and Bradley, J.T. (1986). Early detection and juvenile hormone dependence of cricket *Acheta domestica* vitellogenin. *J. Insect Physiol.* **32**:109–116.

Berzofsky, J.A. (1985). Intrinsic and extrinsic factors in protein antigenic structure. *Science* **229**:932–940.

Beverley, S.M., and Wilson, A.C. (1982). Molecular evolution in *Drosophila* and higher Diptera. I. Microcomplement fixation studies of a larval hemolymph protein. *J. Mol. Evol.* **18**:251–264.

Beverley, S.M., and Wilson, A.C. (1984). Molecular evolution in *Drosophila* and higher Diptera. II. A time scale for fly evolution. *J. Mol. Evol.* **21**:1–13.

Beverley, S.M., and Wilson, A.C. (1985). Ancient origin for Hawaiian Drosophilinae inferred from protein comparisons. *Proc. Natl. Acad. Sci. USA* **82**:4753–4757.

Bog-Hansen, T.C., and Brogren, C.H. (1983). Enzyme characterization in quantitative immunoelectrophoresis—some potentialities and some pitfalls. *Scand. J. Immunol.* **17S10:**325–330.

Bohn, H., and Barwig, B. (1984). Hemolymph clotting in the cockroach *Leucophaea maderae* (Blattaria). Influence of ions and inhibitors; isolation of plasma coagulogen. *J. Comp. Physiol. B* **154:**457–467.

Bohn, H., and Saks, T.M. (1986). Inhibition of cross-linking in clotting of cockroach hemolymph *Leucophaea maderae* Blaberidae Blattaria. *J. Comp. Physiol. B* **156:**625–634.

Borovsky, D., and Whitney, P.L. (1987). Biosynthesis, purification and characterization of *Aedes aegypti* vitellin and vitellogenin. *Arch. Insect Biochem. Physiol.* **4:**81–99.

Brennan, M.D., Weiner, A.J., Goralski, T.J., and Mahowald, A.P. (1982). The follicle cells are a major site of vitellogenin synthesis in *Drosophila melanogaster*. *Dev. Biol.* **89:**225–236.

Brock, H.W., and Roberts, D.B. (1983). An immunological and electrophoretic study of the larval serum proteins of *Drosophila* species. *Insect Biochem.* **13:**57–63.

Brookes, V.J. (1986). The polypeptide structure of vitellogenin and vitellin from the cockroach, *Leucophaea maderae*. *Arch Insect Biochem. Physiol.* **3:**577–591.

Buschor, J., and Lanzrein, B. (1983). Isolation and quantification of vitellogenin in the haemolymph of the ovoviviparous cockroach *Nauphoeta cinerea*. *Comp. Biochem. Physiol.* **76B:**65–72.

Butler, J.E., and Leone, C.A. (1968). Determination of immunologic correspondence for taxonomic studies by densitometric scanning of antigen-antibody precipitates in agar-gel. *Comp. Biochem. Physiol.* **25:**417–426.

Chino, H., and Yazawa, M. (1986). Apolipophorin III in locusts: Purification and characterization. *J. Lipid Res.* **27:**377–385.

Cox, D.L. (1987). Analysis of the cuticular proteins of *Hyalophora cecropia* with polyclonal antibodies. *Insect Biochem.* **17:**485–492.

Darnell, J., Lodish, H., and Baltimore, D. (1986). *Molecular Cell Biology*. Scientific American Books. New York, N.Y.

deBianchi, A.G., and Marinotti, O. (1984). A storage protein in *Rhynchosciara americana* (Diptera, Sciaridae), *Insect Biochem.* **14:**453–461.

deBianchi, A.G., and Marinotti, O. (1985). A specific protein in the genus *Rhynchosciara* (Diptera, Sciaridae). *Experientia* **41:**1330–1332.

deBianchi, A.G., Marinotti, O., Espinoza-Fuente, F.P., and Pereira, S.D. (1983). Purification and characterization of *Musca domestica* storage protein and its developmental profile. *Comp. Biochem. Physiol.* **76B:**861–867.

deKort, C.A.D., and Koopmanschap, A.B. (1987). Isolation and characterization of a larval hemolymph protein in *Locusta migratoria*. *Arch. Insect Biochem. Physiol.* **4:**191–203.

Della-Cioppa, G., and Engelmann, F. (1987). The vitellogenin of *Leucophaea maderae*. Synthesis as a large phosphorylated precursor. *Insect Biochem.* **17:**401–415.

Dillwith, J.W., and Chippendale, G.M. (1984). Purification and properties of a protein that accumulates in the fat body of prediapausing larvae of the soutwestern cornborer, *Diatraea gradiosella*. *Insect Biochem.* **14:**369–381.

Dillwith, J.W., Lenz, C.J., and Chippendale, G.M. (1985). An immunochemical study of the diapause-associated protein of the southwestern corn borer, *Diatrea grandiosella*. *Insect Biochem.* **15:**711–722.

Duhamel, R.C., and Kunkel, J.G. (1978). A molting rhythm for serum proteins of the cockroach, *Blatta orientalis*. *Comp. Biochem. Physiol.* **60B:**333–337.

Duhamel, R.C., and Kunkel, J.G. (1983). Cockroach larval-specific protein, a tyrosine-rich serum protein. *J. Biol. Chem.* **258:**14461–14465.

Duhamel, R.C., and Kunkel, J.G. (1987). Molting cycle regulation of hemolymph protein clearance in cockroaches: Possible size-dependent mechanism. *J. Insect Physiol.* **33:**155–158.

Gellissen, G., Wajc, E., Cohen, E., Emmerich, H., Applebaum, S.W., and Flossdorf, J. (1976). Purification and properties of oocyte vitellin from the migratory locust. *J. Comp. Physiol. B* **108:**287–301.

Graf, R., Raikhel, A.S., Brown, A.O., Lea, A.O., and Briegel, H. (1986). Mosquito trypsin: Immunocytochemical localization in the midgut of blood-fed *Aedes aegypti* (L.). *Cell Tissue Res.* **245:**19–27.

Greenberg, S.L.W., Kunkel, J.G., and Stuart, A.M. (1978). Vitellogenesis in a primitive termite, *Zootermopsis anqusticollis* (Hagen) (Hodotermitidae). *Biol. Bull.* **155:**336–346.

Hagedorn, H.H., Kunkel, J.G., and Wheelock, G. (1978). The specificity of an antiserum against mosquito vitellogenin and its use in a radioimmunological precipitation assay for protein synthesis. *J. Insect Physiol.* **24:**481–489.

Harboe, M., and Closs, O. (1983). Immunization with precipitates obtained by crossed immunoelectrophoresis. *Scand. J. Immunol.* **17S10:**353–359.

Haunerland, N.H., and Bowers, W.S. (1986a). A larval-specific lipoprotein: Purification and characterization of a blue chromoprotein from *Heliothis zea*. *Biochem. Biophys. Res. Commun.* **134:**580–586.

Haunerland, N.H., and Bowers, W.S. (1986b). Arylphorin from the corn earworm, *Heliothis zea*. *Insect Biochem.* **16:**617–625.

Haunerland, N.H., Ryan, R.O., Law, J.H., and Bowers, W.S. (1986). Lipophorin from the grasshopper, *Gastrimargus africanus*. Isolation and properties of apolipoprotein III. *Insect Biochem.* **16**(5):797–802.

Hawkes, R., Niday, E., and Gordon, J. (1982). Dot-immunobinding assay for monoclonal and other antibodies. *Anal. Biochem.* **119:**142–147.

Hayden, A.R., and Becker, E.L. (1960). Antigen antibody reactions in agar. V. percipitate density studies using the Oudin technique. *J. Immunol.* **85:**591–602.

Imboden, H., and Law, J.H. (1983). Heterogeneity of vitellins and vitellogenins of the tobacco hornworm *Manducca sexta* L. Time course of Vg appearance in the hemolymph of the adult female. *Insect Biochem.* **13:**151–162.

Imboden, H., Koenig, R., Ott, P., Lustig, A., Kampfer, U., and Lanzrein, B. (1987). Characterization of the native vitellogenin and vitellin of the cockroach *Nauphoeta cinerea* and comparison with other species. *Insect Biochem.* **17:**353–366.

Irie, K., and Yamashita, O. (1983). Egg-specific protein in the silkworm *Bombyx mori:* Purification, properties, localization and titre changes during oogenesis and embryogenesis. *Insect Biochem.* **13:**71–80.

Isaac, P.G., and Bownes, M. (1982). Ovary and fat body vitellogenin synthesis in *Drosophila melanogaster*. *Eur. J. Biochem.* **123:**527–534.

Izumi, S., Tomino, S., and Chino, H. (1980). Purification and properties of vitellin from the silkworm, *Bombyx mori*. *Insect Biochem.* **10:**199–208.

Jahn, R., Schiebler, W., and Greengard, P. (1984). A quantitative dot-immunobinding assay for proteins using nitrocellulose membrane filters. *Proc. Natl. Acad. Sci. USA* **81:**1684–1687.

Kawooya, J.K., Osir, E.O., and Law, J.H. (1986). Physical and chemical properties of microvitellogenin, a protein from the egg of the tobacco hornworm, *Manduca sexta*. *J. Biol. Chem.* **261**:10844–10849.

Koenig, R., and Lanzrein, B. (1985). Binding of vitellogenin to specific receptors in oocyte membrane preparations of the ovoviviparous cockroach *Nauphoeta cinerea*. *Insect Biochem.* **15**:735–747.

Kozma, R., and Bownes, M. (1986). Identification and comparisons of the yolk polypeptides and the genes which code for them in *Drosophila melanogaster* sibling species. *Mol. Gen. Genet.* **204**:302–309.

Kunkel, J.G. (1975). Larval-specific serum protein in the order Dictyoptera. II. Antagonist effects of ecdysone and regeneration on LSP concentration in the hemolymph of the Oriental Cockroach, *Blatta orientalis*. *Comp. Biochem. Physiol.* **51B**:177–180.

Kunkel, J.G. (1981). A minimal model of metamorphosis: Fat body competence to respond to juvenile hormone. In: *Current Topics In Insect Endocrinology and Nutrition*. G. Bhaskaran, S. Friedman, and J.G. Rodriquez (Eds.). New York: Plenum.

Kunkel, J.G., and Lawler, D.M. (1974). Larval-specific serum protein in the order Dictyoptera. I. Immunological characterization in larval *Blattella germanica* and cross-reaction throughout the order. *Comp. Biochem. Physiol.* **47B**:697–710.

Kunkel, J.G., and Nordin, J.H. (1985). Yolk proteins. In: *Comprehensive Insect Physiology. Biochemistry and Pharmacology*. Chap. 4, Vol. I. G.A. Kerkut and L.I. Gilbert (Eds.). New York: Pergamon Press, pp. 83–111.

Kunkel, J.G., and Pan, M.L. (1976). Selectivity of yolk protein uptake: Comparison of vitellogenins of two insects. *J. Insect Physiol.* **22**:809–818.

Kunkel, J.G., Johnson, M., Haggerty, W., and Sargent, T.D. (1976). Conservation of an active site for oocyte recognition in rapidly evolving vitellogenins. *Am. Zool.* **16**:246.

Kunkel, J.G., Shepard, G., McCarthy, R.A., Ethier, D.B., and Nordin, J.H. (1980). Concanavalin A reactivity and carbohydrate structure of *Blattella germanica* vitellin. *Insect Biochem.* **10**:703–714.

Laurell, C.-B. (1972). Electroimmunoassay. *Scand. J. Clin. Lab. Invest.* (29 Suppl.) **124**:21–37.

Legocki, R.P., and Verma, D.P.S. (1980). Multiple immunoreplica technique: Screening for specific proteins with a series of different antibodies using one polyacrylamide gel. *Anal. Biochem.* **III**:385–392.

Levenbook, L., and Bauer, C. (1984). The fate of the larval storage protein calliphorin during adult development of *Calliphora vicina*. *Insect Biochem.* **14**:77–86.

Lewin, R. (1985). Hawaiian *Drosophila:* Young islands, old flies. *Science* **229**:1072–1074.

Ma, M., He, G., Newton, P.B., and Borkovec, A.B. (1986). Monitoring *Aedes aegypti* vitellogenin production and uptake with hybridoma antibodies. *J. Insect Physiol.* **32**(3):207–214.

Marinotti, O., and deBianchi, A.G. (1983). A larval hemolymph protein in the eggs of *Rhyncosciara americana*. *Insect Biochem.* **13**:647–653.

Masuda, H., and Oliveira, P.L. (1985). Characterization of vitellin and vitellogenin from *Rhodnius prolixus*. Identification of phosphorylated compounds in the molecule. *Insect Biochem.* **15**:543–550.

Mintzas, A.C., and Reboutsicas, D. (1984). Isolation, characterization and immunological properties of ceratitin-4, a major haemolymph protein of the Mediterranean fruit fly *Ceratitis capitata*. *Insect Biochem.* **14**:285–291.

Moore, G.W., and Goodman, M. (1968). A set theoretical approach to immunotaxonomy: Analysis of species comparisons in modified Ouchterlony plates. *Bull. Math. Biophys.* **30**:279–289.

Mundall, E.C., and Law, J.H. (1979). Physical and chemical characterization of vitellogenin from the hemolymph and eggs of the tobacco hornworm, *Manduca sexta*. *Comp. Biochem. Physiol.* **63B**:459–468.

Nakanishi, M., Wilson, A.C., Nolan, R.A., Gorman, G.C., and Bailey, G.S. (1969). Phenoxyethanol: Protein preservative for taxonomists. *Science* **163**:681–683.

O'Brien, S.J., Nash, W.G., Wildt, D.E., Bush, R.E., and Benveniste, M.E. (1985). A molecular solution to the riddle of the giant panda's phylogeny. *Nature* **317**:140–144.

O'Farrell, P.H. (1975). High resolution two-dimensional electrophoresis of proteins. *J. Biol. Chem.* **250**:4007–4021.

Ogawa, K., and Tojo, S. (1981). Quantitative changes of storage proteins and vitellogenin during the pupal-adult development in the silkworm, *Bombyx mori* (Lepid.: Bombycidae), *Appl. Entcomol. Zool.* **16**:288–296.

Oliveira, P.L., Gondim, K.C., Guedes, D.M., and Masuda, H. (1986). Uptake of yolk proteins in *Rhodnius prolixus*. *J. Insect Physiol.* **32**:859–866.

Olmsted, J.B. (1981). Affinity purification of antibodies from diazotized paper blots of heterogeneous protein samples. *J. Biol. Chem.* **256**:11955–11957.

Osir, E.O., Wells, M.A., and Law, J.H. (1986). Studies on vitellogenin from the tobacco hornworm, *Manduca sexta*. *Arch. Insect Biochem.* **3**:217–233.

Ouchterlony, O. (1968). *Handbook of Immunodiffusion and Immunoelectrophoresis*. Ann Arbor, MI: Ann Arbor Science Publishers.

Pan, M.L. (1977). Juvenile hormone and vitellogenin synthesis in the cecropia silkworm. *Biol. Bull.* **153**:336–345.

Pan, M.L., Bell, W.J., and Telfer, W.H. (1969). Vitellogenic blood protein synthesis by insect fat body. *Science* **165**:393–394.

Pan, M.L., and Wyatt, G.R. (1971). Juvenile hormone induces vitellogenin synthesis in the monarch. *Science* **174**:503–505.

Peferoen, M., Delang, I., and DeLoof, A. (1986). Immunocytochemical localization of vitellogenic and non-vitellogenic yolk proteins in the ovary of *Leptinotarsa decemlineata*. *Comp. Biochem. Physiol. B* **83**:347–354.

Peferoen, M., and DeLoof, A. (1986). Synthesis of vitellogenic and non-vitellogenic yolk proteins by the fat body and ovary of *Leptinotarsa decemlineata*. *Comp. Biochem. Physiol. B* **83**:251–254.

Purcell, J.P., Kunkel, J.G., and Nordin, J.H. (1986). Oligosaccharide and polypeptide processing events associated with the programmed degradation of an insect vitellin. *Fed. Proc.* **45**:1712.

Raikhel, A.S. (1984). The accumulation pathway of vitellogenin in the mosquito oocyte: A high-resolution immuno- and cytochemical study. *J. Ultrastruct. Res.* **87**:285–302.

Raikhel, A.S., and Lea, A.O. (1986). Analysis of mosquito yolk by monoclonal antibodies. In: *Molecular Entomology*. J.H. Law (Ed.). Los Angeles: UCLA Symposia on Molecular and Cellular Biology. Alan R Liss Inc., pp. 403–413.

Raikhel, A.S., Pratt, L.H., and Lea, A.O. (1986). Monoclonal antibodies as probes for processing of yolk protein in the mosquito. Production and characterization. *J. Insect Physiol.* **32**:879–890.

Robbs, S.L., Ryan, R.O., Schmidt, J.O., Keim, P.S., and Law, J.H. (1985). Lipophorin of the larval honeybee, *Apis mellifera* L. *J. Lipid Res.* **26**:241–247.

Ryan, R.O., Schmidt, J.O., and Law, J.H. (1984). Arylphorin from the haemolymph of the larval honeybee, *Apis mellifera. Insect Biochem.* **14**:515–520.

Ryan, R.O., Anderson, D.R., Grimes, W.J., and Law, J.H. (1985a). Arylphorin from *Manduca sexta.* Carbohydrate structure and immunological studies. *Arch. Biochem.* **243**:115–124.

Ryan, R.O., Keim, P.S., Wells, M.A., and Law, J.H. (1985b). Purification and properties of a predominantly female specific protein from the hemolymph of the larva of the tobacco hornworm, *Manduca sexta. J. Biol. Chem.* **260**:782–787.

Ryan, R.O., Wang, X., Willott, E., and Law, J.H. (1986). Major hemolymph proteins from larvae of the black swallowtail butterfly, *Papilio polyxenes. Arch. Insect Biochem. Physiol.* **3**:577–591.

Shapiro, D.J., Taylor, J.M., McKnight, G.S., et al. (1974). Isolation of hen oviduct ovalbumin and rat liver albumin plysomes by indirect immunoprecipitation. *J. Biol. Chem.* **249**:3665–3671.

Sharrock, W.J. (1983). Yolk proteins of *Caenorhabditis elegans. Dev. Biol.* **96**:182–188.

Smith, D.E., and Fisher, P.A. (1984). Identification, developmental regulation and response to heat shock of two antigenically related forms of a major nuclear envelope protein in *Drosophila* embryos. *J. Cell Biol.* **99**:20–28.

Storella, J.R., Wojchowski, D.M., and Kunkel, J.G. (1985). Structure and embryonic degradation of two native vitellins in the cockroach, *Periplaneta americana. Insect Biochem.* **15**:259–275.

Tanaka, A. (1973). General accounts on the oocyte growth and the identification of vitellogenin by means of immunospecificity in the cockroach, *Blattella germanica* (L.). *Dev. Growth Diff.* **15**:153–168.

Tanaka, A., and Ishizaki, H. (1974). Immunohistochemical detection of vitellogenin in the ovary and fat body during a reproductive cycle of the cockroach, *Blattella germanica. Dev. Growth Diff.* **16**:247–255.

Telfer, W.H. (1954). Immunological studies of insect metamorphosis. II. The role of a sex-limited blood protein in egg formation by the cecropia silkworm. *J. Gen. Physiol.* **37**:539–558.

Telfer, W.H., Keim, P.S., and Law, J.H. (1983). Arylphorin, a new protein from *Hyalophora cecropia:* Comparisons with calliphorin and manducin. *Insect Biochem.* **13**:601–613.

Telfer, W.H., Rubenstein, E., and Pan, M.L. (1981). How the ovary makes yolk in Hyalophora. In: *Regulation of Insect Development and Behavior.* F. Sehnal, A. Zabza, J.J. Menn, and B. Cymborowski (Eds.). pp. 637–654.

Tillman, P., and Norman, C. (1983). Droperidol-fentanyl as an aid to blood collection in rabbits. *Lab. Anim. Sci.* **33**:181–182.

Tojo, S., Kiguchi, K., and Kimura, S. (1981). Hormonal control of storage protein synthesis and uptake by the fat body in the silkworm, *Bombyx mori. J. Insect Physiol.* **27**:491–497.

Tojo, S., Nagata, M., and Kobayashi, M. (1980). Storage proteins in the silkworm, *Bombyx mori. Insect Biochem.* **10**:289–303.

Towbin, H., Staehelin, T., and Gordin, J. (1979). Electrophoretic transfer of proteins from polyacrylamide gels to nitrocellulose sheets: Procedure and some applications. *Proc. Natl. Acad. Sci. USA* **76**:4350–4354.

Trost, J.T., and Goodman, W.G. (1986). Hemolymph titers of the biliprotein, insectacyanin, during development of *Manduca sexta. Insect Biochem.* **16**(2):353–358

Tung, A.S., Ju, S.-T., Sato, S., and Nisonoff, A. (1976). Production of large amounts of antibodies in idividual mice. *J. Immunol.* **116**:676–681.

Venkatesh, K., and Chippendale, G.M. (1986). Synthesis and release of proteins from cultured larval fat body of the southwestern cornborer, *Diatrea grandiosella. Insect Biochem.* **16**:917–927.

Verbruggen, R. (1975). Quantitative immunoelectrophoretic methods: A literature survey. *Clin. Chem.* **21**:5–43.

Wojchowski, D.M., and Kunkel, J.G. (1987). Purification of two distinct oocyte vitellins and identification of their corresponding vitellogenins in fat body and hemolymph of *Blaberus discoidalis. Insect Biochem.* **17**:189–198.

Wojchowski, D.M., Parsons, P., Nordin, J.H., and Kunkel, J.G. (1986). Processing of pro-vitellogenin in insect fat body: A role for high-mannose oligosaccharide. *Dev. Biol.* **116**:422–430.

Wu, S.J., and Ma, M. (1986). Hybridoma antibodies as specific probes to *Drosophila melanogaster* yolk polypeptides. *Insect Biochem.* **16**:789–796.

Wyatt, G.R., and Pan, M.L. (1979). Insect plasm proteins. *Ann. Rev. Biochem.* **47**:779–817.

Zhu, J., Indrasith, L.S., and Yamamshita, O. (1986). Characterization of vitellin egg-specific protein and 30-kDalton protein from *Bombyx* eggs and their fates during oogenesis and embryogenesis. *Biochim. Biophys. Acta* **882**:427–436.

Chapter 2

ELISA and Monoclonal Antibodies

Michael Ma, Klaus-Peter Sieber, Joanne Ballarino, and Shuenn-Jue Wu

I. Enzyme-Linked Immunosorbent Assay

1. General Overview

Immunochemical methods have been used extensively to identify and quantify biologically important substances. The nature of the "antigen-antibody reaction" enables researchers to selectively monitor molecules of interest among a mixture of potentially interfering materials found in complex biological samples. Classical immunochemical techniques such as immunodiffusion and immunoprecipitation (see Chapter 1) do not in some cases offer sufficient detection sensitivity and are incapable of handling the large number of samples that are currently routinely processed. With the advent of radioimmunoassay (RIA: see Chapter 6), an analytical immunochemical technique has become available for measuring minute amounts of biologically active substances such as hormones, enzymes, and neuropeptides.

Utilizing a combination of radioactive compounds and antibodies (ABs), radioimmunoassays (RIAs) have revolutionized biological research in numerous areas, especially in physiology and endocrinology. In recent years, knowledge in the field of invertebrate endocrinology has exploded through the development and popularization of an RIA that allowed the measurement of ecdysteroid titers (Borst and O'Connor, 1972). This is undoubtedly the single most important reason for the rapid advancement of insect ecdysteroid research. The continuous demand for precision in experimentation has prompted insect biologists to seek more appropriate analytical methods for quantification of proteins and other bioactive compounds.

Despite the advantages of RIA, the cost of liquid scintillation counters and the potential hazards of contamination have limited this method to centralized laboratories where all the procedures are self-contained and closely supervised. The emergence of the enzyme-linked immunosorbent assay (ELISA) has provided biologists with an excellent alternative to the RIA. The principle behind ELISA is essentially the detection of an antigen (Ag) with enzyme-labeled antibodies. The enzyme moiety keys a colorimetric reaction that is proportional to the amount of Ag molecules bound to the enzyme-AB conjugate.

In 1971, Engvall and Perlmann described an ELISA for the quantification of rabbit IgGs by labeling anti-rabbit immunoglobulin with enzyme. In the same year, Van Weemen and Shuurs conjugated an enzyme directly to Ag to compete with nonlabeled Ag for limited AB binding sites, a principle that is similar to competitive RIA. Besides the high sensitivity and specificity inherent in this method, the maintenance cost is relatively low, the setup cost is affordable for most laboratories, the problems associated with radiochemical wastes can be avoided, and the sensitivity of detection approaches that of the RIA.

With the introduction of the 96-well microplate method by Voller et al. (1974), ELISA became extremely popular for the diagnosis of plant and animal diseases (Voller et al., 1979), by allowing the simultaneous processing of a large number of samples. Parallel rapid commercial development of instrumentation for automation of various parts of the ELISA protocol have further contributed to its popularity. Subsequently, as researchers became familiar with the operational aspects of ELISA, there were increased applications for the quantification of compounds with capabilities similar to the RIA. Unfortunately, ELISA methods have not been adopted with quite the same fervor by insect physiologists and biochemists.

In this section, we present a case study that involves the development and application of ELISA as the key analytic method for an insect protein, vitellogenin. Parallel studies and additional innovative applications will surely be developed for many other areas of insect research. We have witnessed and experienced the important role of RIA in advancing insect ecdysteroid research. In a similar way, ELISA will become a valued immunochemical method for insect research that requires accurate measurement of biologic materials.

2. ELISA as a Quantitative Procedure

There are three ELISA protocols that are frequently employed for the analytic measurement of a chemical in biological samples. They are referred to as (1) competitive ELISA, (2) direct double-AB sandwich ELISA, and (3) indirect AB sandwich ELISA (Voller et al., 1979). All three protocols employ a primary or coating AB for capturing Ag molecules.

In the competitive ELISA procedure, enzyme-labeled Ag is used to compete with the nonlabeled Ag present in a sample preparation for available AB sites. It is important to note that this protocol can only be effective if the Ag is stable and can be easily purified in milligram quantities. The amount of bound enzyme-labeled Ag can then be estimated after the addition of enzyme substrate and subsequent colorimetric reading.

The double-AB sandwich ELISA employs an enzyme-labeled AB that is also specific to the Ag. An Ag molecule is "sandwiched" between the coating or primary AB and enzyme-labeled secondary AB. This procedure appears to have gained the most popularity in ELISA literature, since there are relatively few steps involved and the methods for coupling enzyme-secondary AB have become quite routine.

The indirect double-AB sandwich ELISA is a modification of this procedure which allows the use of commercially available conjugates, rather than performing the time-consuming conjugation of extracted ABs. The indirect double-AB sandwich ELISA differs from the direct method in that the Ag is sandwiched between specific ABs developed in two different animal species (such as mouse, rabbit, goat, and guinea pig). If the coating AB is developed from guinea pig, then the secondary AB must be from another species such as rabbit. The amount of secondary AB bound to the Ag is in turn detected by a commercially available enzyme-AB conjugate specific for the immunoglobulin of the secondary AB; in this example enzyme-labeled goat anti-rabbit will act as the tertiary AB. Depending on the choice of secondary ABs, this tertiary AB can be goat anti-mouse, anti-rabbit or anti-guinea pig. This protocol has been applied successfully for analysis of mosquito vitellogenin (Ma et al., 1984b, 1986) and the gypsy moth nuclear polyhedrosis virus (Ma et al., 1984a).

In the process of developing a reliable quantitative ELISA procedure, each step of the protocol has to be optimized and standardized. In the following paragraphs, we will attempt to point out the important precautionary measures necessary for a successful quantitative ELISA.

A. Preparation of Specific ABs

The most important component for the success of any immunochemical technique is the preparation of a high-titered antiserum. To generate a quality antiserum, the immunogen should be highly purified with methods such as gel electrophoresis, conventional column, or high-pressure liquid chromatography procedures. Purification reduces the possibility that ABs to impurities might produce undesirable cross-reactivity. Also, biological samples may contain many different proteins that could potentially interfere with the binding of ABs to their target molecules. For these reasons, it is important to obtain antisera that are free of ABs to nontarget proteins in the test samples.

In practice, it is not always possible to obtain an ideal antiserum. In many cases, an absolutely pure Ag preparation is extremely difficult to obtain owing to its low concentration and instability in biological extracts. Also, chemicals that are commonly used in purification procedures might alter the structure or stability of the Ag thus affecting the quality of the antiserum. To reach a compromise, the Ag can be obtained in a highly purified form while retaining some of the host or nontarget proteins as a minor impurity.

In the event that it is possible to obtain a preparation of Ag-free host protein, ABs that have an affinity to the epitopes of nontarget proteins can be removed by immunoprecipitation. Another approach to remove such cross-reacting ABs is the use of immunoaffinity chromatography using CNBr-activated Sepharose 4B (Pharmacia) to which host nontarget proteins are coupled as ligand. The activation of Sepharose 4B with cyanogen bromide results in a reactive product which can be coupled to proteins or nucleic acids via primary amino groups or other nucleophilic groups under mild conditions. Any ABs that might have high affinity to nontarget or host proteins will be adsorbed by the ligands on the column support. Bound ABs can then be removed by either high-salt or low-pH buffer solutions. The antiserum can be eluted through such an affinity column as many times as needed to remove all the undesirable immunoreactive ABs, resulting in a preparation that is highly specific to the Ag.

B. Coating ABs

Primary ABs or coating ABs are commonly used in ELISA protocols that require accurate quantification of Ag in sample preparations. The first step is to sensitize the surface of a solid phase with ABs. The immobilized ABs will capture and retain Ag molecules among interfering proteins in a test sample. The solid phases commonly employed are polystyrene and polyvinyl microplates and polystyrene beads. The plastic tubes developed for RIAs are also applicable.

Since immunoglobulins bind to the polymer surfaces by hydrophobic-hydrophilic type interactions, it is advantageous to have the coating AB preparation as pure as possible, because other serum proteins such as albumin would compete with ABs for available active sites on the solid phase. The immunoglobulin subclass G (IgG) ABs can be purified by adsorption on protein A–Sepharose CL-4B (Pharmacia), followed by elution with 0.1 N glycine-HCl, pH 3.0. This extra step assures maximum coverage of the solid phase by high-affinity immunoglobulins and improves capture efficiency for the Ag as a result. The concentration of the coating AB and the length of this passive sensitization period should be determined experimentally.

C. Blocking

The step that usually follows coating is the addition of blocking solution to cover unoccupied active sites on the solid phase before the addition of biological samples. While the usual 1% bovine serum albumin solution is adequate, we have found that a milk powder-based cocktail known as BLOTTO (bovine lacto transfer technique optimizer) consistently provided the most effective blocking (Johnson et al., 1984). BLOTTO consists of 5% (w/v) nonfat dry milk and 0.0001% methiolate in phosphate-buffered saline. The optimal blocking time should also be determined experimentally with different antigenic preparations.

D. Preparing Biological Samples for ELISA

The purity, concentration and chemical nature of the Ag will dictate how samples should be prepared for ELISA. This is an important step if the ELISA is used routinely as a quantitative procedure. In cases where the Ag is abundant in the test sample, proper dilution series are needed to ensure that the optical readings at the end of an ELISA run will fall within the linear range of the standard curve. If the Ag is present in small amounts, appropriate concentration procedures might be necessary to increase the Ag concentration to a level well within the range of detectability. It is important to note that such a procedure must be easy to standardize and cause minimal degradation to the Ag.

There are instances where the Ags are lipophilic molecules that have high affinity for the inside walls of the plastic wells. These molecules may compete and exchange more frequently with the coated ABs for active sites. There are two ways to minimize this type of nonspecific binding: (1) increase the blocking time with BLOTTO or other blocking buffers, and (2) employ BLOTTO as a diluent for preparing test samples. These measures have worked well with phosphoglycoproteins such as insect vitellogenin (Ma et al., 1984b, 1986).

E. Enzyme Conjugates and Substrates

Protein A–purified immunoglobulin can be conjugated to enzymes such as alkaline phosphatase, horseradish peroxidase, or galactosidase as the secondary antibody for ELISA. We have found that the two-step glutaraldehyde method (Voller et al., 1979) consistently yielded more sensitive assays (Arameas, 1969). This protocol involves the conjugation of an equal ratio of enzyme and AB molecules, with dilutions up to 1 : 2,000 commonly employed. In cases where a large number of samples are processed, conjugates can be reused up to four times. These reused conjugates might have higher specificity, because nonspecific ABs would have been absorbed in previous assays. Substrates such as *p*-nitrophenyl phosphate yield a sensitivity of 10–50 ng/ml. We have also experimented with fluorogenic methylumbelliferin substrates (Kato et al., 1976; Ishikawa and Kata, 1978). Although the fluorogenic substrate is more sensitive, the background readings are also considerably higher. With all the improvements made in the quality and affordability of ELISA instruments in recent years, fluorogenic ELISA does not appear to offer a substantial advantage over the standard colorimetric reading protocols.

3. Monitoring Gypsy Moth Vitellogenin: A Case Study

Vitellogenesis is the process of yolk synthesis, translocation, and deposition in developing oocytes and is common to all egg-laying animals (Engelmann, 1970). In insects, vitellogenin is usually synthesized in the fat body and transported to the ovaries via the hemolymph. Vitellogenin (Vg) is then selectively incorporated into the oocytes by a process of receptor-mediated endocytosis.

In elucidating the details of this physiological process, it is important to have a specific and sensitive monitor for vitellogenin. An analytical procedure such as ELISA is valuable for establishing the normal time course of vitellogenin production, and hormonal regulation can be accurately monitored with this method. In developing an ELISA procedure, several steps have to be undertaken to ensure the production of an antiserum that is highly specific to Vg. It is conceivable that the same type of approach can be used for monitoring other important hemolymph proteins.

A. Purification of Vitellin

Since Vg and vitellin are immunologically and biochemically similar, most researchers have collected soluble yolk protein from homogenized mature ovaries as the primary source of immunogen. Depending on the insect species, the soluble yolk protein extract contains anywhere from 60% to 90% vitellin (Englemann, 1979). Milligrams of partially purified vitellin can be obtained with this method.

In the gypsy moth, we carried out a more extensive purification protocol. Ovarian proteins were solubilized in 0.15 M NaCl, followed by precipitation with 50% ammonium sulfate solution. After dialysis, the ovary extracts were subjected to three successive chromatography procedures: (1) Sephacryl S-300, (2) DEAE ion-exchange column, and (3) a hydroxyapatite column. We had previously identified the native molecular weight of gypsy moth vitellin to be 400 kD with polyacrylamide gel electrophoresis (PAGE) and further identified two major subunits of approximately 180 and 160 kD by SDS-PAGE. The latter method was used to monitor the elution of vitellin from the different column procedures. These purification steps yielded highly purified vitellin suitable for the production of a specific polyclonal antibody.

B. Production of Antiserum to Purified Vitellin

Rabbit antiserum to the purified vitellin was obtained through four injections using the immunization protocol described by Kabat and Mayer (1961). The AB titer between booster injections was monitored periodically by the indirect ELISA method (Voller et al., 1979). In our laboratory, antisera with an end point of 1:100,000 or better have been found to be adequate for ELISA procedures.

C. Production of a Female-Specific Polyclonal Antibody

The rabbit antiserum was repeatedly adsorbed with male pupal hemolymph so that the AB solution would be rendered female-specific. Further removal of cross-reacting ABs was performed by immunoaffinity chromatography using CNBr-activated Sepharose 4B to which purified vitellin was coupled as ligand. The immunoaffinity purified IgGs were subsequently isolated by protein A Sepharose affinity chromatography. The specificity of the ABs to vitellin was verified successively by Ouchterlony immunodiffusion (see Chapter 1), indirect ELISA (Voller et al., 1979), and Western blotting followed by immunochemical staining (Towbin et al., 1979). These purified ABs were used for coating microplates as well as for the preparation of AB-enzyme conjugate.

The double-AB sandwich ELISA (Voller et al., 1979) developed with the above reagents was found to be quite sensitive and specific. Vg can be detected in concentrations as low as 10 ng/ml as shown in a typical standard curve (Fig. 2.1). We found this curve has a usable range of 100 ×. With samples that have high Vg concentrations, the hemolymph samples have to be diluted in excess of 500,000 × for the readings to fall within the usable range. In practice, a

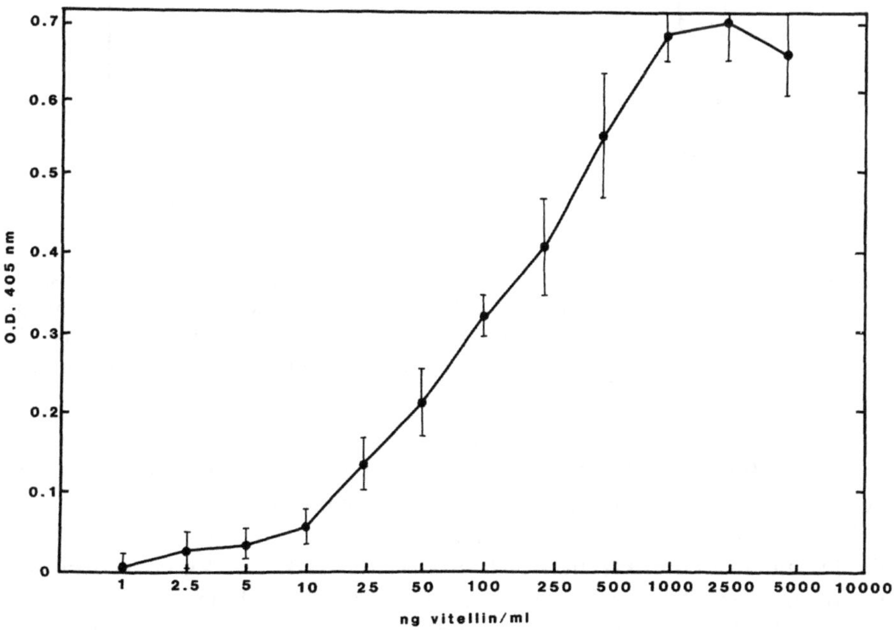

Figure 2.1. Typical double-AB sandwich ELISA standard curve using affinity-purified rabbit ABs raised against ovarian extract as the coating AB and alkaline phosphatase-labeled affinity-purified rabbit extract AB raised against ovarian extract as the indicator AB.

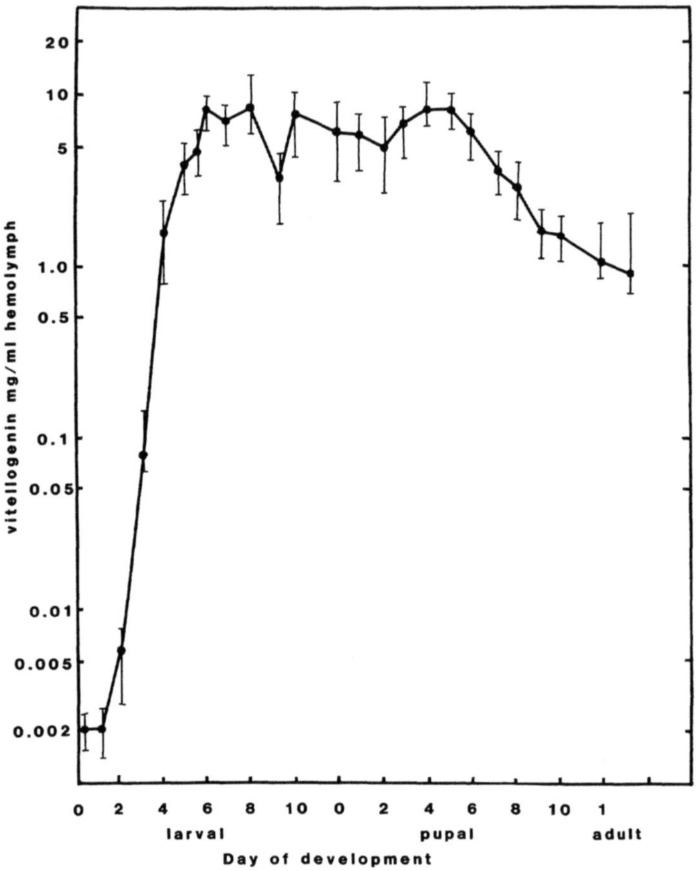

Figure 2.2. Vitellogenin concentration in the female hemolymph of last larval instar, pupa and adult gypsy moth measured by the double-AB sandwich ELISA. Each point represents the mean of five replicates per insect hemolymph sample, and at least 10 insects were sampled per day. The bars represent the standard deviation around the mean.

1 : 500,000 dilution of 10 μl of the gypsy moth hemolymph was necessary for detection of Vg during the regular production periods.

Vg was first detected in the hemolymph on day 2 of the last larval instar of the female at a concentration of 6 μg/ml (Fig. 2.2). The Vg concentration rapidly increased to a maximum of 8.25 mg/ml on day 6 of the last larval instar and remained constant through the stadium. Around day 4 of the pupal stage, the Vg concentration gradually began to decline. This can be attributed to the initiation of patency in follicular epithelial cells of the ovary, followed by the endocytotic uptake of Vg (Fig. 2.3). Most research in lepidopteran

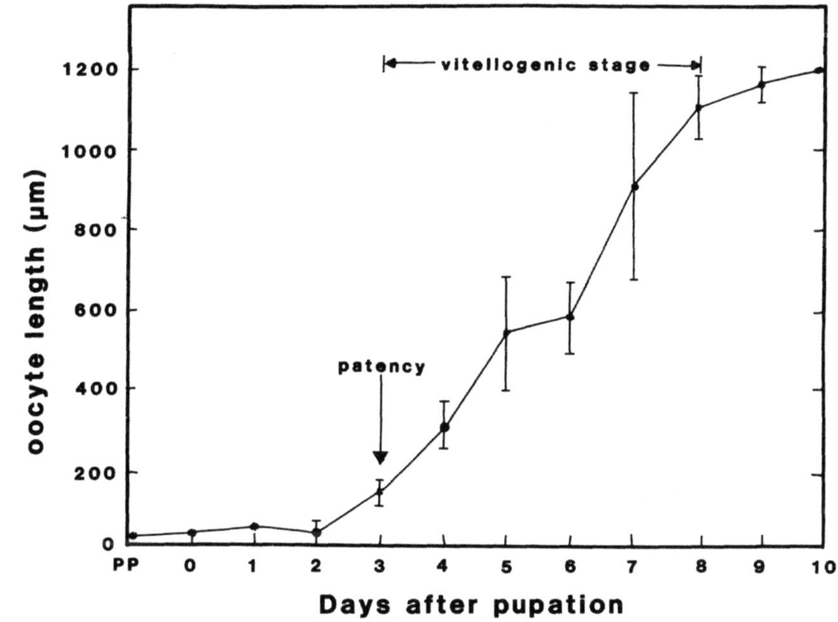

Figure 2.3. Daily measurement of the length of the terminal (T) oocyte in the pupal stage. Each point represents the mean of at least three ovarioles from at least five different insects. Bars represent the standard deviation around the mean.

vitellogenesis has centered around the uptake of yolk proteins by the ovaries during the pupal-adult stage as the initial event of vitellogenesis. It is evident from our gypsy moth data that vitellogenesis is not necessarily an exclusively pupal or adult event. The immunochemical method permitted the accurate measurement of hemolymph Vg at all stages of development. The ELISA was shown to be quite specific in that no Vg was detected in any of the male hemolymph samples collected from larvae, pupae, or adults.

In conclusion, this ELISA method has been shown to be useful in establishing the normal time course and temporal changes in Vg levels. The sensitivity of this technique will permit the use of experimental designs that require the accurate measurement of changes of Vg levels due to surgical manipulation or hormonal or chemical treatments. Since microliter hemolymph samples are sufficient for analysis, multiple bleedings can conceivably be performed on the same individual at different times before and after experimental manipulations. In summary, the ability to monitor Vg accurately in individual animals offers many advantages in experimental designs, and this approach can easily be adopted for other hemolymph proteins such as enzymes, hormones, storage and binding proteins, and the like.

II. Monoclonal Antibodies

1. Overview

In 1965, Harris and Watkins from Cambridge University reported that the hybridization of mouse and human cells was successfully induced by inactivated Sendai virus, a member of the parainfluenza group of myxoviruses, resulting in hybrid cells that were viable. Somatic cell hybrids have since been used extensively in studies of gene mapping, gene expression, and differentiation (Harris, 1970; Ephrussi, 1972; Ringertz and Savage, 1976).

Köhler and Milstein (1975) subsequently published a historic paper on the fusion of mouse splenocytes and myeloma cells. The hybrids resulting from

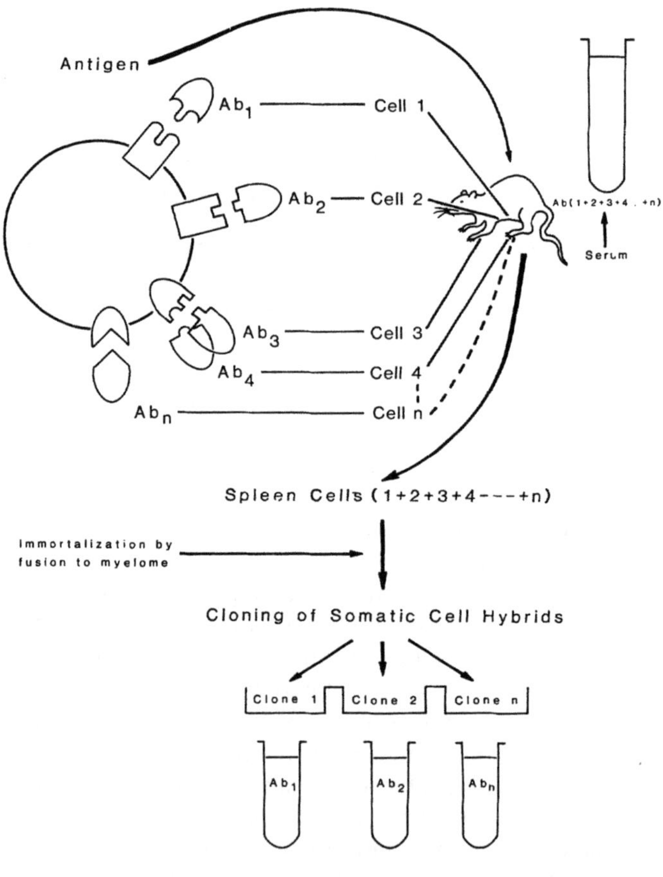

Figure 2.4. Diagram describing hybridoma technology. Spleen cells are removed from a mouse previously injected with the Ag. Each spleen cell produces one species of AB with specificity to one antigenic determinant. Fusion of these spleen cells with myeloma cells will produce hybridomas that continue to produce AB and that can be propagated indefinitely in culture. (Redrawn from Milstein and Lennox, 1980.)

this fusion could be propagated in culture and were capable of producing ABs with predefined specificity to the Ag. ABs produced by these hybrid cells, or hybridomas, are directed against different antigenic determinants of the immunogen (Fig. 2.4). Individual hybridomas can be cloned to establish different cell lines. Each cell line produces a homogeneous population of antibodies or monoclonal antibodies (mABs) directed against a specific antigenic determinant. Cell culture and cryogenic storage techniques can immortalize these cell lines and produce an inexhaustible supply of standard AB reagents for experimental or therapeutic purposes. The mAB offers a whole new dimension in analyses that has revolutionized the biological sciences.

Laboratory practices of hybridoma technology have been reviewed extensively (Kennett et al., 1980; Milstein and Lenox, 1980; Galfre and Milstein, 1981; McMichael and Fabre, 1982). The protocols for developing mABs can be carried out in laboratories that routinely use cell culture and immunological techniques (Fig. 2.5). A brief summary of the procedures involved in hybridoma technology follows.

1. Immunization. The first step involves the immunization of BALB/c mice, the mouse strain that myeloma cells were originally isolated from, with the immunogen.
2. Cell fusion. After a period of immunization, the spleen cells are released into suspension along with myeloma cells grown in culture media. The mixture of cells is pelleted, and the supernatants are removed. Polyethylene glycol (PEG) is introduced slowly under controlled temperature conditions. PEG, previously identified as a reagent for plant protoplast fusion, also promotes membrane fusion of mammalian cells via a mechanism that remains unknown. After only minutes of incubation with PEG, the fusant is replaced with culture medium containing HAT (hypoxanthine, aminopterin, and thymidine), and the cells are distributed into microplates. Since the myeloma cell line has been previously selected to be HAT-sensitive and the splenocytes cannot be propagated in culture, the only cells that can grow in this selective medium are hybridomas.
3. Screening. After colonies of hybridomas develop in microplate wells, aliquots of media from those wells are removed to test for the presence of antibody and its affinity to the antigen. Indirect ELISA (Voller et al., 1979) and solid-phase RIA (Kennett et al., 1980) are techniques that are employed frequently for such screening.
4. Cloning. The hybridomas that test positive in the screening procedure must then be isolated and cloned by limiting dilution or other cloning methods (Kennett et al., 1980). This step ensures that a hybridoma cell line is derived from a single cell and thus guarantees genetic homogeneity.
5. Mass propagation. The production of large amounts of mABs can be achieved by either mass culture or injection of hybridomas into the intraperitoneal cavity of BALB/c mice to induce ascites production. Given all the modern developments in instrumentation for concentrating and purifying mABs, researchers can choose either approach, depending on their laboratory budget and experimental needs.

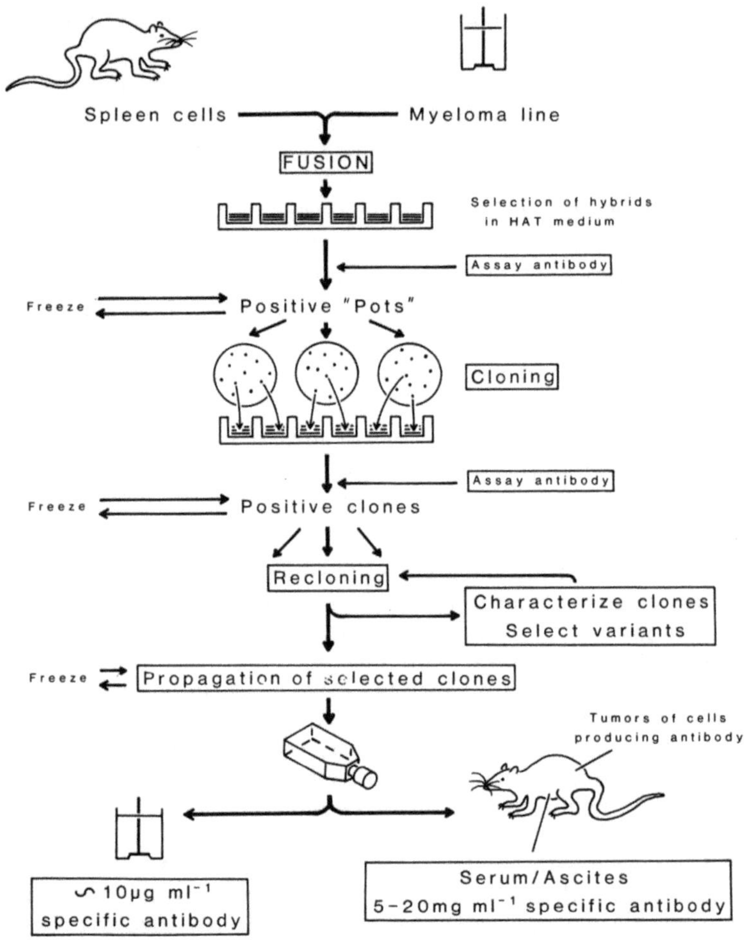

Figure 2.5. Procedure involved in the production of monoclonal ABs. (Redrawn from Galfre and Milstein, 1981.)

In insect physiology and biochemistry, we often encounter research problems that require the specific monitoring and quantification of a compound. If an antiserum can be developed to that compound, we can employ methods such as RIA or ELISA for quantification, and immunocytochemistry to localize the antigen in tissue sections (see Chapter 4). Unfortunately, the identification, isolation, and purification of an Ag are not always as straightforward as for the gypsy moth Vg previously described.

In the course of our efforts to develop an ELISA for quantifying the Vg or yolk polypeptides of small insects such as *Aedes aegypti* and *Drosophila melanogaster*, we experienced considerable difficulty in obtaining a high-purity immunogen. Since one of the advantages of hybridoma technology is that specific ABs can be developed from impure immunogens, we proceeded to generate monoclonal antibodies to Vg for both species with the hope of obtaining specific reagents for monitoring the dynamics of these proteins at different stages of ovarian development. Some of the data will be presented in the next section to illustrate the application of mABs for quantifying a protein Ag. We will also discuss a strategy to obtain mABs for insect neurohormones which presents an interesting challenge, because they are usually present in low concentrations in biological fluids. Another case study will be presented in which mABs are used as markers to monitor the process of spermatophore assembly.

2. Endocrine Control of Dipteran Vitellogenesis

Dipterans are known to have extremely intricate hormonal regulation of vitellogenesis (Hagedorn, 1985). *Aedes aegypti* and *Drosophila melanogaster* are two species that have been studied intensively. The yellow fever mosquito has been used as a model for the reproductive biology of mosquitoes that are important disease vectors, while the fruit fly offers great potential for studying the genetic aspects of vitellogenesis. Despite an extensive background in morphological and endocrine aspects of these two insect species, numerous control pathways in the scheme of hormonal regulation have remained controversial (Postlethwait and Jowett, 1981; Fuchs and Kang, 1981; Hagedorn, 1985; Bownes, 1986). Most researchers employed the radioactive amino acid incorporation method to monitor the synthetic rates of Vg (Hagedorn, 1985). However, an analytic method like the ELISA developed for gypsy moth Vg was needed for studies that require accurate quantification of Vg.

A. In Vivo Studies

Aedes aegypti and *Drosophila melanogaster* are very small insects compared to the gypsy moth. To measure Vg in individual hemolymph samples, all the steps in the ELISA protocol must be optimized. Attempts were made to incorporate the use of mABs in an ELISA for the detection of Vg in both insects (Ma et al., 1986; Wu and Ma, 1986). Initially, a panel of mABs was developed to the soluble yolk proteins of *Aedes aegypti*. Hybridoma clones that produced mABs specific to Vg were identified by indirect ELISA, Western blotting and subsequent immunohistochemical staining procedures (Towbin et al., 1979). Vg and vitellin of *A. aegypti* consist of a large subunit of about 200 kD and a small one of about 66 kD (Fig. 2.6). Monoclonal antibodies that specifically bind to those polypeptides were mass-produced to be used as coating ABs in an indirect double-AB sandwich ELISA procedure (Voller et al., 1979; Ma et al., 1986).

A specific rabbit polyclonal AB directed at Vg was used as the secondary AB, and alkaline phosphatase-labeled goat anti-rabbit conjugate (Kirkegaard

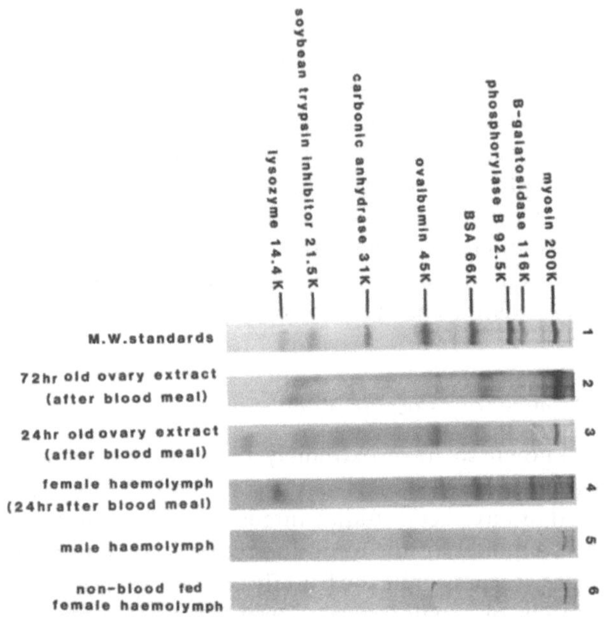

Figure 2.6. SDS-PAGE gel analysis of *Aedes aegypti* ovarian and hemolymph proteins. Track 1, molecular weight standard. Track 2, ovarian protein 72 h after blood meal. Track 3, ovarian protein 24 h after blood meal. Track 4, pooled hemolymph sample from five females after blood meal. Track 5, pooled hemolymph sample from five males. Track 6, pooled hemolymph sample from five nonblood-fed females.

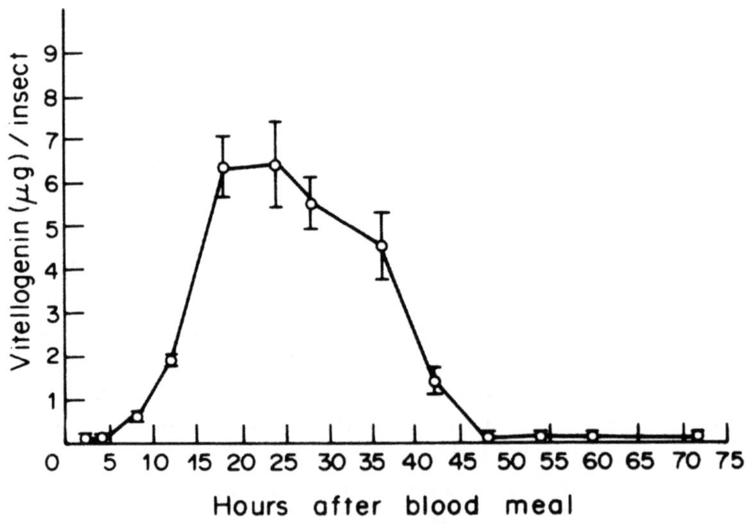

Figure 2.7. Hemolymph Vg levels of *Aedes aegypti* at different times after a blood meal. Each point represent the mean of 10–15 females, and bars represent standard error.

and Perry Laboratories) was introduced as the tertiary AB. The sensitivity of this indirect double-AB sandwich allowed us to analyze the hemolymph Vg level of individual insects at different times after blood feeding (Fig. 2.7). Vg was found to appear in the hemolymph at about 5 h and reached its peak at 24 h. The concentration dropped to about 1 μg/ml around 42 h and became virtually undetectable after 50 h.

Another version of the ELISA was developed by Wu and Ma (1986) for the quantification of hemolymph Vg in an insect even smaller than the mosquito, the *Drosophila* female. Monoclonal antibodies that are specific to the *Drosophila* Vg, which consists of three yolk polypeptides (YP1, YP2, YP3), were used as primary coating ABs (Fig. 2.8). An enzyme-AB conjugate was prepared by coupling alkaline phosphastase to an affinity-purified rabbit polyclonal AB directed at the Ag. Hemolymph levels of YP were determined

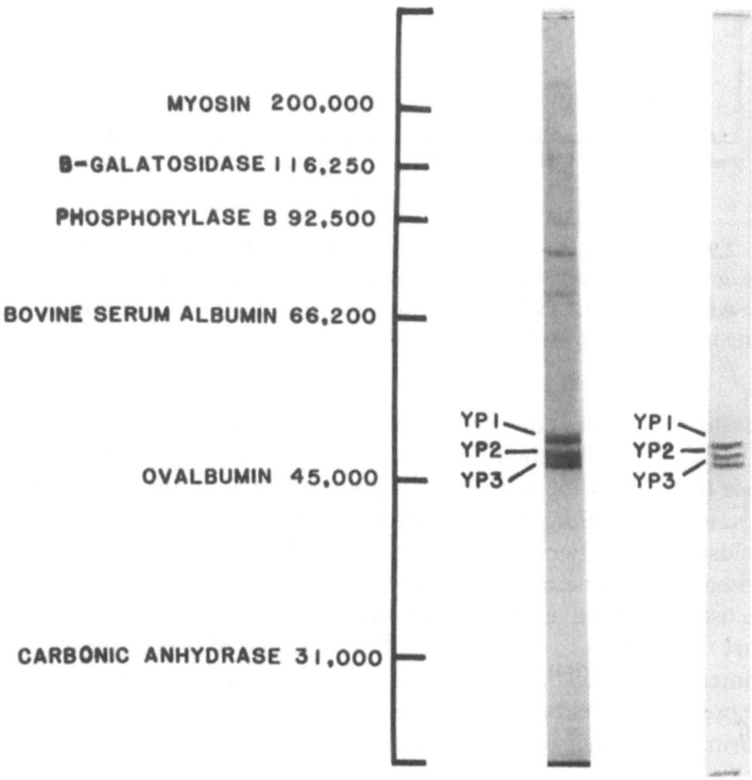

Figure 2.8. Three major yolk polypeptides of *D. melanogaster*. (A) 8% SDS-PAGE analysis of the partially purified soluble yolk proteins with molecular weight standards indicated on the left. (B) Western blot pattern produced after protein transfer to nitrocellulose paper and incubation with female-specific rabbit polyclonal antibodies (or monoclonal antibodies).

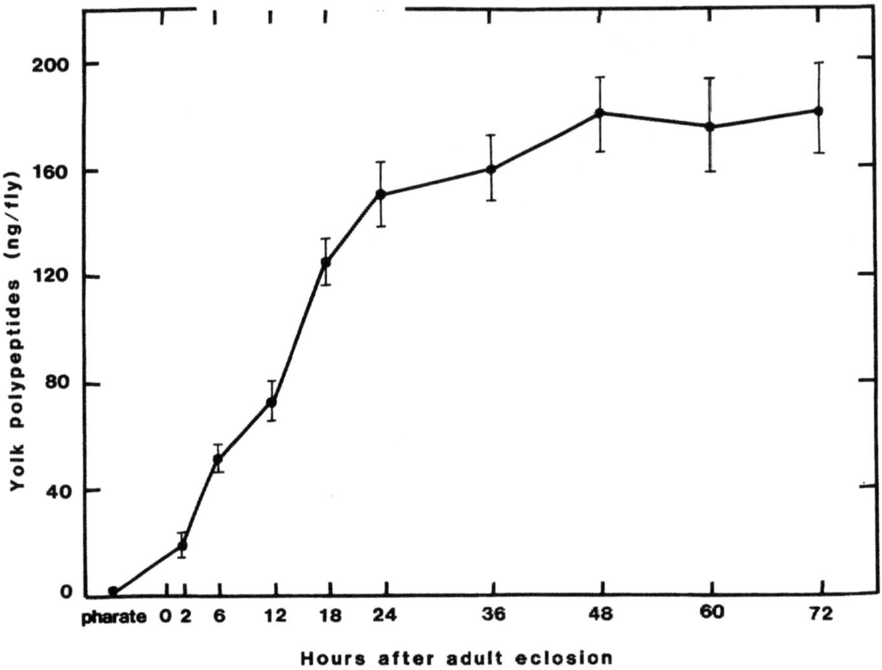

Figure 2.9. Yolk polypeptide concentration in the hemolymph of individual *D. melanogaster* females collected at different times after adult eclosion measured by the double-AB sandwich ELISA. Each point represents the mean of at least five insects, and bars represent standard error around the mean.

by measuring samples collected from each female at different times posteclosion (Fig. 2.9). Within the first day, the YP concentration exhibited a rapid increase from 20 to 150 ng per fly, beginning 2 h after emergence. The peak was found to be 180 ng per fly at 48 h, and it stayed at that level up to 72 h posteclosion. Hemolymph samples, obtained from pharate males and females processed in the same manner, were found to have no detectable YPs.

The use of a sensitive ELISA with mABs as the coating AB has enabled us to chart the changes in Vg titers at different stages of development. In future experiments, the effects of chemicals, microsurgery, and physiological stress on the process of vitellogenesis can be determined, because any deviation from the norm can be easily detected with ELISA.

B. In Vitro Studies

Another application of ELISA in Vg research is the monitoring of fat-body secretory activity in vitro. Organ culture is extremely useful in endocrinological research for demonstrating the direct relationship between a hormone and its target tissue at various stages of readiness or competence. In mosquitoes, fat body can be dissected and cultured in medium (Hagedorn and Judson, 1972), and synthetic hormones or biological extracts can be administered precisely into the medium. The production of Vg was monitored at different times by taking aliquots of culture medium for ELISA analysis. This approach was particularly useful in investigating the response of fat body to 20-hydroxyecdysone (20-HE) which is a subject of controversy in dipteran vitellogenesis research (Fuchs and Kang, 1981). Our studies have shown conclusively that cultured fat body from 3-day-old non-blood-fed females can be activated by a physiological dose of 20-HE (Ma et al., 1986). After an initial period of hormonal stimulation, the fat body Vg secretory activity rises and declines with a specific time course when incubated in either hormone-supplemented or hormone-free medium (Fig. 2.10). It is interesting to note that the temporal changes of secretory activity of fat body in vitro coincides with the hemolymph Vg titer in vivo. The patterns of response have led us to believe that the production of Vg in response to 20-HE is programmed.

In a similar study, *Drosophila melanogaster* fat body was incubated with physiological amounts of 20-HE to establish the hormone and target tissue relationship. We found that 20-HE stimulates the fat body to produce YPs in

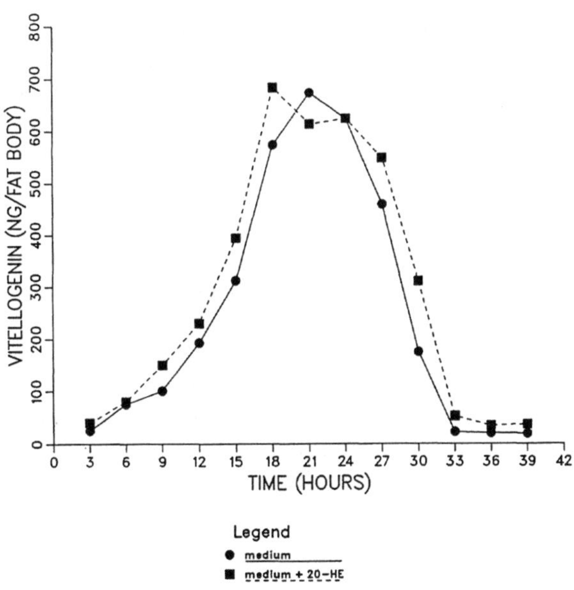

Figure 2.10. Time course response of in vitro fat body of non-blood-fed female *Aedes aegypti* to 10^{-6} M of 20-HE. After 5 h of incubation with 20-HE, one group of five fat bodies was cultured in 50 μl of medium in the presence of hormone and another group in hormone-free medium. Each point represents the mean of three groups of five fat bodies.

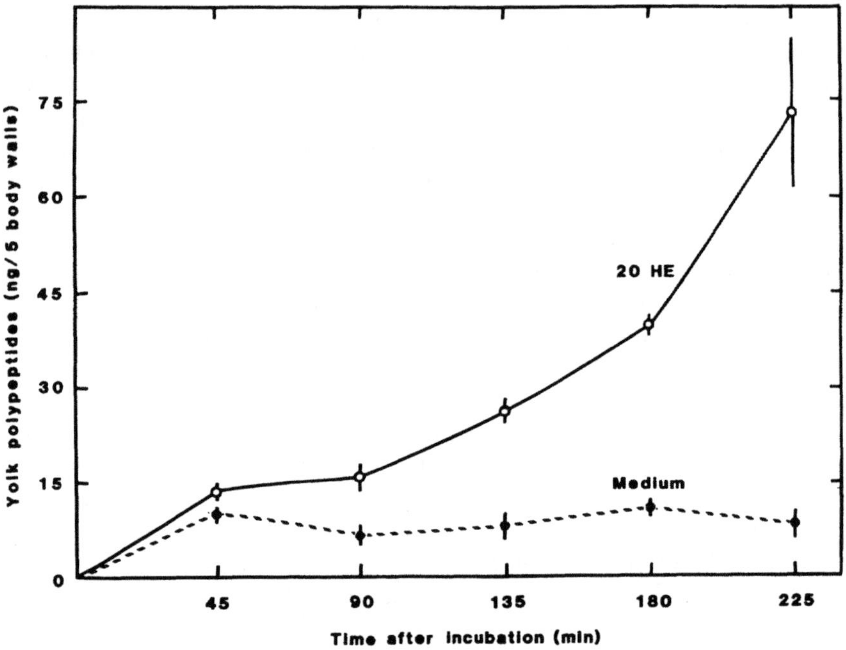

Figure 2.11. Time course of the response of in vitro fat body of *Drosphila melanogaster* to 10^{-6}M 20-HE. A group of five mesenteric preparations in 50 μl medium with or without hormone as control respectively. Each point represents the mean of three groups of five preparations from decapitated nonvitellogenic females. The error bars represent the standard error of the mean.

the absence of other endogenous factors (Fig. 2.11). Although the fat body is small in these insects, five pairs of fat bodies per preparation can produce Vg well within the range of detectability. This is the first direct evidence on the role of 20-HE in *Drosophila* vitellogenesis. Using this method, it will be possible to investigate how other factors are involved in preparing the fat body and ovarian cells for the production of yolk polypeptides. Furthermore, we believe that this method can be adopted quite easily for studying the in vitro secretory behavior of other insect cells.

3. Localization of Insect Brain Peptides

The role of neurohormones such as prothoracicotropins and allatatropins has been well documented (Doane, 1973; Gilbert et al., 1981; Raabe, 1982; Bollenbacher and Granger, 1985). Despite the fact that insects have provided excellent systems for in vivo as well as in vitro studies of biologically active factors, research on insect neurohormones lags behind similar studies of mammalian systems. One of the major difficulties has been the inability to obtain enough purified material for chemical characterization. Painstaking tasks to purify thousands of carefully dissected insect brains are often diminished by activity loss during chromatographic separation procedures. Another difficulty associated with purification is that bioassays used to monitor active fractions can be extremely time-consuming and complex (Tobe and Pratt, 1975; Hagedorn et al., 1975).

The distribution of neurosecretory sites of synthesis and release has been described in the central nervous system of numerous insect species on the basis of histological staining of neurosecretory cells (NSCs) by either paraldehyde fuchsin or chrome hematoxylin (Panov, 1980; Rowell, 1976). These stains, however, do not provide information on the nature of the products of the neurosecretory cells. To localize the origin of a particular neurohormone, a specific histochemical probe is required for identifying the synthesis, translocation, and release sites of the hormone. In recent years, ABs raised against vertebrate neurohormones have been found to produce specific staining patterns in insect secretory cells (Proux et al., 1982; El-Salhy et al., 1983). We have to be cautious in interpreting these findings, because positive immunoreactivity may only point to structural homology rather than functional similarities. Although many of the results are intriguing, ABs to vertebrate neuropeptides have limited applications in insect neurohormone studies unless accompanied by conventional biochemical analysis and bioassay. Unfortunately, there has been little progress on the development of specific ABs to insect neurohormones such as eclosion hormone or prothoracicotropin (see Chapter 8).

Since it is very difficult to obtain sufficient amounts (in the microgram range) of highly purified hormone for the production of a polyclonal AB, hybridoma technology is the logical approach, because specific mABs can be obtained with only partially purified immunogen. Monoclonal antibodies have many potential applications in insect neurohormone research: (1) A quantification method such as an immunoassay (ELISA/RIA) can be developed to obtain hormone titers at different developmental stages. (2) Specific AB probes can be used to monitor the distribution of the hormone in the brain with immunocytochemical techniques (see Chapter 4). This would be especially useful for studies of insect neuronal development where, owing to the dramatic changes in gross morphology of the insect CNS during metamorphosis, mABs would be invaluable as cell markers. (3) Affinity chromatography with mABs as ligands can be used for the selective purification or concentration of a neurohormone in biological extracts containing a complex mixture of proteins.

The problems with purification of insect neurohormones dictate a special strategy for mAB development. The procedures are outlined in a comprehensive flow chart in Figure 2.12. After immunization and cell fusion procedures are carried out, screening with indirect ELISA or RIA using crude Ag is

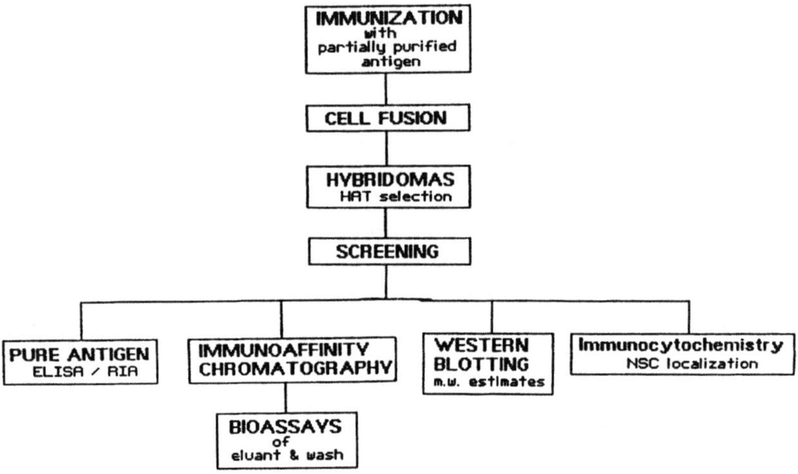

Figure 2.12. A comprehensive flow chart on the development of mABs to insect neurohormones.

commonly employed, since highly purifed Ag is usually not available or is difficult to obtain.

Once ELISA-positive clones have been identified, it is necessary to single out those clones that produce ABs to the neurohormone and not to other contaminating proteins. Initial screening procedures must therefore be followed by a battery of techniques that when used in combination, can ensure specificity. These techniques include (1) immunocytochemistry, (2) Western blotting, and (3) immunoaffinity chromatography coupled with bioassay. We have found it profitable to first screen and characterize ABs with the less time-consuming procedures of immunocytochemistry and Western blotting, and then to perform the final characterization using immunoaffinity chromatography coupled with bioassay.

One alternative in the screening path is to characterize mABs with immunocytochemical techniques which allow the localization of the Ag. Caution is advised when drawing conclusions from immunocytochemical screening, because the lack of staining in areas of expected cross-reactivity does not always indicate a lack of response to Ag, since the antigen may be sensitive to fixation procedures. Moreover, a positive response even in cells that are thought to be the site of synthesis or release may be inconclusive, because neurosecretory cells may contain other products which have structural homology to the neurohormone or which are present in the hormone preparation used in the immunization procedures. However, a localized specific staining pattern on tissue sections can give the first indication of a possible hormone-specific AB.

Information on the molecular weight of the protein bound by mABs can be revealed by gel electrophoretic separation of a partially purified Ag followed by Western blotting (Towbin et al., 1979). Incubations are performed with mABs from the ELISA-positive clones as primary ABs. This technique provides the molecular weight of proteins bound by the ABs, but one must bear in mind that this technique can only detect polypeptides within the molecular weight range of the electrophoretic separation. Although not conclusive, AB staining of a polypeptide with a molecular weight similar to that of a neurohormone can indicate a possible hormone-specific AB.

A final characterization of the AB can be performed using immunoaffinity chromatography coupled with a bioassay to correlate immunoreactivity to bioactivity of the neurohormone. A bioactive sample, prepared from brain extracts, is allowed to pass through a column prepared by coupling the mAB from one clone to CNBr-activated Sepharose. The wash and subsequent elution are monitored separately for bioactivity. If the mAB is directed against the neurohormone, the bioactive molecules retained by the column support can be dissociated by a high-salt or low-pH buffer. The bioassay activity should be detected only in the eluant and not the initial wash. Although this approach can yield conclusive results, it is too time-consuming for the screening of a large number of clones. In screening for mABs to neurohormones, it is our belief that all the procedures previously discussed are necessary to ensure specificity.

This screening strategy was used to characterize mABs against partially purified *Manduca* head extract that contained 4,400-fold purified prothoracicotropin (PTTH) (Kingan, 1982). To illustrate our screening strategy, we will describe the specific screening steps used in this investigation and interpret their results. After the initial screening for clones with ELISA, we used mABs that were found to cross-react in immunocytochemistry.

A description of the immunocytochemical technique used for these studies follows. Tissue sections were fixed in PBS containing 2% paraformaldehyde and 0.15% picric acid prior to sectioning on a cryostat. After brief postfixation, the sections were incubated with mABs. Subsequent development of tissue sections with an immunoperoxidase technique involving a secondary biotinylated AB and avidin (Hsu et al., 1981) with diaminobenzidine (DAB) as the substrate yielded a highly sensitive brown stain. By using a blue counterstain (toluidine blue), we could increase the contrast of the immunostain. This technique, based on a avidin biotin complex (ABC), is somewhat more sensitive than immunofluorescence and yields permanently stained tissue sections.

In the brain of *Manduca sexta* pupae, immunoreactive staining was localized in several anteromedial neurosecretory cells (NSCs) (Fig. 2.13a), which correspond to paraldehyde fuchsin-positive (PF) median NSCs found in the last larval instar (Bhaskaran and Jones, 1980). Staining patterns reveal differing amounts of immunoreactivity in these cells. Similar observations by Hiruma and Agui (1977) in *Mamestra brassicae* led to the distinction of two cell types in nondiapausing pupae, NSC I and NSC II, with the latter con-

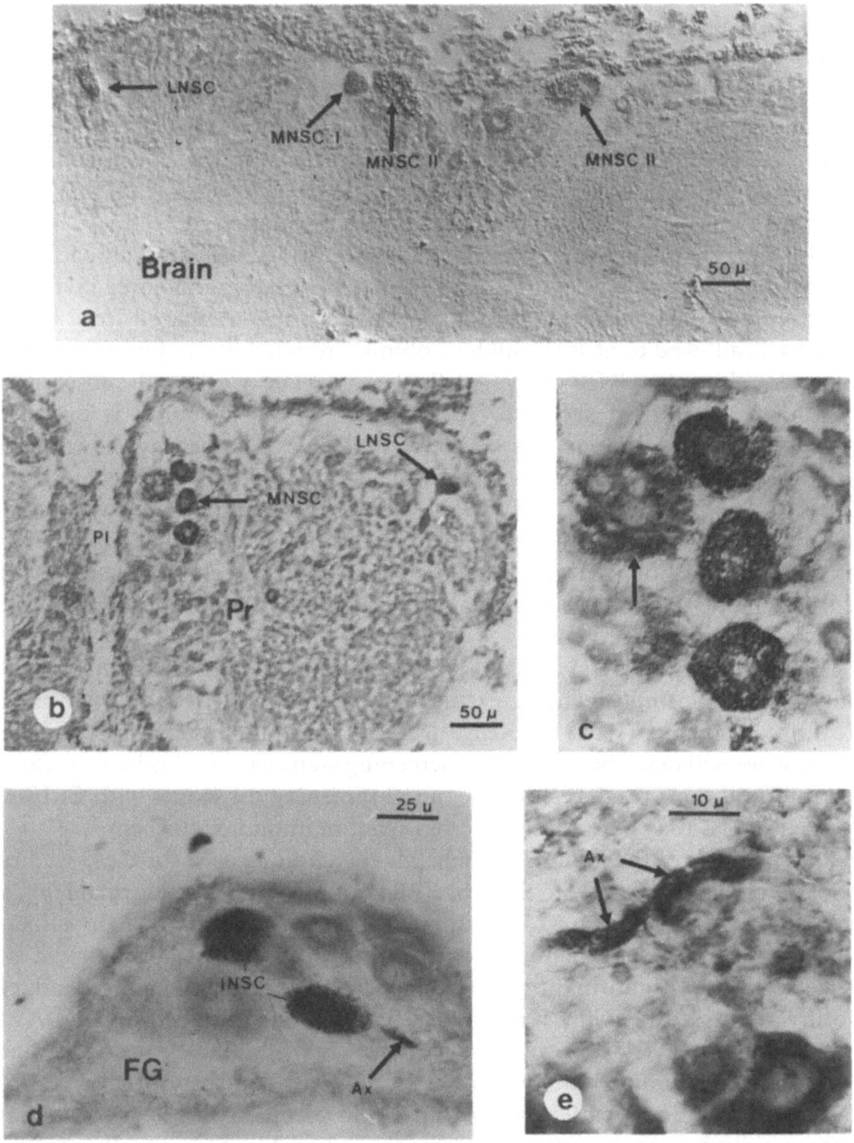

Figure 2.13. Micrographs of AB-positive neurosecretory cells in frozen sections of brain and retrocerebral complex of *Manduca sexta* pupa. (a) Cross section of brain showing selective staining of two different intensities of stain (MNSC I and MNSC II) and lateral cells (LNSC). (b) Longitudinal section of protocerebrum (Pr) with NSCs that contain immunoreactivity. (c) Higher magnification of a portion of (b) illustrates the granular structure of the immunoreactive stain in MNSC. Arrow indicates a cell with less immunoreactivity. (d) Frontal ganglion (FG) of the last larval instar with two immunoreactive intrinsic neurosecretory cells (INSC). Ax: Axonal process containing neurosecretion. (e) Axons (Ax) with immunoreactivity in frontal ganglion (FG) of 3-day-old pupa projecting to the recurrent nerve (RN). INSC on bottom of micrograph is counterstained with toluidine blue.

taining more PF stain. These authors attributed control of molting to median NSC II. As shown in Figure 2.13a, one lateral NSC was also stained when incubated with mAB 01D9. In *Manduca sexta*, one of two lateral NSC in each hemisphere were found to contain PTTH activity (Agui et al., 1979). Further characterization of the mAB with the electrophoretic method, however, revealed specific cross-reactivity to a single protein (36 kD) in brain homogenates when assayed with the Western blot technique (Fig. 2.14). PTTH in *Manduca sexta* has been reported to have a big form with a molecular weight ranging from 25 (Kingan, 1982) to 29 kD (Bollenbacher et al., 1984). These findings suggest that the mAB is not directed against PTTH, as was indicated by the staining pattern in pupal brain sections.

Findings in *Manduca sexta* larval brains, however, open up intriguing possibilities for the nature of the immunoreactive protein. Brain preparations from 4-day-old fifth-instar larvae contain very little or no immunoreactivity in the median NSC and lateral NSC. This corresponds to findings in 2- to 5-day-old fifth-instar larvae that report no or very little PF-positive neurosecretory material in NSC (Bhaskaran and Jones, 1980). Significant and strong immunoreactvity in brains of the last larval instar was found only in two NSCs of the frontal ganglion (Fig. 2.13d), first identified by Bell et al. (1974) as intrinsic PF-positive NSC. As yet no PTTH activity has been linked to the frontal ganglion. Although these findings confirm the dissimilarity of the 01D9-immunoreactive protein from PTTH, these results gain potential significance, since it may soon be possible to attribute a function to these two NSCs.

In comparative immunohistochemcial studies, we applied mAB 01D9 to a dipteran species with a somewhat different neurohormonal organization. Localization of identified neurohormones in mosquitoes has not been reported, but an ecdysiotropic hormone that controls egg development has been attributed to the (posterior) median NSC through ablation experiments (Lea, 1967; 1972). A factor from adult *Aedes aegypti* with a molecular weight different from the *M. sexta* Ag (Fig. 2.14) was found to cross-react strongly with mAB 01D9 in tissue sections.

Axons containing immunoreactivity and having a possible origin in the subesophageal ganglion (Fig. 2.15d) project via the circumesophageal connectives to the protocerebrum (Fig. 2.15b, c). Immunoreactive arborization within the neuropile (Fig. 2.15a) was found in the vicinity of the posterior NSC, which contained no immunoreactivity. The origin of accumulated granules with immunoreactivity around the gut and in the subesophageal ganglion (SOG) was traced to cells located in the anterior part of the SOG (Fig. 2.15d). The different immunocytochemical staining pattern in the mosquito cerebral neurosecretory system may be based on a different functional organization of the cephalic ganglia-neurohemal organ complex, as has been shown by several authors (Meola and Lea, 1972; Normann, 1983).

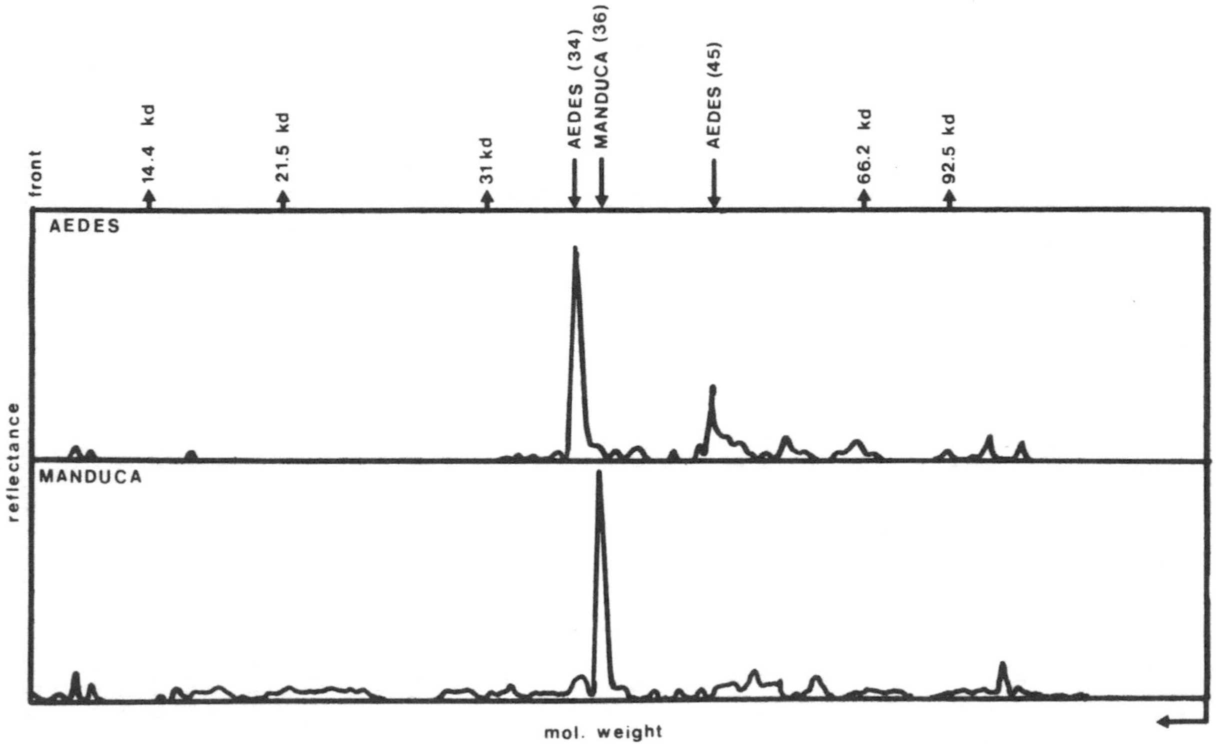

Figure 2.14. Electropherogram of boiled pupal brain extract from *Manduca sexta* and of boiled extract from *Aedes aegypti* adult heads. Brain extracts were separated on 15% SDS-PAGE, transferred on nitrocellulose paper, and subsequently incubated with mAB against partially purified *Manduca* brain extract. Immunoreactivity was detected with an avidin-biotin peroxidase complex that reacted with diaminobenzidine (DAB) as substrate. Diagram illustrates immunoreactivity at 36 kD (*Manduca*) and 34 KD (*Aedes*). The AB also reacted with a minor protein band of *Aedes* at 45 KD. The blot was scanned with a Videolab (Biorad) densitometer in the reflective mode.

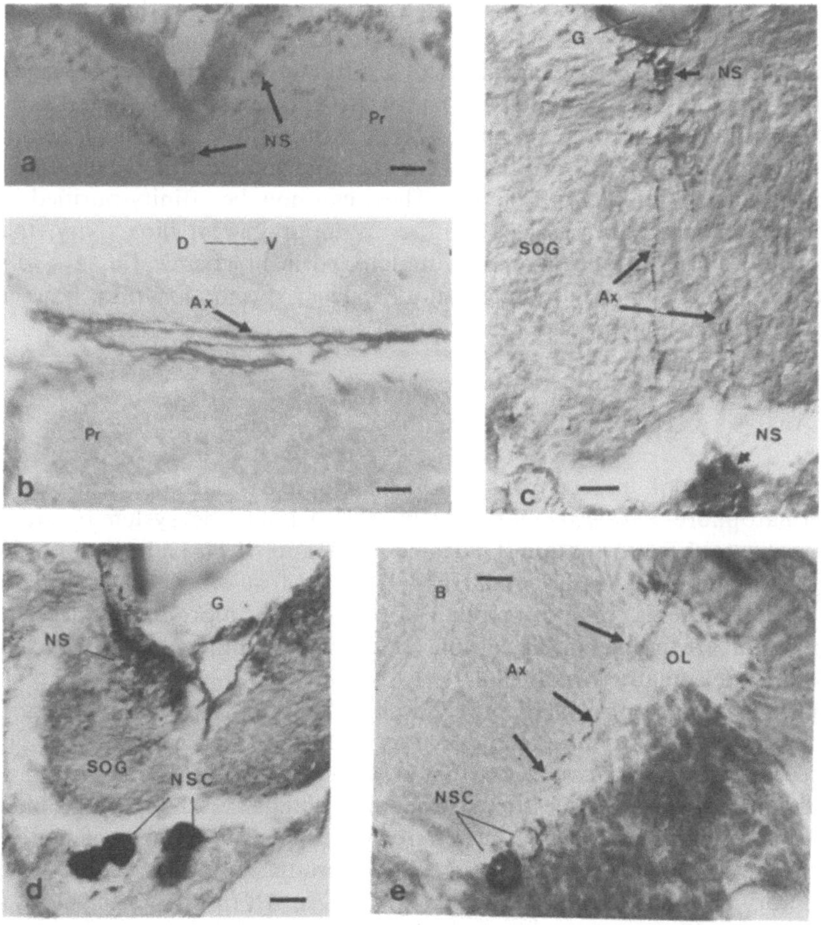

Figure 2.15. Micrographs of AB-positive stain in frozen brain sections of adult *Aedes aegypti.* (a) Immunoreactivity in arborizations in the protocerebrum (Pr). (b) Antibody-positive axons shown in (a) can be followed along the midline of the pars intercerebralis. (c) These immunoreactive axons originate in the subesophageal ganglion (SOG). (d) Accumulated immunoreactive material is found in the cell bodies of the SOG as well as the arborized axons. (e) Two cells and their axonal processes on each side of the brain in close vicinity to the optic lobe (OL) are the only cells that stained with another mAB (02B10). Axonal processes that lead to the lateral protocerebrum are not shown. Bar is 10 μm.

Immunocytochemistry as a method for screening antigens that are present in small quantities was demonstrated in an additional example. One pair of neurosecretory cells in each hemisphere of adult mosquito brain cross-reacted with ABs produced by another cell line (02B10). Located in the lateral part of the brain, these cells project axons with neurosecretory material to the optic lobes (Fig. 2.15e). This screening technique thus enables the localization of factors with yet unknown function. These can now be affinity-purified and assayed for their physiological nature. Without known bioactivity, these putative neurohormones would go undetected in biochemical assays, since they are present in only minute amounts. Their specific location, however, suggests a role in neuroendocrine control.

4. Analysis of Spermatophore Assembly

Many species of insects transfer semen within a gelatinous capsule, the spermatophore. *Tenebrio molitor* had been used as a model system for studies of spermatophore formation (Gadzama and Happ, 1974; Dailey et al., 1980; Happ et al., 1982; Dailey and Happ, 1983). Most of the spermatophore structural proteins come from a pair of bean-shaped reproductive accessory glands and are molded into a capsule containing a multilayered wall and core within the ejaculatory duct.

Grimes and Happ (1986) have developed mABs as specific probes for monitoring the biosynthesis and export of these structural proteins. They succeeded in isolating two clones that produced ABs to two proteins found in both the bean-shaped accessory reproductive glands and the spermatophore. With immunohistochemistry, they were able to localize the sites of synthesis within these structures. One clone, PL 3.4, produced mABs specific for the secretory products of cell type 3 (Fig. 2.16A). An AB from another clone, PL 6.3, was specific to the secretions of cell type 7 (Fig. 2.16B). The movements of the Ags from these two cell types were traced from the accessory glands to the prespermatophore mass in the ejaculatory duct.

Immunohistochemical examination of spermatophore sections also helped to elucidate the role of these proteins as building blocks in the outer wall and inner core of the spermatophore. The secretion product from cell type 3 can be seen at the perimeter of each secretory granule in the packed mass of the outer wall of the spermatophore (Fig. 2.16C), while those from cell type 7 constitute the innermost layer of the spermatophore wall (Fig. 2.16D). It is certainly possible that in the future, additional mABs can be used as biochemical tools for the characterization of spermatophore proteins at different stages of synthesis, transport, and assembly. The information generated with this approach will certainly contribute to the understanding of the architecture and the mechanism of spermatophore assembly in the male *Tenebrio molitor*.

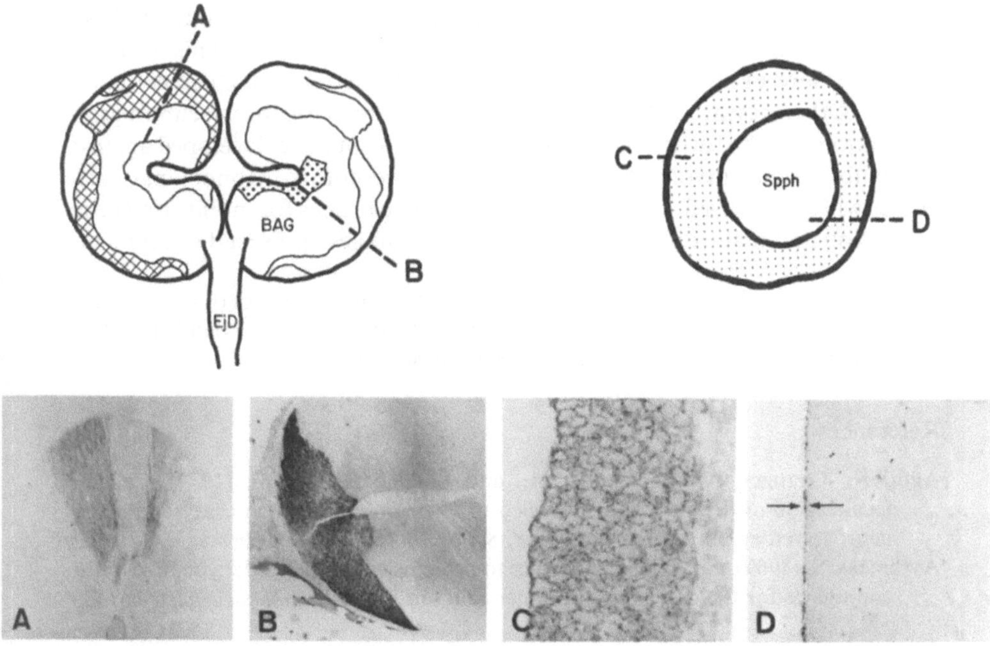

Figure 2.16. Illustration of assessory glands of *Tenebrio molitor*. On the left gland of the pair, cell type 3 is indicated by the cross-hatching. On the right gland of the pair, cell type 7 is indicated by dots. (× 30). (A) A paraffin section along the line indicated on the diagram. It has been treated with mAB PL 3.4, which binds only to the secretory products of cell type 3 and not to surrounding cells. (B) A paraffin section along the line indicated in the diagram. This section has been treated with PL 6.3, an mAB specific for the secretions on cell type 7 (× 200). The diagram on the right shows the spermaophore in cross section. The thick wall has been cut for electron microscopy at C and D. (× 1200). (C) An oblique Epon section along the line indicated, that has been treated with PL 3.4 AB. The Ag from cell type 3 can be seen at the perimeter of each secretory granule in the packed mass at the outside of the spermatophore. (× 10,000). (D) Epon section along the line indicated, that has been treated with PL 6.3 AB. The innermost layer of the wall of the spermatophore (between arrows) contains Ag from cell type 7. (× 10,000). Paraffin sections were treated successively with primary AB, rabbit anti-mouse IgG conjugated with peroxidase, and diaminobenzidine-hydrogen peroxide. Vectastain was used with the Epon sections for electron microscopy. (Courtesy of Drs. G. Happ and K. Grimes)

III. Conclusion

It is evident that ELISAs with polyclonal and monoclonal ABs are invaluable for quantitative assessment of physiologic processes in individual insects and organ cultures. Monoclonal antibodies represent a new class of standardized analytic reagents for linking biochemical characteristics to morphologic and functional properties. We should see an increase in application of these technologies in insect physiology and endocrinology research. This is only a small measure of the excitement provided by these immunochemical techniques. Other research directions could evolve with further technological innovations. In the future, antiidiotype antibodies could be developed against mABs for studies involving Vg and hormonal receptors. Monoclonal antibodies will also be useful in obtaining specific gene probes for monitoring gene expression and tissue differentiation in different stages of insect development.

Acknowledgment. The research was supported by National Institutes of Health grant AI21227; College of Agriculture and Life Sciences, University of Maryland at College Park; and the Maryland Agricultural Experiment Station.

References

Agui, N., Granger, N.A., Gilbert, L.I., and Bollenbacher, W.E. (1979). Cellular localization of the insect prothoracicotropic hormone: in vitro assay of a single neurosecretory cell. *Proc. Natl. Acad. Sci. USA* **76**:5694–5698.

Avrameas, S. (1969). Coupling of enzymes to proteins with glutaraldehyde. Use of the conjugates for the detection of antigen and antibodies. *Immunochemistry* **6**:43–52.

Bell, R.A., Borg, T.K., and Ittycheria, P.I. (1974). Neurosecretory cells in the frontal ganglion of the tobacco hornworm, *Manduca sexta. J. Insect Physiol.* **20**:669–678.

Bhaskaran, G., and Jones, G. (1980) Neuroendocrine regulation of corpus allatum activity in *Manduca sexta.* The endocrine basis for starvation-induced supernumerary larval moult. *J. Insect Physiol.* **26**:431–440.

Bollenbacher, W.E., and Granger, N.A. (1985). Endocrinology of the prothoracicotropic hormone. In: *Comprehensive Insect Physiology, Biochemistry and Pharmacology*, Vol. 7; *Endocrinology I* G.A. Kerkut and L.I. Gilbert (Eds). New York: Pergamon Press, pp. 109–151.

Bollenbacher, W.E., Katahira, E.J., O'Brian, M.A., et al. (1984). Insect prothoracicotropic hormone: Evidence for two molecular forms. *Science* **224**:1243–1245.

Borst, D.W., and O'Connor, J.D. (1972). Arthropod molting hormone: Radioimmune assay. *Science* **178**:418–419.

Bownes, M. (1982). Hormonal and genetic regulation of vitellogenesis in *Drosophila. Q. Rev. Biol.* **57**:247–274.

Bownes, M. (1986). Expression of the gene coding for vitellogenin (yolk protein). *Annu. Rev. Entomol.* **31**:507–531.

Dailey, P.J., Gadzama, N.M., and Happ, G.M. (1980). Cytodifferentiation in the accessory glands of *Tenebrio molitor*. VI. A congruent map of cells and their secretions in the layered elastic product of the male bean-shaped gland. *J. Morphol.* **166**:289–322.

Dailey, P.J., and Happ, G.M. (1983). Cytodifferentiation in the accessory glands of *Tenebrio molitor*. XI. Transitional cell types during establishment of pattern. *J. Morphol.* **178**:139–154.

Doane, W.W. (1973). Role of hormones in insect development. In: *Developmental Systems: Insects* S.J. Counce and C.H. Waddington (Eds.). New York: Academic Press, Vol. 2, pp. 291–497.

El-Salhy, M., Falkmer, S., Kramer, K.J., and Speirs, R.D. (1983). Immunohistochemical investigations of neuropeptides in the brain, corpora cardiaca and corpora allata of an adult lepidoteran insect, *Manduca sexta* (L.) *Cell Tissue. Res.* **232**:295–317.

Engelmann, F. (1970). *The Physiology of Insect Reproduction.* Oxford: Pergamon Press.

Engelmann, F. (1979). Insect vitellogenin identification, biosynthesis and role in vitellogenesis. In: *Advances in Insect Physiology*, Vol 14. J.E. Treherne, M.J. Berridge, and V.W. Wigglesworth (Eds.). London: Academic Press.

Engvall, E., and Perlmann, P. (1971). Enzyme-linked immunosorbent assay (ELISA). Quantitative assay of Immunoglobulin G. *Immunochemistry* **8**:871–874.

Ephrussi, B. (1972). *Hybridization of Somatic Cells.* Princeton, NJ: Princeton University Press.

Fuchs, M.S., and Kang, S.H. (1981). Edysone and mosquito vitellogenesis: A critical appraisal. *Insect Biochem.* **11**:627–633.

Gadzama, N.M., and Happ, G.M. (1974). The structure and evaluation of the spermatophore of *Tenebrio molitor* L. (Coleoptera: Tenebrionidae). *Tissue Cell* **6**:95–108.

Galfre, G., and Milstein, C. (1981). Preparation of monoclonal antibodies: Strategies and procedures. *Methods Enzymol.* **73B**:3–46.

Gilbert, L.I., Bollenbacher, W.E., Agui, N., et al. (1981). The prothoracicotropes: Source of the prothoracicotropic hormone. *Am. Zool.* **21**:641–653.

Grimes, K.A., and Happ, G.M. (1986). A monoclonal antibody against a structural protein in the spermatophore of *Tenebrio molitor* (Coleoptera). *Insect Biochem.* **16**:635–643.

Hagedorn, H.H. (1985). The role of ecdysteroids in reproduction. In: *Comprehensive Insect Physiology, Biochemistry and Pharmacology* G.A. Kerkut and L.I. Gilbert (Eds.). Oxford: Pergamon Press, pp. 205–262.

Hagedorn, H.H., and Judson, C.L. (1972). Purification and synthesis of *Aedes aegypti* yolk proteins. *J. Exp. Zool.* **182**:367–377.

Hagedorn, H.H., O'Connor, J.D., Fuchs, M.S., Sage, B., Schlaeger, D.A., and Bohm, M.K. (1975). The ovary as a source of α-ecdysone in an adult mosquito. *Proc. Natl. Acad. Sci. USA* **72**:3255–3259.

Happ, G.M., Yuncker, C. and Dailey, P.J. (1982). Cytodifferentiation in the accessory glands of *Tenebrio moliter* VII. Patterns of leucine incorporation by the bean-shaped glands of males. *J. Exp. Zool.* **220**:82–91.

Harris, H. (1970). Cell Fusion. Harvard University Press, Cambridge, Mass.

Harris, H., and Watkins, J.F. (1965). Hybrid cells derived from mouse and man: Artificial heterokaryons of mammalian cells from different species. *Nature* **205**:640.

Hiruma, K., and Agui, N. (1977). Relationship between histological changes and function of neurosecretory cells in the brain of the cabbage armyworm *Mamestra brassicae* L. *Appl. Entomol. Zool.* **12**:42–49.

Hsu, S.M., Raine, L., and Fanger, H. (1981). A comparative study of the peroxidase and anti-peroxidase method and an avidine-biotin complex method for studying

polypeptide hormone with radioimmunoassay antibodies. *Am. J. Clin. Pathol.* **75**:734.
Ishikawa, E., and Kato K. (1978). Ultrasensitive enzyme immunoassay. *Scand. J. Immunol.* **8** (Suppl. 7):43–55.
Johnson, D.A., Gautsch, J.W., Sportsman, J.R., and Elder, J.H. (1984). Improved technique utilizing nonfat dry milk for analysis of proteins and nucleic acids transferred to nitrocellulose. *Gene Anal. Technol.* **1**:3–8.
Kabat, E.A., and Mayer, M.M. (1961). Preparation of antigens for immunization and injection schedules. In: *Experimental Immunochemistry*, 2d Ed. Springfield, IL: Thomas, pp. 871–872.
Kato, K., Fukui, H., Hamaguchi, Y., and Ishikawa, E. (1976). Enzyme-linked immunoassay: Conjugation of the FAB′ fragment of rabbit IgG with B-d-galactosidase from *E. coli* and its use for immunoassay. *J. Immunol.* **116**:1554–1560.
Kennett, R.H., Mckearn, T.J., and Bechtol, K.B. (1980). Monoclonal antibodies. In: *Hybridoma: A New Dimension in Biological Analyses*. New York: Plenum Press.
Kingan, T.G. (1982). Purification of the prothoracicotropic hormone from the tobacco hornworm *Manduca sexta*. *Life Sci.* **28**:2585–2594.
Köhler, G., and Milstein, C. (1975). Continuous cultures of fused cells secreting antibody of predefined specificity. *Nature* **256**:495–497.
Lea, A.O. (1967). The medial neurosecretory cells and egg maturation in mosquitoes. *J. Insect Physiol.* **13**:419–429.
Lea, A.O. (1972). Regulation in egg maturation in the mosquito by the neurosecretory system: The role of the corpus cardiacum, *Gen. Comp. Endocrinol.* (Suppl.) **3**:602–608.
Ma, M., Burkholder, J.K., Webb, R.E., and Hsu, H.T. (1984a). Plastic bead ELISA: An inexpensive epidemiological tool for detecting gypsy moth (Lepidoptera: Lymantriidae) nuclear polyhedrosis virus. *J. Econ. Entomol.* **77**:537–540.
Ma, M., Newton, P.E., Gong, H., et al. (1984b). Development of monoclonal antibodies for monitoring *Aedes atropalpus* vitellogenesis. *J. Insect Physiol.* **30**:529–536.
Ma, M., Gong, H., Newton, P.E., and Borkovec, A.B. (1986). Monitoring *Aedes aegypti* vitellogenin production and uptake with hybridoma antibodies. *J. Insect Physiol.* **32**:207–213.
McMichael, A.J., and Fabre, J.W. (1982). *Monoclonal Antibodies in Clinical Medicine*. London: Academic Press, 663 pp.
Meola, S.M., and Lea, A.O. (1972). The ultrastructure of the corpus cardiacum of *Aedes sollicitans* and the histology of the cerebral neurosecretory system of mosquitoes. *Gen. Comp. Endocrinol.* **18**:210–234.
Milstein, C., and Lennox, E. (1980). In: *Current Topics in Developmental Biology: Developmental Immunology*, Vol. 14. M. Friedlander (Ed.). London: Academic Press, pp. 1–32.
Normann, T.C. (1983). Cephalic neurahemal organs in adult diptera. In: *Neurohemal Organs of Arthropods. Their Development, Evolution, Structures and Functions*. A.P. Gupta (Ed.). Springfield, IL: Charles C. Thomas.
Panov, A.A. (1980). Demonstration of neurosecretory cells in the insect nervous system. In: *Neuroanatomical Techniques*. N.J. Straussfeld and T.A. Miller (Eds.). New York: Springer-Verlag, pp. 26–51.

Postlewait, J., and Jowett, T. (1981). Regulation of vitellogenesis in *Drosophila*. In: *Regulation of Insect Development and Behavior*. F. Sehnal, A. Zabza, J.J. Meen, and B. Cymborowski (Eds.). Wroclaw, Poland: Wroclaw Technical University Press, pp. 591–627.

Proux, J., Rougon, G., and Cupo, A. (1982). Enhancement of excretion across locust Malphigian tubules by a diuretic vasopressin-like hormone. *Gen. Comp. Endocrinol.* **47**:449–457.

Raabe, M. (1982) *Insect Neurohormones*. New York: Plenum Press.

Ringertz, N.R., and Savage, R.E. (1976). In: *Cell Hybrids*. London: Academic Press.

Rowell, H.F. (1976). The cells of the insect neurosecretory system: Constancy, variability and the concept of the unique identifiable neuron. *Adv. Insect Physiol.* **12**:63–123.

Tobe, S.S. and Pratt, G.E. (1975). Corpus allatum activity in vitro during ovarian maturation in the desert locust, *Schistocerca gregaria*. *J. Exp. Biol.* **62**:611–627.

Towbin, H., Staeheln, R., and Gordon, J. (1979). Electrophoretic transfer of proteins from polyacrylamide gels to nitrocellulose sheets. Procedure and some applications. *Proc. Natl. Acad. Sci. USA* **76**:4350–4354.

Van Weemen, B.K., and Schuurs, A.H.M.W. (1971). Immunoassay using antigen-enzyme conjugates. *FEBS Lett.* **15**:232–236.

Voller, A., Bidwell, D.C., and Bartlett, A. (1979). The enzyme-linked immunosorbent assay (ELISA). A guide with abstracts of microplate applications. Chantilly, VA: Dynatech Laboratories.

Voller, A., Bidwell, D.C., Huldt, G., and Engvall, E. (1974). A microplate method of enzyme-linked immunosorbent assay and its application to malaria. *Bull. WHO* **51**:209–211.

Wu, S.-J., and Ma, M. (1986). Hybridoma antibodies as specific probes to *Drosophila melanogaster* yolk polypeptides. *Insect Biochem.* **16**:789–795.

Chapter 3

Time-Resolved Fluorescence Immunoassay

Rob C.H.M. Oudejans, Hans Voshol, Thomas K.F. Schulz, and Ad M.T. Beenakkers

I. Introduction

For more than 20 years, immunological techniques have been applied extensively to measure hormone titers quantitatively, based on the highly specific and sensitive binding of antigens to antibodies. Originally, radioisotopes were used for labeling the hormone or the antibody to measure the hormone titers via radioimmunoassay (RIA) or immunoradiometric assay (IRMA). Although this approach can still be of great value, applications of radioisotopes involve several disadvantages like potential health and environmental hazards and particularly also the often relatively short half-life of the radiolabel itself and the process of radiolysis causing deterioration of the labeled compounds.

Because of these disadvantages, research has been directed to replace the radioactive label with nonradioactive ones. The search for nonisotopic labels has resulted in the application of very divergent labels such as fluorescent, electron spin, or enzyme labels. In principle, any kind of label can be used if the requirements of sensitivity and ease of detection are met. However, the nonradioisotopic labels seldom match those of the commonly used radioisotopes.

One of the most promising fluorescent probes seems to be europium chelates used in time-resolved fluorescence immunoassays (Hemmilä, 1985). Although the technique has been in development for many years (Dakubu et al., 1984; Lövgren et al., 1984), only very recently have all the basic problems been overcome, and this has resulted in the introduction of a commercially available measuring system known as the ARCUS fluorometer

(LKB/Wallac). Europium as a label has become a unique tool with high potency and a sensitivity, often being more than 1,000 times greater than that of ^{125}I. Parenthetically, it has to be stressed that, whatever the label, the ultimate success of developing any type of immunoassay for a particular hormone depends largely on the availability and/or development of suitable antibodies (ABs).

In the present chapter we describe our continuing efforts to develop a time-resolved fluorescence assay for locust adipokinetic hormone (AKH). Currently, only laborious and relatively inaccurate bioassays are available for this hormone, based on measuring the elevation of lipid concentration in the hemolymph of resting locusts upon treatment with the unknown fraction or extract (Stone and Mordue, 1980). Because of variation between individual animals, a large series of measurements have to be conducted.

Several authors have stressed the need for a reliable and sensitive assay for AKH, and although several papers have been published on the synthesis of radiolabeled AKHs and analogues (Hardy and Sheppard, 1983; Hardy et al., 1983; Muramoto et al., 1984; Yamashiro et al., 1984), as far as we know, no (radio)immunoassay procedure is available. [See note added in the proof, p. 88] A portion of our results regarding the time-resolved fluorescence immunoassay have been presented in a preliminary paper (Oudejans et al., 1986).

II. Locust Adipokinetic Hormones: Blocked Peptides

AKH plays a crucial role in locusts by acting in the regulation of the mobilization and utilization of metabolic fuels during flight (Beenakkers et al., 1985a,b; O'Shea and Schaffer, 1985). Originally it was reported as a lipid-mobilizing factor in the corpora cardiaca of locust species (Beenakkers, 1969; Mayer and Candy, 1969). Later, its structure was elucidated by Stone et al. (1976) as a blocked decapeptide (pGlu-Leu-Asn-Phe-Thr-Pro-Asn-Trp-Gly-Thr-NH_2) (AKH I) for the locust species *Locusta migratoria* and *Schistocerca gregaria*. Subsequently, a second adipokinetic hormone was found by Carlsen et al. (1979), of which the primary structure recently was identified by Siegert et al. (1985) as a blocked octapeptide (pGlu-Leu-Asn-Phe-Ser-Thr-Gly-Trp-NH_2 for *Schistocerca gregaria* and pGlu-Leu-Asn-Phe-Ser-Ala-Gly-Trp-NH_2 for *Locusta migratoria*) (AKH II).

Although their main function appears to be the mobilization of metabolic substrates from fat body, several other effects have been attributed to these hormones (cf. Beenakkers et al., 1985a). Moreover, it is far from clear why two structurally similar hormones also elicit rather identical effects (Mordue and Stone, 1977, 1978; Stone et al., 1978; Goldsworthy and Wheeler, 1984; Schooneveld et al., 1985; Goldsworthy et al., 1986a,b; Siegert and Mordue, 1986).

The AKHs of locusts are representatives of a large peptide family, all characterized by their blocked nature and a pyroglutamate, (pGlu), as the N-terminal amino acid. The structures of several members of this so-called AKH/RPCH (red pigment concentrating hormone) family (Greenberg et al., 1983; O'Shea and Schaffer, 1985) are shown in Table 3.1. Their rather similar structure presents difficulties when developing immunoassays for comparing hormones of this family.

III. Time-Resolved Fluorometry of Lanthanide Chelates

Chelates of lanthanides are a potential alternative to radioisotopic and other nonisotopic labels in immunological methods, particularly in immunoassays (Soini and Kojola, 1983; Dakubu et al., 1984; Soini and Lövgren, 1987). Many lanthanide ions are fluorescent with a high quantum yield, an exceptionally large Stokes shift, and narrow emission peaks (Fig. 3.1). The detection limit is down to 10^{-18} mol (Déchaud et al., 1986).

Since all biological samples contain other fluorescent molecules (Fig. 3.1A), background interference is the most important problem with any fluorometric determination. However, the fluorescence of lanthanide chelates is a long-lasting process (with a decay time in the millisecond range), whereas the fluorescence of interfering substances is very short (the decay time being in the nanosecond range). By activating the detector after a delay, during which all interfering fluorescence has ceased, measuring fluorescence originating from the lanthanide chelate alone is easily possible. Consequently, this so-called time-resolved measurement of fluorescence is potentially several orders of magnitude more sensitive than the measuring procedure with conventional fluorometers. Figure 3.2 shows that after an excitation pulse at 340 nm produced by a pulsed light source, the background fluorescence of non-lanthanide compounds ceases after about 10 nsec. After a delay of 400 μsec, the counting of fluorescence of the lanthanide chelates is initiated and continued until 800 μsec. The procedure can be repeated after 1 msec, etc.

Several remarks can be made regarding the use of the proper lanthanide. Fluorescence of interfering substances falls in the range of 400–600 nm (Fig. 3.1A). Using interference filters, the sharp emission peak of europium at 614 nm (Fig. 3.1D) can easily be separated from scattering caused by excitation (340 nm) and also from fluorescence of interfering compounds.

Of the lanthanides, europium and terbium have been used most extensively. They have different excitation and emission wavelengths and also different fluorescent properties in solution. The strongest fluorescence emission peaks of terbium and europium are at 490/545 and 590/613 nm, respectively, whereas their excitation maxima are below 300 nm and at 340 nm, respectively. Although terbium forms more stable chelates in water and no protection against quenching in water is needed, europium has been chosen as the

Table 3.1. Chemical Structure of Known Members of the AKH/RPCH Family

Hormone[a]	Organism	Structure	Ref.[b]
AKH I	*Locusta migratoria, Schistocerca gregaria*	pGlu-Leu-Asn-Phe-Thr-Pro-Asn-Trp-Gly-Thr-NH_2	1
AKH II	*Locusta migratoria*	pGlu-Leu-Asn-Phe-Ser-Ala-Gly-Trp-NH_2	2
AKH II	*Schistocerca gregaria, S. nitans*	pGlu-Leu-Asn-Phe-Ser-Thr-Gly-Trp-NH_2	2,3
AKH	*Manduca sexta*	pGlu-Leu-Thr-Phe-Thr-Ser-Ser-Trp-Gly-NH_2	5
AKH	*Heliothis zea*	pGlu-Leu-Thr-Phe-Thr-Ser-Ser-Trp-Gly-NH_2	6
RPCH	Crustacea species	pGlu-Leu-Asn-Phe-Ser-Pro-Gly-Trp-NH_2	4
HGH I	*Periplaneta americana*	pGlu-Val-Asn-Phe-Ser-Pro-Asn-Trp-NH_2	10
(= CC I)			7
(= M I)			8,9
HGH II	*Periplaneta americana*	pGlu-Leu-Thr-Phe-Thr-Pro-Asn-Trp-NH_2	10
(= CC II)			7
(= M II)			8,9

[a] AKH, adipokinetic hormone; CC, cardioaccelatory peptide; HGH, hyperglycemic hormone; RPCH, red pigment concentrating hormone; M, myotropic peptide.
[b] 1, Stone et al. (1976); 2, Siegert et al. (1985); 3, Gäde et al. (1986); 4, Fernlund (1974); 5, Ziegler et al. (1985); 6, Jaffe et al. (1986); 7, Scarborough et al. (1984); 8, Witten et al. (1984); 9, O'Shea et al. (1984); 10, Siegert et al. (1986).

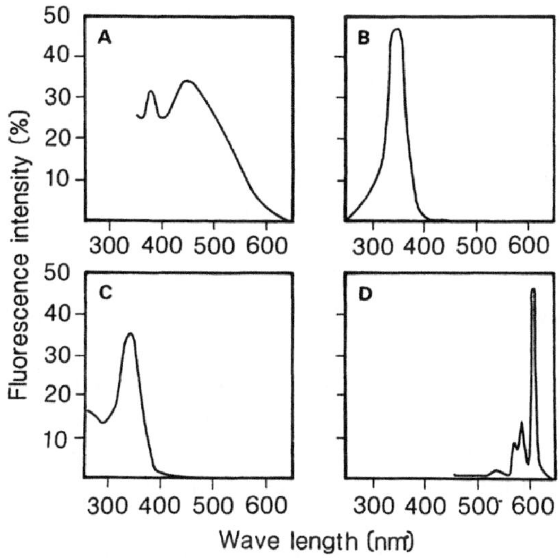

Figure 3.1. Absorption, excitation, and emission spectra. (A) Fluorescence emission of serum; (B) excitation of europium; (C) absorption of europium; (D) emission of europium. Redrawn after Soini (1984).

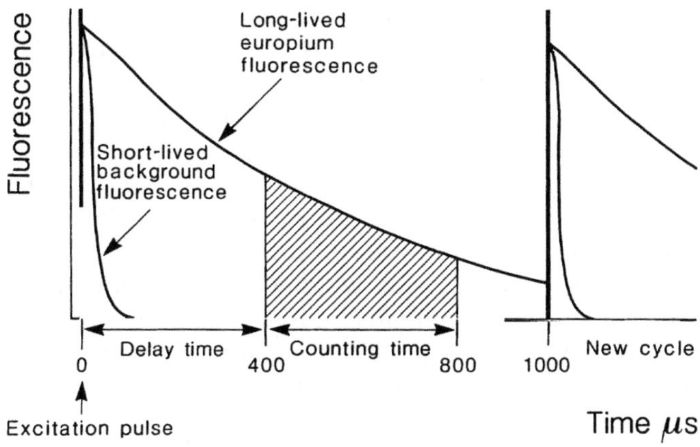

Figure 3.2. Time course of counting europium fluorescence. Upon an excitation pulse of light of 340 nm at t=0, the short-lived background fluorescence ceases rapidly, whereas the long-lived fluorescence of europium is measured after a delay. Counting time is the shaded area. After Soini (1984).

best candidate because of its excitation wavelength, which requires only plastic or optical glass for measurement of samples, whereas the very low excitation wavelength of terbium requires the use of quartz.

Theoretical and methodological considerations of the fluorometric determination of europium ion concentrations can be found in Soini and Hemmilä (1979) and in Dakubu and Ekins (1985). The latter authors also give full equations for determination of total fluorescence reaching the detector.

IV. Description of the LKB 1230 ARCUS Fluorometer

The first commercially available automatic time-resolved fluorometer has been developed by LKB/Wallac. It was introduced as the 1230 ARCUS fluorometer in 1984 and is equipped with a sample changer and a built-in microprocessor to measure large series of samples (Soini, 1984). The light source, a xenon lamp, is used with a flash duration of a few microseconds and a repetition rate of about 1,000 flashes per second. Using an interference filter, the optimal excitation wavelength of 340 nm for europium chelates can be achieved, but with other filters different excitation wavelengths are also possible.

Usually, samples are prepared in standard 12-well microtitration strips which can be placed in racks for direct counting in the ARCUS apparatus: up to 30 racks can be loaded on the conveyer table. Also, cuvette racks for 10 sample tubes (80 × 12 mm OD) can be used, and if necessary, the two rack types can be measured mixed together.

Excitation of the samples takes place through the side of the well, whereas the generated fluorescence is measured through the bottom. The most suitable wavelength for the fluorescence can be chosen from five different emission filters controlled by software, so measuring of double-labeled samples at two different wavelengths using consecutive counting periods is possible. The photons are measured with a side window-type photomultiplicator tube equipped with a red-sensitive multialkali cathode, usually with good response at longer wavelengths. The photomultiplier output pulses are amplified and fed into the main scaler after dividing them by 100 to lower the pulse frequency.

At the end of the counting period, the microprocessor reads the scaler contents and stores the accumulated counts in the memory. An external printer and a double floppy disk drive are connected by an RS-232C interface. All data obtained with the ARCUS can be edited with the software to calculate the concentrations in any form for the compound of interest. Also, several plots or comparisons with standard curves are possible.

A soft-key concept is used to operate the ARCUS by means of four touch keypads: the software generates a new meaning for each key every time a new program phase is entered by using one of these keys. Each sample is measured about 1,000 times within 1 sec, resulting in a maximal throughput of samples of about 1,500 per hour!

V. Labeling of Antibodies with Lanthanide Chelates

The procedure for labeling ABs with europium can be found in Hemmilä et al. (1984) and Lövgren et al. (1984), based on the paper by Sundberg et al. (1974). For our tests two antisera were available: anti-AKH[1–4] No. 433, specific for AKH I and AKH II (Schooneveld et al., 1986), and anti-[Tyr^1]AKH No. 241, specific for AKH I (Schooneveld et al., 1983). Labeling of both antisera was done by Dr. Timo Lövgren (Wallac Biochemical Laboratory, Turku, Finland).

Antiserum 241 appeared to contain 24 mol europium/mol IgG, and antiserum 433 16 mol europium/mol IgG. Bovine serum albumin (1 mg/ml) was added to the labeled AB solutions. For measurement of europium concentrations, see also Lyle and Rahman (1963).

VI. Applications

1. Kits Available for Human Peptide and Steroid Hormones

Since the introduction of the DELFIA (dissociation-enhanced lanthanide fluoroimmunoassay; see Appendix) method, a series of ready-to-use kits is available for several human hormones and other compounds of general interest. Some are based on a solid-phase, two-site fluoroimmunoassay (direct sandwich technique) like human thyrotropin (hTSH), human luteinizing hormone (hLH), human prolactin (hPRL), human chorionic gonadotropin (hCG), human alphafetoprotein (hAFP), hepatitis B surface antigen (HB_sAg), and ferritin. Others are based on a competitive reaction—for example, kits for cortisol, digoxin, and thyroxin (T4). Moreover, Bertoft et al. (1984) have published a competitive DELFIA method for testosterone, whereas Déchaud et al. (1986) have developed a time-resolved fluorometer and have presented a method for measuring prolactin. Paterson et al. (1986) have evaluated the assay for hTSH.

In general, the detection limits obtained with time-resolved fluorescence are at least equal to, or better than, those for the most sensitive RIAs.

2. Development of an Assay for Locust Adipokinetic Hormone

A. Hormones and Hormone Derivatives

Although AKHs can easily be isolated and purified from insect corpora cardiaca (Stone and Mordue, 1980), quantities are usually too small for subsequent chemical modification and coupling to carrier proteins. However, several synthetic AKHs and also [Tyr[1]]AKH I are available from Peninsula Laboratories, Inc. (Belmont, CA). AKH[1–4] was prepared via custom synthesis by Cambridge Research Biochemicals (Harnston, U.K.).

B. Synthesis of Antigens

AKH per se cannot be used as an Ag owing to its small size. Moreover, its structure does not allow a direct chemical attack to bind the hormone to a carrier protein which as a complex can be used as the Ag. So, in order to couple AKH, one has to replace at least one amino acid in the hormone by another—for instance pGlu by tyrosine ([Tyr[1]]AKH). By using bifunctional reagents (Tesser et al., 1975; Leung and Meares, 1977; Samokhin and Filimonov, 1985), the tyrosine analogue of AKH can easily be coupled to larger carrier proteins which can be used as Ags and also for the coating of microtitration strips with "immobilized" hormone. We have synthesized conjugates of [Tyr[1]]AKH with thyroglobulin and bovine serum albumin, but other conjugates are possible as well—for instance, with ovalbumin and keyhole limpet (*Megathura crenulata*) hemocyanin.

The necessary chemical modification of the hormone prior to its coupling to a carrier protein may result in less appropriate structures to raise suitable ABs. Moreover ABs against the carrier protein itself will be generated, so polyclonal ABs have to be purified to obtain AKH-specific ABs. To circumvent these difficulties, we have also tried to raise ABs against [Tyr[1]]AKH coupled to Affigel-10 (BioRad). Results are being evaluated. A direct "coupling" of the native hormone to fumed silica to be used as the Ag may be possible as well (Van Ness et al., 1984).

a. Conjugation of [Tyr[1]]AKH with a carrier protein. Typically a conjugation with thyroglobulin is described, but other carrier proteins can be used as well. Fifty microliters of a solution of 0.75 mg bis-[2-(succinimidooxycarbonyloxyl)ethyl] sulfone (BSOCOES; Pierce, Rockford, CA) (Tesser et al., 1975) was added to 1 mg [Tyr[1]]AKH and incubated for 15 min. The reaction mixture was subsequently added to 17 mg thyroglobulin (Sigma) and dissolved in 500 μl bidistilled water. During the addition the reaction vial was shaken vigorously to assure a homogenous coupling reaction. After 15 min, the reactants were run over a Sephadex G-25 column to separate the resulting thyroglobulin-BSOCOES-[Tyr[1]]AKH complex from other reaction products. The coupling was checked by thin-layer chromatography on silica F_{254} (Fig. 3.3).

b. Coupling of [Tyr[1]]AKH to Affigel-10. [Tyr[1]]AKH (250 μg) was dissolved in 250 μl dimethylformamide and added to 0.5 ml Affigel-10 (BioRad), which was washed previously with three bed volumes of isopropyl alcohol

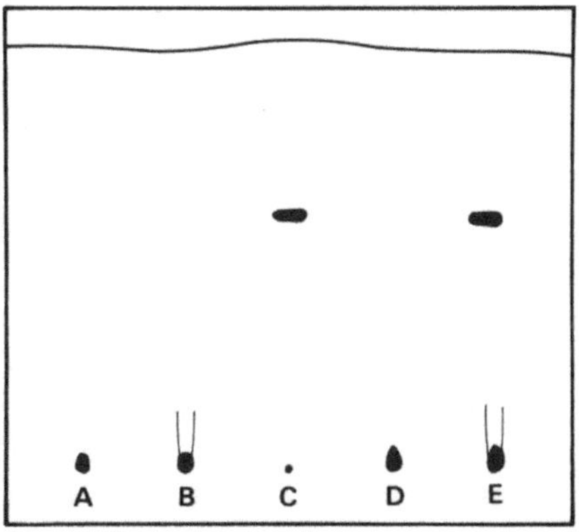

Figure 3.3. Thin-layer chromatography on silicagel F_{254} (Merck, Darmstadt, Germany), using n-butanol : acetic acid : ethyl acetate : water (1 : 1 : 1 : 1 by volume) as the solvent. (A) Thyroglobulin; (B) conjugate of [Tyr[1]]AKH and thyroglobulin; (C) [Tyr[1]]AKH; (D) thyroglobulin treated with 0.1N NaOH; (E) conjugate of [Tyr[1]]AKH and thyroglobulin treated with 0.1 N NaOH.

followed by three bed volumes of cold (4°C) deionized water. The reaction mixture was agitated gently for 1 h at room temperature. Any remaining ester was then blocked for 1 h by adding 50 μl 1 M ethanolamine HCl (pH 8). After blocking, the gel was washed with phosphate-buffered saline (PBS).

C. Production of Monoclonal Antibodies

To obtain ABs specific for AKH I, BALB/c mice were immunized by subcutaneous injection at four sites on the back of the animals with 50–100 μg [Tyr[1]]AKH-thyroglobulin in Freund's complete adjuvant mixed in a 1 : 1 ratio (by volume) with PBS. Four days before cell fusion, mice were boostered with 50–100 μg antigen in PBS (Schulz et al., 1987; see also Chapter 2). Fusion of spleen lymphocytes with the nonproducing mouse myeloma strain Sp2/0 was carried out using a standard fusion protocol (Galfré and Milstein, 1981; see Chapter 2).

For intrasplenic immunization (Spitz et al., 1984), 20–40 μg of [Tyr[1]]AKH coupled to Affigel-10 was used. Four to five days after immunization, spleen cells were fused, and the culture medium was screened for ABs specific for AKH by an ELISA. For standard ELISA, multiwell disposotrays (Linbro, Flowlabs) were coated by incubating the wells with 100 μl of a solution containing 1–5 μg [Tyr[1]]AKH-bovine serum albumin per milliliter PBS for 2 h at 37°C. After coating, the wells were incubated with 100 μl of the culture media. ABs bound to the immobilized hormone were visualized by peroxidase-conjugated goat anti-mouse (GAM-PO; Tago Inc.) with o-phenyldiamine reagent essentially as described by Engvall (1980).

D. Assays for Adipokinetic Hormone

In principle, fluorescence immunoassays in which the fluorescent group is a europium chelate can be applied to all types of immunoassays already in use. We have tried to develop both a sandwich-type assay (an immunofluorometric assay, IFMA) and a competitive assay (a fluoroimmunoassay, FIA).

a. Immunometric assay (IFMA). In the sandwich-type assay (essentially according to Lövgren et al., 1984), two ABs against the hormone of interest are used. The procedure starts with the binding of the first (unlabeled) AB onto the wells of a microtitration strip. The next step is the incubation of the wells with the samples (standards and unknowns) and subsequently with an excess of the second AB labeled with europium. After thorough washing to remove the nonbound labeled AB, the europium from the bound AB is dissociated with enhancement solution and the fluorescence is measured with the ARCUS fluorometer. The fluorescence obtained is proportional to the concentration of the hormone. In order to achieve the highest sensitivity for this type of assay, two different ABs specific for different determinants are to be used. The procedure is as follows.

Polystyrene microtitration strips (12 wells, Eflab, Helsinki, Finland) were coated by adsorption with [Tyr^1]AKH antiserum 241. Individual wells were filled with 200 μl of a solution of the antiserum (5 μg IgG/ml) in 0.1 M phosphate buffer, pH 4.9. After shaking the strips for 20 h at room temperature with a Dynatech shaker, the wells were washed three times with saline (0.9% sodium chloride and 0.05% sodium azide) using the Nuncwash system and saturated for 2 h by treatment with 300 μl of a 0.05 M TRIS-HCl buffer, pH 7.4, containing 0.9% sodium chloride, 0.05% sodium azide, and 0.5% bovine serum albumin. After washing twice with saline, the coated strips can be stored at 4°C in a humid environment.

Prior to further use, the precoated wells were washed with DELFIA washing solution and then incubated for 2 h with a solution consisting of 100 μl AKH samples (known or unknown) and 150 μl DELFIA assay buffer per well. The incubation was terminated by washing three times with washing solution. The second incubation with 50 ng of europium-labeled antiserum 433 in 200 μl assay buffer for 1 h was ended by washing the wells six times with washing solution. Subsequently, the europium of the bound labeled AB was dissociated by shaking for 5 min with 200 μl enhancement solution per well (Lövgren et al., 1984). After covering, the strips were left for 10 min and fluorescence was measured with the ARCUS fluorometer using counting settings according to the specifications of the manufacturer.

All efforts to obtain results with the immunometric assay as previously described were unsuccessful: no fluorescence correlating with the quantity of AKH could be measured. Several considerations (i.e., the small hormone molecule and less specific polyclonal antisera) gave reason to increase the incubation time, particularly that of the second AB. However, by increasing the incubation time of the second AB to an overnight period, the results obtained were not significantly different from those obtained after standard conditions. Also, an increase of the first incubation with the sample to 4 h gave no improvement. Increasing the amount of the labeled second AB only showed an increase in fluorescence (both in controls and unknowns). Evidently the number of washings is not enough to remove the excess non-bound label. By increasing the number of washings, this can be overcome.

In conclusion, using standard conditions with increased incubation times (up to 4 h for the first and up to overnight for the second) results on the concentration of AKH present (1 to 100 pmol) could not be obtained. Even 1 nmol AKH was ineffective. We conclude that the AKH molecule is probably too small to bind two ABs, because both the coating of the first AB (checked by an ELISA procedure) and the binding of the labeled AB to AKH (checked by a spot test) were correct. Moreover, the europium on the AB did not prohibit the binding of the labeled AB with AKH, since immunogold staining of AKH in AKH-containing granules of corpora cardiaca was not affected as checked by electron microscopy. Evidently the binding of the hormone to two ABs at the same time is the limiting step. Even by selecting appropriate ABs specific for regions of the molecule distinct from one another, this assay type may be not successful.

b. Competitive assay (FIA). A competitive assay (essentially according to Bertoft et al., 1984) is based on the competition of a limiting quantity of labeled AB between the free hormone to be measured and the same hapten bound directly or indirectly via a larger "carrier" protein to the wells. The procedure starts with the binding of the hormone-carrier protein complex to the wells of the microtitration strips. The samples to be measured together with the europium-labeled AB are then incubated in the wells. Depending on the concentrations of the competing substances the labeled AB will partly bind to the free hormone and partly to its immobilized form. The complexes formed between hormone and labeled AB are removed by a washing procedure, and the fluorescence of the europium from the AB bound to the coated fraction of the hormone can be measured. The more fluorescence obtained, the less hormone present in the sample. The procedure is as follows.

Polystyrene microtitration strips were coated by adsorption with [Tyr^1]-AKH conjugated with thyroglobulin as the carrier protein by filling individual wells with 250 μl of a solution of 0.1 μg thyroglobulin-hormone conjugate in 0.1 M sodium carbonate buffer, pH 9.3, and shaking for 20 h at room temperature. After washing the wells twice with saline, the strips can be stored at 4°C in a humid environment.

Prior to further use, the precoated wells were washed with DELFIA washing solution. The incubation was done with 50 μl of the hormone solution and 200 μl of assay buffer, containing 50 ng of europium-labeled anti-AKH 241. The incubation was terminated by washing six times with washing solution. If higher doses of tracer have to be used, the wells must be washed up to 10 times. Fluorescence can be measured as described for the IFMA assay type.

In principle, different combinations of immobilized hormone and labeled AB can be used: coating with [Tyr^1]AKH-thyroglobulin or AKH[1–4]-thyroglobulin and incubation with labeled antiserum 241 or 433. Unfortunately, the AKH[1–4]-thyroglobulin could not be used, because the available batch appeared to be insoluble and therefore coating was impossible. Experiments with [Tyr^1]AKH-thyroglobulin revealed that antiserum 433 seems to bind better to the immobilized hormone than antiserum 241, although 241 was raised specifically against that Ag. However, since both antisera are polyclonal, a direct comparison of binding capacities is valid only if identical titers of the binding ABs are used.

Effects of AKH in assays using either labeled antiserum 241 or 433 could not be observed. An obvious explanation would be that the antisera also contain ABs against the thyroglobulin portion of the Ag. To test this, wells were coated with [Tyr^1]AKH-thyroglobulin or thyroglobulin. From the fluorescence obtained after incubation with the labeled antisera, it can be concluded that one-fourth to one-third of the binding of the antisera is due to nonspecific binding to the thyroglobulin part of the immobilized hormone. Maximal binding of the labeled antiserum 241 was only two times higher than the nonspecific binding, leaving a range too small for reliable measurements. A purified antiserum can be obtained after preincubation of the antiserum with thyroglobulin. A sample of 100 μg labeled antiserum 241 was incubated in 1 ml of a 1 μM aqueous solution of thyroglobulin for 1 h at 37°C and subsequently overnight at room temperature. The thyroglobulin-AB complex formed did not precipitate.

Free ABs and thyroglobulin-bound ABs were separated by gel filtration over Ultrogel AcA 22 with 50 mM TRIS-HCl buffer, pH 7.65, containing 0.05% BSA. Elution was monitored by measuring the europium fluorescence with the ARCUS fluorometer (Fig. 3.4). Complexed ABs eluted at about 50 ml, and free ABs at about 70 ml. The labeled antiserum purified in this way was used in an assay, using wells coated with thyroglobulin or [Tyr^1]AKH-thyroglobulin. Binding of the purified labeled antiserum was lowered four- to sixfold, but the range left between nonspecific and maximal binding was smaller than with the nonpurified antiserum.

In conclusion, antiserum 241 is not appropriate to be used in competitive assays owing to its low specificity for AKH. To overcome these difficulties, ABs more specific for AKH must be generated. Monoclonal ABs specific for AKH are currently produced in our laboratory, and we believe they possess high enough specificity to develop an AKH time-resolved fluorescence assay. Moreover, using carrier proteins for coating that differ from those used for raising ABs should be of great value.

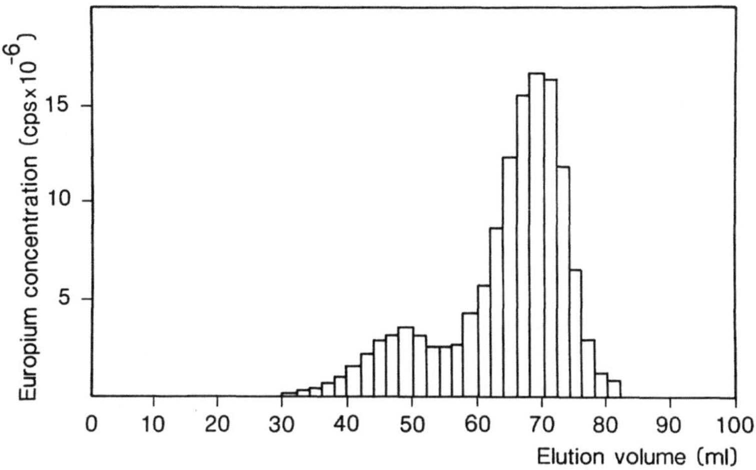

Figure 3.4. Elution profile of gel filtration of 100 μg of europium-labeled antiserum 241 preincubated with thyroglobulin over Ultrogel AcA 22. Europium fluorescence was measured in only 5-μl samples of eluate!

VII. Future Trends

The excellent specificity and ultrasensitivity of the measurement of fluorescence of europium as a label applied in time-resolved fluorescence immunoassays is a reliable method to replace part of the radioisotopic labels currently in use in immunoassays. Apart from several standard kits already available for the measurement of human hormones, the method is of course also appropriate to be used in developing assays for invertebrate hormones. Although our efforts to develop such an assay for locust AKH are unsuccessful, we believe that the problems encountered can be overcome by developing more specific monoclonal ABs. In developing an immunoassay, the right ABs are so crucial that without their availability any immunoassay is impossible. The most promising approach to develop an AKH assay seems to be a competitive assay.

Apart from the low detection limit and wide dynamic range and linearity for measuring europium labels, other advantages of europium are the high specific activity of the label and the stability of labeled ABs or hormones, and therefore avoiding repetitive labeling and the deterioration of stored labeled compounds. Moreover, several other applications for fluorescent lanthanides make this method a rather universally applicable one. Soini and Lövgren (1987) survey the use of lanthanides in serological assays, virus antigen assays, screening assays from dry blood spots, labeling nucleic acids for recombinant DNA techniques, conformation studies, and labeling of phospholipids. Currently, in our laboratory europium-labeled ABs are also used to quantify lipoproteins in extrasts of locust flight muscles. The proteins are separated by gel electrophoresis and transferred onto nitrocellulose paper by electroblotting. Once immunopositive bands have been identified (e.g., by peroxidase staining), corresponding areas are excised and incubated with specific primary ABs and europium-labeled secondary ABs, after which the bound europium is dissociated and counted (H.G.A.M. Van Antwerpen and M.C. Van Heusden, personal communication). In this way it is also possible to study the binding characteristics of Ag-AB interactions in concentration- or time-dependent assays or to determine competition between various ABs.

Acknowledgments. We thank Dr. Huug Schooneveld (Agricultural University, Wageningen, The Netherlands) for generously providing the anti-AKH sera, and Dr. Timo Lövgren (Wallac Biochemical Laboratory, Turku, Finland) for labeling the antibodies and for sending information and a copy of a manuscript of a review on time-resolved fluorescence.

Note added in Proof

Recently a sensitive radioimmunoassay for locust AKH was described by P. Moshitzky, D. Yamashiro, L. Stuve, J. Ramachandran, and S.W. Applebaum (1987). Determination of locust AKH-I by radioimmunoassay and the identification of an AKH-like factor in the locust brain. Insect Biochem. 17:765–769.

References

Beenakkers, A.M.T. (1969). The influence of corpus allatum and corpus cardiacum on lipid metabolism in *Locusta migratoria*. *Gen. Comp. Endocrinol.* **13:**492.

Beenakkers, A.M.T., Bloemen, R.E.B., De Vlieger, T.A., Van Der Horst, D.J., and Van Marrewijk, W.J.A. (1985a). Insect adipokinetic hormones. *Peptides* **6** (Suppl. 3): 437–444.

Beenakkers, A.M.T., Van Der Horst, D.J., and Van Marrewijk, W.J.A. (1985b). Insect lipids and lipoproteins, and their role in physiological processes. *Prog. Lipid Res.* **24:** 19–67.

Bertoft, E., Eskola, J.U., Näntö, V., and Lövgren, T. (1984). Competitive solid-phase immunoassay of testosterone using time-resolved fluorescence. *FEBS Lett.* **173:**213–216.

Carlsen, J., Herman, W.S., Christensen, M., and Josefsson, L. (1979). Characterization of a second peptide with adipokinetic and red pigment-concentrating activity from the locust corpora cardiaca. *Insect Biochem.* **9:**497–501.

Dakubu, S., and Ekins, R.P. (1985). The fluorometric determination of europium ion concentration as used in time-resolved fluoroimmunoassay. *Anal. Biochem.* **144:**20–26.

Dakubu, S., Ekins, R., Jackson, T., and Marshall, N.J. (1984). High sensitivity, pulsed-light time-resolved fluoroimmunoassay. In: *Practical Immunoassay*. W. Butt (Ed.). New York: Dekker, pp. 71–101.

Déchaud, H., Bador, R., Claustrat, F., and Desuzinges, C. (1986). Laser-excited immunofluorometric assay of prolactin, with use of antibodies coupled to lanthanide-labelled diethyleneaminepentaacetic acid. *Clin. Chem.* **32**: 1323–1327.

Engval, E. (1980). Enzyme immunoassay ELISA and EMIT. *Methods Enzymol.* **70**: 419–439.

Fernlund, P. (1974). Structure of red-pigment-concentrating hormone of the shrimp, *Pandalus borealis*. *Biochim. Biophys. Acta* **371**: 304–311.

Gäde G., Goldsworthy, G.J., Schaffer, M.H., Cook, J.C. and Rinehart, K.L. Jr. (1986). Sequence analyses of adipokinetic hormones II from corpora cardiaca of *Schistocerca nitans*, *Schistocerca gregaria*, and *Locusta migratoria* by fast atom bombardment mass spectrometry. *Biochem. Biophys. Res. Commun.* **134**: 723–730.

Galfré, G., and Milstein, C. (1981). Preparation of monoclonal antibodies: Strategies and procedures. *Methods Enzymol.* **73**: 3–46.

Goldsworthy, G.J., and Wheeler, C.H. (1984). Adipokinetic hormones in locusts. In: *Biosynthesis, Metabolism and Mode of Action of Invertebrate Hormones.* J. Hoffmann and M. Porchet (Eds.). Berlin: Springer, pp. 126–135.

Goldsworthy, G.J., Mallison, K., and Wheeler, C.H. (1986a). The relative potencies of two known locust adipokinetic hormones. *J. Insect. Physiol.* **32**: 95–101.

Goldsworthy, G.J., Mallison, K., Wheeler, C.H., and Gäde, G. (1986b). Relative adipokinetic activities of members of the adipokinetic hormone/red pigment concentrating hormone family. *J. Insect Physiol.* **32**: 433–438.

Greenberg, M.J., and Price, D.A. (1983). Invertebrate neuropeptides: Native and naturalized. *Annu. Rev. Physiol.* **45**: 271–288.

Hardy, P.M., and Sheppard, P.W. (1983). Synthesis of [4,5-dehydro Leu^2]- and [3,4-dehydro Pro^6]-locust adipokinetic hormone. *J. Chem. Soc. Perkin Trans.* **1**: 723–729.

Hardy, P.M., Sheppard, P.W., Brundish, D.E., and Wade, R. (1983). Tritiated peptides. 13. Synthesis of [4,5-3H-Leu2]- and [3,4-3H-Pro6]-locust adipokinetic hormone. *J. Chem. Soc. Perkin Trans.* **1**: 731–734.

Hemmilä, I. (1985). Fluoroimmunoassays and immunofluorometric assays. *Clin. Chem.* **31**: 359–370.

Hemmilä, I., Dakubu, S., Mukkala, V.-M., Siitari, H., and Lövgren, T. (1984). Europium as a label in time-resolved immunofluorometric assays. *Anal. Biochem.* **137**: 335–343.

Jaffe, H., Raina, A.K., Riley, C.T., et al. (1986). Isolation and primary structure of a peptide from the corpora cardiaca of *Heliothis zea* with adipokinetic activity. *Biochem. Biophys. Res. Commun.* **135**: 622–628.

Leung, C.S.-H., and Meares, C.F. (1977). Attachment of fluorescent metal chelates to macromolecules using "bifunctional"chelating agents. *Biochem. Biophys. Res. Commun.* **75**: 149–155.

Lövgren, T., Hemmilä, I., Petterson, K., Eskola, J.U., and Bertoft, E. (1984). Determination of hormones by time-resolved fluoroimmunoassay. *Talanta* **31**: 909–916.

Lyle, S.J., and Rahman, M.M. (1963). Complexometric titration of yttrium and the lanthanons. I. A comparison of direct methods. *Talanta* **10**: 1177–1182.

Mayer, R.J., and Candy, D.J. (1969). Control of haemolymph lipid concentration during locust flight: An adipokinetic hormone from the corpora cardiaca. *J. Insect Physiol.* **15:**611–620.

Mordue, W., and Stone, J.V. (1977). Relative potencies of locust adipokinetic hormone and prawn red pigment-concentrating hormone in insect and crustacean systems. *Gen. Comp. Endocrinol.* **33:**103–108.

Mordue, W., and Stone, J.V. (1978). Structure and metabolism of adipokinetic hormone. In: *Comparative Endocrinology.* P.J. Gaillard and H.H. Boer (Eds.): Amsterdam: Elsevier/North-Holland, pp. 487–490.

Muramoto, K., Ramashandran, J., Moshitzky, P., and Applebaum, S.W. (1984). Preparation of specifically tritiated locust adipokinetic hormone analog with full biological activity. *Int. J. Peptide Protein Res.* **23:**443–446.

O'Shea, M., and Schaffer, M. (1985). Neuropeptide function: The invertebrate contribution. *Annu. Rev. Neurosci.* **8:**171–178.

O'Shea, M., Witten, J., and Schaffer, M. (1984). Isolation and characterization of two myoactive neuropeptides: Further evidence of an invertebrate peptide family. *J. Neurosci.* **4:**521–529.

Oudejans, R.C.H.M., Voshol, H., and Beenakkers, A.M.T. (1986). Development of a time-resolved fluorescence immunoassay for locust adipokinetic hormone. *Bull. Soc. Zool. Fr.* **111:**44–45.

Paterson, N., Biggart, E.M., Chapman, R.S., and Beastall, G.H. (1985). Evaluation of a time-resolved immunofluorometric assay for serum thyroid stimulating hormone. *Ann. Clin. Biochem.* **22:**606–611.

Samokhin, G.P., and Filimonov, I.N. (1985). Coupling of peptides to protein carriers by mixed anhydride procedure. *Anal. Biochem.* **145:**311–314.

Scarborough, R.M., Jamieson, G.C., Kalish, F., et al. (1984). Isolation and primary structure of two peptides with cardioacceletory and hyperglycemic activity from the corpora cardiaca of *Periplaneta americana. Proc. Natl. Acad. Sci. USA* **81:**5575–5579.

Schooneveld, H., Tesser, G.I., Veenstra, J.A., and Romberg-Privee, H.M. (1983). Adipokinetic hormone and AKH-like peptide demonstrated in the corpora cardiaca and nervous system of *Locusta migratoria* by immunocytochemistry. *Cell Tissue Res.* **230:**67–76.

Schooneveld, H., Romberg-Privee, H.M., and Veenstra, J.A. (1985). Adipokinetic hormone-immunoreactive peptide in the endocrine and central nervous system of several insect species: A comparative immunological approach. *Gen. Comp. Endocrinol.* **57:**184–194.

Schooneveld, H., Romberg-Privee, H.M., and Veenstra, J.A. (1986). Immunocytochemical differentiation between adipokinetic hormone (AKH)-like peptides in neurons and glandular cells in the corpus cardiacum of *Locusta migratoria* and *Periplaneta americana* with C-terminal and N-terminal specific antisera to AKH. *Cell Tissue Res.* **243:**9–14.

Schulz, T.K.F., Van der Horst, D.J., Amesz, H., Voorma, H.O., and Beenakkers, A.M.T. (1987). Monoclonal antibodies specific for apoproteins of lipophorins from the migratory locust. *Arch. Insect Biochem. Physiol.* **6:**97–107.

Siegert, K.J., and Mordue, W. (1986). Quantification of adipokinetic hormones I and II in the corpora cardiaca of *Schistocerca gregaria* and *Locusta migratoria. Comp. Biochem. Physiol.* **84A:**279–284.

Siegert, K., Morgan, P., and Mordue, W. (1985). Primary structures of locust adipokinetic hormones II. *Biol. Chem. Hoppe-Seyler* **366:**723–727.

Siegert, K.J., Morgan, P.J., and Mordue, W. (1986). Isolation of hyperglycaemic peptides from the corpus cardiacum of the American cockroach, *Periplaneta americana. Insect Biochem.* **16:**365–371.

Soini, E. (1984). Pulsed light, time-resolved fluorometric immunoassay. In: *Monoclonal Antibodies and New Trends in Immunoassays.* C.A. Bizollon (Ed.). Amsterdam: Elsevier, pp. 197–208.

Soini, E., and Hemmilä, I. (1979). Fluoroimmunoassay: Present status and key problems. *Clin. Chem.* **25:**353–361.

Soini, E., and Kojola, H. (1983). Time-resolved fluorometer for lanthanide chelates—new generation of nonisotopic immunoassays. *Clin. Chem.* **29:**65–68.

Soini, E., and Lövgren, T. (1987). Time-resolved fluorescence of lanthanide probes and applications to biotechnology. *CRC Crit. Rev. Anal. Chem.* **18:**105–154.

Spitz, M., Spitz, L., Thorpe, R., and Eugui, E. (1984). Intrasplenic primary immunization for the production of monoclonal antibodies. *J. Immunol. Methods* **70:**39–43.

Stone, J.V., and Mordue, W. (1980). Adipokinetic hormone. In: *Neurohormonal Techniques in Insects.* T.A. Miller (ed.). New York: Springer, pp. 31–80.

Stone, J.V., Mordue, W., Batley, K.E., and Morris, H.R. (1976). Structure of locust adipokinetic hormone, a neurohormone that regulates lipid utilisation during flight. *Nature* **263:**207–211.

Stone, J.V., Mordue, W., Broomfield, C.E., and Hardy, P.M. (1978). Structure-activity relationships for the lipid-mobilising action of locust adipokinetic hormone. Synthesis and activity of a series of hormone analogues. *Eur. J. Biochem.* **89:**195–202.

Sundberg, M.W., Meares, C.F., Goodwin, D.A., and Diamanti, C.I. (1974). Selective binding of metal ions to macromolecules using bifunctional analogs of EDTA. *J. Med. Chem.* **17:**1304–1307.

Tesser. G.I., De Hoog-Declerck, R.A.O.M.M., and Westerhuis, L.W. (1975). Some new bifunctional reagents for the conjugation of a protein with an amino compound in water. *Hoppe-Seyler's Z. Physiol. Chem.* **356:**1625–1629.

Van Ness, J., Laemmli, U.K., and Pettijohn, D.E. (1984). Immunization in vitro and production of monoclonal antibodies specific to insoluble and weakly immunogenic proteins. *Proc. Natl. Acad. Sci. USA* **81:**7897–7901.

Witten, J.L., Schaffer, M.H., O'Shea, M., Cook, J.C., Hemling, M.E., and Rinehart, K.L. Jr. (1984). Structures of two cockroach neuropeptides assigned by fast atom bombardment mass spectrometry. *Biochem. Biophys. Res. Commun.* **124:**350–358.

Yamashiro, D., Applebaum, S.W., and Li, C.H. (1984). Synthesis of shrimp red pigment-concentrating hormone analogs and their biological activity in locusts. *Int. J. Peptide Protein Res.* **23:**39–41.

Ziegler, R., Eckart, K., Schwarz, H., and Keller, R. (1985). Amino acid sequence of *Manduca sexta* adipokinetic hormone elucidated by combined fast atom bombardment (FAB)/tandem mass spectrometry. *Biochem. Biophys. Res. Commun.* **133:**337–342.

APPENDIX: Composition of various solutions used

Phosphate-buffered saline (PBS):
0.8% sodium chloride, 0.02% potassium chloride, 0.115% disodium hydrogen phosphate, and 0.021% sodium dihydrogen phosphate

DELFIA washing solution:
0.9% sodium chloride, 0.1% Germall II, 0.02% Tween 20

DELFIA enhancement solution:
0.1 M acetate buffer, adjusted to pH 3.2 with potassium hydrogen phthalate and containing 15 μM 2-naphthoyltrifluoroacetone, 50 μM tri-n-octylphosphine oxide, and 0.1% Triton X-100

Assay buffer:
0.05 M TRIS-HCl, pH 7.7, containing 0.9% sodium chloride, 0.05% sodium azide, 0.5% bovine serum albumin, 0.05% bovine globulin, 0.01% Tween 40, and 20 μM diethylenetriaminepentaacetic acid (DTPA)

Chapter 4

Immunocytochemistry

Hugo Schooneveld and Jan A. Veenstra

I. Introduction

Recently, immunocytochemical (ICC) techniques have enjoyed increasing popularity for the rapid and convenient qualitative detection of antigens (Ags) in tissues and cells. The techniques are basically rather simple. As most of the necessary reagents are commercially available, every laboratory is now able to carry out standard ICC experiments, provided the specific antibodies (ABs) are (made) available to recognize the Ag of interest. This chapter gives a number of ICC procedures that have been applied routinely in entomological laboratories of the authors and of others who were so kind to provide their data to increase the value of this chapter.

In essence, all methods rely on the visualization of binding of ABs against the Ags (usually fixed) in the tissue. Apart from the fact that the substances are sometimes unique for insects, the methodology is similar for all biological material. Excellent reviews have already been published by Sternberger (1979), Picket (1981), Cuello (1983), Van Noorden and Polak (1983), and Polak and Varndell (1984). Some of the topics related to insects have been treated in various ways by Eckert and Ude (1983) and by Duve and Thorpe (1983, 1984). In the interval, a number of technical improvements have become available, and the scope of ICC applications has been expanded considerably. We selected those data which, in our opinion, are worth considering at the moment. Where possible, we chose examples of certain applications of interest to those working with insects.

ICC methods are considered more versatile than conventional histological and histochemical methods for the demonstration of cell constituents. Since the great majority of the biological substances are suitable, in principle, for the production of antisera, the range of Ags to be detected is almost unlimited. Some of the procedures for antiserum production will be given below; for details one is referred to standard textbooks on immunology (e.g. Hurrell, 1982; Tijssen, 1985).

One of the prerequisites for success is that the Ag is prevented from leaving the cell and is configured so that it is recognized by the antisera. Much attention will therefore be given to fixation procedures.

Depending on the expected distribution of the Ag, it may be advisable to study Ag-AB reactions in wholemounts. Large and three-dimensional structures such as neurons and their branching axons may then be much easier to analyze. However, the size of the tissue or the solubility of the Ag may require that the tissue (fresh or fixed) be sectioned, either by cryotome or vibratome or after paraffin embedding. The routine of dehydration, embedding, and deparaffination does not affect the immunostainability of the Ags in a surprising number of cases.

The next step is to apply ABs to the section or tissue and to locate the Ag-AB complexes by one of several methods. The choice is sometimes dictated by the quality of the antiserum and hence the lowest dilution that yields good labeling without too much background staining, by the concentration of Ag in the cell, the required speed of assay, the required storage period of the slides, and the required contrast (signal-to-noise ratio) for photomicrography. The procedures for light microscopy and electron microscopy are also different in some respects. Relevant details on methodological aspects in particular will be given.

II. Pretreatment of Insects

If one is interested in the presence and distribution of an antigen, as is often the case in neuroanatomy, it is of obvious importance that the insects to be studied be in a condition where the Ags are indeed available. For instance, neurochemical mediators (neurotransmitters, neuromodulators, and neurohormones) often need to be artificially concentrated for successful observation. It is common practice in our laboratory to treat the insects with a critical dose of colchicine, which blocks axonal transport by the disruption of microtubules, 2 days before dissection of the specimen. Optimal effects are obtained with dosages fully blocking locomotion of the animals in the absence of undue mortality in that population over that period. The smaller insects such as the Colorado potato beetle are injected with 1.5 μg of colchicine dissolved in 2 μl Ringer; the larger American cockroach requires 150 μg in 50 μl (Bishop and O'Shea, 1982). Optimal doses for each species should be determined by experimentation and may differ, depending on the physiological state of the animals.

One other variable factor influencing Ag accumulation is the developmental stage and physiological condition. We usually find more immunoreactive material in mature and old specimens than in those just after ecdysis, and more in reproducing Colorado potato beetles than in diapausing specimens. It may therefore be worthwile to include specimens of a wide variety of conditions in the first series of experiments. We found that peptide concentrations in certain neurons increased considerably if reproducing specimens were starved for several days.

III. Fixation

1. General Considerations

The correct preservation of immunoreactivity of the substance of interest implies that it is anchored to the cell matrix to prevent it from being washed out and configured so that immunorecognition can take place reproducibly. Success further depends on the availability of ABs recognizing the Ag in the fixed condition. These prerequisites may cause problems of a different nature. One important factor is the choice of the Ag used for immunization of the donor animal (see Chapter 1). The immune system may, for certain reasons, not produce the ABs one was hoping for. The other factor is that once ABs have been obtained, they do not work in ICC. It is not uncommon to have an antiserum that works well in radioimmunoassay (RIA) but poorly or not at all in ICC—e.g. anti-proctolin (Bishop et al., 1981). This could be because of the inaccessibility of the antigenic determinants for the antibodies after fixation, due to steric hindrance caused by cross-linking of the antigenic determinants to the cell matrix by the fixative. Another, theoretical possibility, which has hardly been explored, is that the configuration of the fixed Ag in solution (i.e., during immunization) triggers the defense system of the host to make ABs that cannot recognize the configuration of the Ag in sections, however carefully preserved. The configuration of the Ag in situ is always different from the soluble condition because of the association of the Ag with the cell matrix or secretion granules, among other reasons. It is well known that sequences of short peptides have configurations dictated by the nature of the protein in which they occur (Wilson et al., 1985), and association with cell substrates may well have the same effect.

A highly uncertain factor is the configuration of the Ag during the immunization process. Small molecules like steroids, small peptides, biogenic amines, neurotransmitters, etc. must be conjugated with some carrier (BSA, thyroglobulin, etc.) to be immunogenic. The same carrier probably changes the configuration and limits the molecular flexibility of the Ag, thus restricting the number of ABs produced, as the reactivity of the ABs is a function of the atomic mobility of sites in a protein (Tainer et al., 1984). A recently recognized complication for peptides in solution is that ABs are formed against antigenic determinants arranged discontinuously in the total molecule, depending on the tertiary structure. Amino acids arranged in different fixed conformations but in the same linear sequence may not be recognized (Barlow et al., 1986). The results may again be poor recognition of the Ag in the fixed state. Sometimes the Ag of interest is too small to function as an antigenic determinant. In such cases, ABs are obtained which recognize the Ag plus the cross-linking agent (e.g., antisera to 5HT and other neurotransmitters: see section III.B). It is obvious that such Ags in the tissue should be coupled to the cell matrix in order to obtain an immunoreaction and that the same cross-linking agent must be used. The composition of the fixation mixture may be very critical in these cases.

These considerations are not applicable only to ICC but also to other procedures involving Ag-AB interaction. The ultimate consequence is that the Ag should preferably be in a condition that mimics the condition in the section as closely as possible. That means chemical fixation of the Ag prior to the immunization process. Some success has thus been achieved with immunization protocols for the production of monoclonal ABs to homogenates of ganglia of *Periplaneta americana* that had been "fixed" by 4% paraformaldehyde (Denburg et al., 1986; Verhaert et al., 1986). The resulting ABs may perform poorly in other immunological assays.

2. Fixation of Nonpeptidergic Neurotransmitters/Modulators

By virtue of the absence of better methods, it has until recently been common practice for neuroanatomists to localize and trace specific neurons with neurotransmitter-specific antisera raised to the enzymes involved in the biosynthesis of the neurotransmitters of interest (see Salvaterra et al., 1985). The techniques for these marker enzymes are not different from routine methods treated elsewhere and will not be treated here.

Special procedures for the direct detection of nonpeptidergic neurotransmitters have been worked out only recently. Problems arise easily owing to the relatively small size of the transmitter substances, their solubility, and (sometimes, as for acetylcholine) their lack of reactive groups. The methods available will therefore be treated in some detail.

Antisera to acetylcholine (ACh) have become available after the derivatization of ACh and its complexing to a protein (Geffard et al., 1985b). The resulting ABs were used to localize ACh-containing nerve terminals in certain neuropile areas—that is, after the fixation of locust brains in 0.1 M cacodylate buffer pH 11 containing 12% formol and 1% sodium metabisulfite for 2 min at 4°C in the dark, which allowed the deacetylation of ACh and simultaneous transesterification (Geffard et al., 1985a). The brains were postfixed in Bouin-Hollande without acetic acid but containing 12% formol, 1% sodium metabisulfite, and a saturated solution of mercuric chloride, at 4°C for 36 h in the dark.

For GABA staining in neurons of *Manduca sexta*, a method has been worked out by Hoskins et al. (1986) based on the procedure of Storm-Mathisen et al., (1983). For paraffin sectioning, tissue is fixed for 90 min in a mixture of 0.5% paraformol and 1% glutaraldehyde, followed by paraplast embedding. For vibratome sectioning, the brains were fixed overnight in 0.5% paraformol and 2.5% glutaraldehyde in a 0.1% sodium phosphate buffer, pH 7.4.

Glutamate reactions (in goldfish) require a similar fixation (Storm-Mathisen et al., 1983); the sections were treated first with 1 M ethanolamide to neutralize remaining free aldehyde groups from the fixative. This procedure has not yet been reported to be successful for insects.

Serotonin (5-hydroxytryptamine, 5HT) is preserved by the fixation for a few hours in 4% formol (freshly prepared from paraformol) in 0.1% phosphate buffer pH 7.2 for 2–4 h. The location of 5HT-containing neuronal cell bodies and axons can be studied in situ after incubation of the entire ganglion in the antiserum (Nässel and Elekes, 1984). It is advantageous that the ganglia are desheathed for better penetration of the ABs (Bishop and O'Shea, 1983; Davis, 1985) (see also section IV). Detection of the AB in whole mounts can occur with the PAP method (see Bishop and O'Shea, 1983) or with the fluorescence method (see Davis, 1985). Sections of fixed tissue are preferably made with the cryotome (Schürmann and Klemm, 1984; Nässel and Cantera, 1985).

Paraffin sections can be obtained in two ways. The best way is probably freeze-drying of frozen fresh tissue, fixation in paraformol vapor at 80°C, and vacuum infiltration with paraffin, followed by routine ICC procedures (Klemm et al., 1984). The other method involves embedding of fixed tissue in paraffin and processing the slides through the normal procedures with the PAP method (Nishiitsutsuji-Uwo et al., 1984). The latter method seems to give more background than the in toto fluorescent method of Davis (1985).

For dopamine, a technique has been worked out that works well on locust brain if stringent precautions are fulfilled (Vieillemaringe et al., 1984). Two methods are available. For immunofluorescence, it is sufficient to fix brain tissue for 1.5 h in 4% buffered paraformol, followed by cryostat sectioning and routine analysis with FITC fluorescence. For paraffin sections, tissue is fixed in Bouin-Hollande with 12% formaldehyde and a saturated solution of sublimate. The tissue is then processed quickly. All steps must be performed in the dark and using certain precautions to avoid oxidation of the catechol moiety and photolability of the Formol-dopamine Ag association (Vieillemaringe et al., 1984).

3. Fixation of Proteins and Neuropeptides

Fixation of neuropeptides (and other peptides and proteins) appears to present no problems characteristic of insect tissue. Possible complications are that the structure of the Ag is changed by the fixation procedure used, rendering it unrecognizable by the antisera raised to the native, unfixed Ag. For this reason, different fixation schedules should be tried in case of initial negative results. A good example is the suboptimal fixation of proctolin-containing cells in the terminal abdominal ganglion of *Periplaneta* (Agricola et al., 1985). Nine neuron types became immunoreactive after fixation in a mixture of glutaraldehyde, picric acid, and acetic acid, whereas only two types were revealed after Bouin's fixative. The reverse may also occur; nonspecific immunoreactions may occur after certain fixation and by the use of an antiserum raised to a specific cell constituent but reacting with a completely different molecule.

Similar findings have been observed in insect immunocytochemistry. Numerous antisera against vertebrate peptides have been "borrowed" for demonstrating similar substances in insects (Thorpe and Duve, 1984; Veenstra et al., 1985b). From our experience, we know that immunoreactions with uncharacterized antisera are occasionally obtained if fixation and other experimental conditions are varied. The preabsorption of the antisera with the original Ag may even be effective in that further staining with the serum is prevented, but no information on the immunoreactive substance is obtained other than that it contains perhaps a single recognizable segment.

Substances known to be present are sometimes difficult to identify immunocytochemically. For instance, the structure of AKH-II-S in the corpora cardiaca of *Schistocerca* has recently been elucidated (Siegert et al., 1985; see Chapter 3), but antisera to the four C-terminal amino acids of this peptide fail to react with it, unless, of the several formulas tried, the Karnovski mixture was used (Schooneveld, unpublished). Here, too, the configuration of this small peptide is probably influencing immune recognition.

The same type of problem may arise during tissue preparation for electron microscopy (EM). A 2% glutaraldehyde solution in 0.1% buffer pH 7.2 is used most widely. For the routine PAP and immunogold detection method, a satisfactory stain density is usually obtained, though not always. In our hands, this fixation usually gives very poor results for light microscope (LM) staining purposes with anti-AKH and several other antisera. Results may improve if Formol is added such as in the method of Diederen et al. (1987), based on the formula of Karnovsky (1965).

There are cases where formaldehyde fixation renders Ags unrecognizable by ABs. A "remedy" applied by some is to subject the tissue on the slide to enzymatic digestion by pronase or 0.1% trypsin in TRIS buffer pH 7.8, both after embedding in paraffin (Huang et al., 1976) and in resin (Van Noorden and Polak, 1983; Hermanns et al., 1986). The fundamentals of these treatments are not yet properly understood, and no applications have so far been reported for insects.

4. Miscellaneous Antigens

This chapter would be incomplete without mentioning the possibility of steroid ICC. For details on the methods, we refer to the papers on ecdysteroid ICC by Birkenbeil and Eckert (1982) and Loeb (1986). Furthermore, we would like to mention the possibility of the ICC demonstration of pteridines, some of which are important cofactors in the biosynthesis of biogenic amines (e.g., tetrahydrobiopterin) and may thus be useful markers for aminergic neurons (Matsuura et al., 1983).

IV. Preembedding ICC

It may be advantageous to carry out immunoreactions with intact tissues and to study whole mounts rather than slides: (1) the technique is simple; (2) soluble or labile Ags leak out of the tissue less easily; (3) the immunoreactive cells can be studied more easily in their relative positions or their position in relation to neurons revealed by other means (see Duve et al., 1983) so that a topographical map can be constructed; and (4) embedding for LM and EM can follow to study further details of interest at high power in sectioned material.

The fixation of tissue is not different from the procedures to be described, but special precautions must sometimes be taken to enhance the penetration of the relatively large immunoglobulin molecules if larger distances are to be covered. Saponin (0.5%) has been used to lower the surface tension of the solution and as a wetting agent. Diffusion pathways may be shortened by fragmenting the tissue manually or by use of the Vibratome. The neurilemma may present a diffusion barrier to the ABs. A remedy is to peel off this sheath. Successful staining of peptidergic neurons with antiproctolin has thus been obtained for the ventral ganglia in *Periplaneta* by Bishop and O'Shea (1982) and with antiserotonin by Bishop and O'Shea (1983) and Davis (1985). It is questionable whether desheathing is necessary for all species, as Nässel and Elekes (1984) stained neurons in the CNS of *Calliphora* successfully with antiserotonin AB without this tedious trimming procedure.

For several tissues, AB penetration for preembedding staining is sufficiently rapid without special preparation. In the prothoracic glands of *Galleria*, the location of ecdysone could be determined provided the tissue was UV-irradiated to immobilize the steroid. The antibody was next visualized in Vibratome sections (Birkenbeil and Eckert, 1982). The label indicating bound antibody must obviously remain in place during subsequent processing. Therefore, for both LM and EM, diaminobenzidine (DAB) (which is brown and can be electron-densified by osmium tetroxide) is the label of choice. One of the advantages of preembedding labeling is that postfixation procedures are quite possible to increase visibility of ultrastructure without affecting immunoreactivity. Details on preembedding EM-ICC are given by Priestley (1984).

V. Embedding Materials

For materials that must be sectioned for ICC inspection and when cryostat sectioning is not necessary for maintaining sufficient immunoreactivity, the tissue should be embedded in a solid medium for support during sectioning. All media have their specific advantages. For LM, paraffin is still most widely used, but owing to the use of aggressive solvents and high embedding temperatures, the structural preservation is usually rather poor, although acceptable for most LM applications. The fact that the tissue can be deparaffinized completely ensures that the total tissue in a particular section contributes to the ICC reaction. This gives a high reaction intensity, depending on the thickness of the section.

A far better structural preservation is obtained after embedding in polyethylene glycol (see Wolosewick, 1985) or in one of the synthetic resins available in different types (see Causton, 1984; Newman and Jasani, 1984). Semithin sections made with a glass or steel knife can be stained by a few routine ICC methods and inspected under the light microscope. Best results are usually obtained if postfixation in osmium tetroxide is omitted or applied in much lower concentrations to retain immunoreactivity of the Ag.

The plastic of the epoxy resins should be removed to expose all immunoreactive sites in the tissue. This can be done by soaking the sections for 10–45 min (depending on the thickness of the section) in a solution containing 1% sodium hydroxide in 100% methanol or ethanol. Some use saturated sodium hydroxide solutions in alcohol in the dark and rinse thoroughly afterward in 100% methanol or ethanol. For Epon and related epoxy resins such as Durcupan, Spurr, Araldite, etc., good LM observations can thus be made. However, our experience reveals that staining reactions with the PAP technique are less intense than with paraffin-embedded tissue, but it may be worth testing any of the following intensification and digestion methods for optimizing the results. The optic resolution of labeled compounds is very good and may be further improved by viewing tiny deposits of label in such sections under an interference contrast microscope.

Relatively few reports have appeared on the use of resins of the acrylic type in LM-ICC. This resin cannot be dissolved, but hydrophilic groups can be built in to enhance Ag-AB interactions. Excellent staining of glycol methacrylate-embedded rat tissue by a two-step peroxidase or the ABC method has been reported by Hermanns et al. (1986). It proved essential to uncover immunoreactive groups of surface Ags by a mild protease treatment.

Some find it desirable to adopt protein digestion as a standard procedure for ICC assays (Springall et al., 1984). Several proteases have been used for that purpose—trypsin (Van Noorden and Polak, 1983), pronase, and other enzymes. It is useful to use only sections that have been attached firmly to the glass, not by albumen-glycerin but by poly-L-lysine (PLL) (Sigma; MW 150,000), applied as a thin film by spreading a drop of a 0.01% solution in water on the glass, followed by drying for at least 1 day but no longer than a week or as long as the glass remains sticky.

One of the important features of ICC staining of plastic sections is the possibility of comparing LM sections with neighbouring ultrathin unlabelled sections in the EM. As demonstrated for instance in insects by Veenstra and Schooneveld (1984) and in crustaceans by Jaros et al. (1985), certain relatively small immunoreactive structures, such as neurons and even axons, can be easily pinpointed in semithin LM sections. Ultrathin sections can next be screened efficiently by comparison. This method may be faster than the inspection of immunogold-labeled ultrathin sections (see below), especially if the structure of interest is relatively small. A further advantage of this method is that it allows the direct comparison of the immunoreactivity of structures processed for EM or for LM. The results may be quite different, as was shown with Colorado potato beetle tissues treated with antivasopressin antisera for both types of microscopy (Veenstra et al., 1984)

For ICC at the EM level, several resins are available for both pre- and postembedding staining (see Causton, 1984). The epoxy resins Epon, Araldite, and Spurr are used most often. Satisfactory labeling depends largely on a certain degree of "etching" of the resin surface with, for instance, 10% hydrogen peroxide for a few minutes. Acrylate-based resins cannot be etched, but immune reactions are unhampered by the hydrophilic nature of the resin and may permeate the resin matrix (Anonymous, 1982). Two types of acrylate resins are receiving increasing interest because of (1) their low viscosity even at subzero temperatures and (2) their hydrophilic character, which makes them miscible with water and requires less stringent dehydration procedures. Polymerization can be achieved by heat, by chemical catalysts, and by ultraviolet light. One of these relatively new products is London resin white (LR white, London Resin Company, U.K.) (see Newman and Jasani, 1984). Immunoreactive groups are well preserved, but no reports on insect tissues have been published. The other resin is Lowicryl (Chemische Werke Lowi, Waldkraiburg, FRG), which is offered in two modifications—K4M and HM20 (Carlemalm et al., 1985; Armbruster et al., 1984). The first type is becoming especially popular, and polymerization is usually carried out by UV light, whereby dehydration steps are performed under progressively lower temperatures, all well below 0°C. A well worked out recipe developed for corpora cardiaca staining of *Locusta migratoria* (Diederen et al., 1987) is given in Appendix 9.

VI. Acquisition of Antisera

It is beyond the scope of this article to treat the production of antisera in detail; one is referred to Hudson and Hay (1980), Tijssen (1985), Springer (1985), and other standard treatises on this subject (see Chapter 1).

Countless antisera are available commercially, and several of these are of potential interest for entomologists. They cover a wide range of specificities and will work satisfactorily where Ags in insects and vertebrates are identical. For example, monoclonal ABs to neuronal substances in *Drosophila* have shown the ability to cross-react with several neuronal substances in the human nervous system, suggesting a surprising degree of similarity in Ags in both species (Miller and Benzer, 1983). If ABs to certain cell products are required without the possibility of rigorous Ag purification, two routes can be chosen to prepare antisera. One is to purchase or (partially) purify the substance and use this for raising polyclonal antisera. The latter procedure was followed several years ago by Eckert and Gersch (1978) to demonstrate the site of synthesis of neurosecretory substances in the brain of *Periplaneta americana*, and the activation factor II. After immunoprecipitation of brain homogenates and subsequent SDS gel separation of the dissociated Ags, certain chemical data about the substance could be collected.

The other route is to prepare monoclonal ABs recognizing the substance or the particular cell (neuron) constituents by immunizing mice with the Ag of interest. The basic technique for monoclonal AB production is beyond the scope of this chapter and should be found in other treatises (Hurrell, 1982; Zola and Brooks, 1982; Langone and Van Vunakis, 1983; Milstein and Cuello, 1983; Springer, 1985; Tijssen, 1985; Chapter 2). Data on the use of insect antigens are given by Salvaterra et al. (1985). These authors were interested in the localization of choline acetyltransferase purified from *Drosophila* heads. One of their aims was to map cholinergic neurons in the nervous system, and several neuropile areas in the brain were stained with high specificity.

A different approach is to use homogenates of selected whole tissues (in casu brains) for immunization and to differentiate between neurons (see Barnstable, 1985). Sternberger et al. (1982) showed how subtle changes between rat neurons could be made visible using the monoclonal AB technique. Similar work with the *Drosophila* nervous system was reported by Fujita et al. (1982). A panel of 148 different monoclonal ABs was obtained. Whereas some reacted with widely dispersed Ags, others were highly specific to nerve fibers, neuropile, photoreceptors, and other structures. Denburg et al. (1986) obtained monoclonal ABs, one of which specifically recognized an as yet unidentified substance in the axon terminals in the coxal depressor muscles, originating in the dorsal unpaired medium cells in the thoracic ganglion of the cockroach. After immunizing mice with brain homogenates from *Locusta*, Verhaert et al. (1986) obtained monoclonal ABs recognizing glandular cells in the corpus cardiacum. An interesting technique was developed for the ultrarapid screening of supernatants on tissue sections: single sections were simply glued to the bottom of wells in the 96-well plate, deparaffinized, and incubated under standard conditions.

The preparation of immunogens requires special care when substances other than large molecules like proteins are involved. Compounds with a molecular weight of less than 2,000 are usually not immunogenic and should be conjugated with a carrier (usually a protein) to challenge the defense system of the host animal. Such conjugates are easy to prepare if the Ag possesses chemically reactive groups. These can be used to couple the Ag to the carrier by covalent bonds, and several such coupling reactions, some making use of a spacer molecule and bifunctional reagents such as diethyl or diphenyl carbodiimide, have been described for a wide range of chemical substances (see several chapters in Skowsky and Fischer, 1972; Van Vanakis and Langone, 1980; Springer, 1985). Recent data on the preparation of small peptides and the selection of the proper carrier protein are given by Szelke (1983).

Problems may be encountered for the smaller molecules if no reactive groups are available for conjugation. This is the case with several peptides of the AKH family that have blocked N-terminals (pyroglutamate), amidated C terminals, and a lack of basic or acid amino acids residues suitable for binding to other molecules (see Chapter 3). In these cases chemical analogues must be synthesized that have built-in reactive groups. Thus, for AKH a derivative with a tyrosine residue instead of pGlu was synthesized and was used successfully for raising antisera (Schooneveld et al., 1983). Moreover, an AKH fragment only containing residues 1–4 (pGlu-Leu-Asn-Phe) and a spacer group was used to identify AKH/RPCH-related substances in different tissues and in a wide variety of insects (Schooneveld et al., 1985, 1987a,b) and other invertebrates (Schooneveld, 1987).

Even smaller peptides than these (e.g., the pentapeptide proctolin) can be conjugated with the aid of glutaraldehyde (Eckert et al., 1981; Bishop et al., 1981) and the tetrapeptide FMRFamide (Boer et al., 1980) with the aid of bifunctional reagents (Skowsky and Fischer, 1972; Tesser et al., 1975; Reichlin, 1980; Bauminger and Wilchek, 1980) without problems. It is clear, however, that the chemical reaction should be understood and the molecular structure taken into account.

VII. Postembedding Staining Methods

A variety of ICC staining procedures have been developed for different purposes. Some are rapid but insensitive, others are tedious but give specific staining with high signal-to-noise ratios (low background). Figure 4.1 gives the principles of some interesting methods. The sensitivity of some of the methods is compared in section VII.6.

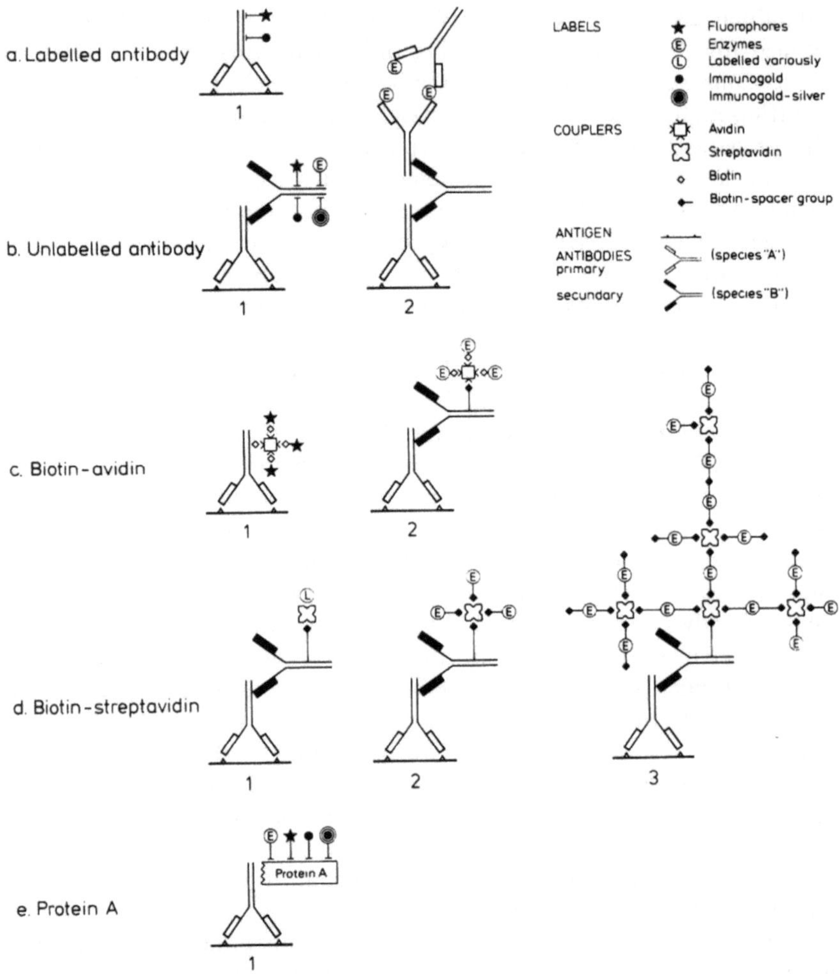

Figure 4.1. Staining methods regularly used. (a) Labeled antibody technique. (b) Unlabeled antibody techniques comprising (1) the two-step method with labeled second antibody, and (2) the three-step PAP method comprising the use of a second antibody linking the third antibody-PAP complex with the primary antibody. (c) Biotin-avidin methods, comprising (1) biotinylated first antibody binding label-bearing avidin, and (2) the ABC method involving the use of biotinylated second antibody associating with an avidin-enzyme complex. (d) Biotin-streptavidin methods, comprising biotinylated second antibody associating with (1) variously labeled streptavidin or streptavidin-enzyme complexes containing (2) three or (3) many more enzyme molecules. (e) Protein A methods comprising the nonimmune recognition of the Fc portion of several IgG molecules by protein A, labeled in various ways. Note that the proportions of the molecules are not drawn to scale. In a1, b1, and e1, the label possibilities, not combinations, are given.

Another point of concern is the cost of reagents, which may vary considerably. Using commercial preparations of second antibodies and employing the techniques as given in Appendices 2–7, we roughly estimated the price per slide (200 μl of solvent to cover the sections on each slide). Prices vary, of course, with the supplier and with the desired degree of specificity and purity. We come to the following prices (in 1986 U.S. dollars):

Fluorescent method (swine anti-rabbit gG/FITC): $0.08
Two-step peroxidase, method (goat anti-rabbit peroxidase, Nordic Immunological Laboratories, Tilburg, The Netherlands): $0.10
PAP method (swine anti-rabbit + rabbit PAP, Nordic): $0.22
Biotin-avidin method (ABC kit, Vectastain, Vector Laboratories, Burlingame, CA): $0.34
Biotin-streptavidin method (biotinylated donkey anti-rabbit IgG + streptavidin complex as depicted in Fig. 4.1d; Amersham International, Amersham, U.K.): $0.30
Immunogold-silver method (goat anti-rabbit IgG with 20 nm colloidal gold particles, LM grade, Auroprobe LM, Janssen Pharmaceutica, Beerse, Belgium), followed by Timm's silver intensification: $0.72

There are several other manufacturers of immunochemicals. We express no preference. Batches of reagents change from time to time, however. In case of irreproducible reactions with a particular reagent, it may be worthwile trying to obtain a better lot of reagent from another supplier. When testing the reactivity of monoclonal ABs with the two-step peroxidase staining, we found that rabbit anti-mouse peroxidase IgG from Dakopatts (Glostrup, Denmark) was far superior to that of some other suppliers.

1. Labeled Antibody Techniques

In these methods, the first AB is visualized directly, either by providing the IgG molecules with a microscopically visible label or with an enzyme. For visible labels one can choose between fluorescent dyes such as FITC or TRITC; some labeled primary ABs are available commercially. Alternatively, gold particles can be used, especially for EM work, as was shown by Myers et al. (1986) for the demonstration of hemoglobin in embryos of *Chironomus thummi*. These reagents must be prepared separately for each antiserum, and this direct approach has found more applications in clinical diagnoses than in research. The affinity of the first AB is reduced drastically by chemical treatments involving the covalent binding of tracers (Sternberger, 1977).

2. Unlabeled Antibody Techniques

These are the most widely applied. The first AB, raised in an animal referred to as species A in Figure 4.1b, is detected with an antiserum raised in species B. Two ways are now available to detect the binding of this second AB.

A. Two-Step Techniques

The fastest staining method uses the second AB in a prelabeled form. Commercially available ABs carry labels such as fluorescent groups, enzymes, or gold particles. Each of these labels can be visualized in a matter of minutes to hours, and each has advantages as well as disadvantages. The availability of different labels makes it possible to perform double or even multiple staining procedures for the simultaneous detection of more than one Ag in the same section.

a. Fluorescent labels. Fluorescent labels have been used from the very start of ICC procedures (see Coons and Kaplan, 1950). Several fluorophores are available, but FITC, TRITC, and rhodamine are the most widely used. Using fluorescent ABs, one has the advantage that slides are ready for inspection directly after completion of this incubation step. The disadvantages are (1) the relative insensitivity of this method (see section VII.F), (2) the deterioration of fluorescence upon UV illumination, although certain mounting media, for instance those containing n-propylgallate (Giloh and Sedat, 1982) and paraphenylene diamide (100 mg/100 ml), retard this effect, and (3) the necessity of mounting and storing the finished preparations in an aqueous medium such as glycerin jelly, which gives suboptimal optical resolution. Fluorescent labels are still among the most widely used methods in ICC.

b. Enzyme labels. Enzyme labels convert a colorless substrate molecule into an intensely stained reaction product. The staining intensity depends on (1) the amount of bound AB, (2) the duration of the enzymatic reaction, and (3) the nature of the substrate. Conjugates with several enzymes are available, but horseradish peroxidase is most widely used. The staining intensities of the different substrates for the peroxidase enzyme have been compared, and the best substrate proved to be 3,3′-diaminobenzidine (DAB) (De Jong et al., 1985), which gives a dark brown, insoluble reaction product. The disadvantage of DAB is its potential carcinogenicity.

The intensity of the DAB reaction may be enhanced if necessary, even when very little reaction product has been generated over the section. A special DAB silver enhancement kit has been developed for this purpose by Amersham International. Intensification is also possible by adding metal ions to the DAB reagent mixture, such as nickel or cobalt chloride in a 1% concentration (Hermanns et al. (1986), or by the addition of 0.1 M imidazol at pH 7.6 (De Jong et al. 1985). Different colors develop and are stable if sections are mounted in DePeX.

Another acceptable peroxidase substrate is 4-chloro-1-naphthol (CN) (Sigma), which yields a dark blue reaction product. This product is soluble in alcohol, and sections should therefore be mounted in aqueous medium. The color fades, even in a refrigerator, but slides may be stored frozen for several months without loss of color.

There are several other enzymes converting certain substrates into stained products (see Polak and Van Noorden, 1983), but their use in this two-step procedure seems to be limited largely to alkaline phosphatase. The best chromogen protocol for this enzyme includes the use of 5-bromo-4-chloro-3-indolyl phosphate-tetranitro blue tetrazonium salt (De Jong et al., 1985).

c. Colloidal gold labels. The immunogold techniques for LM and EM ICC are becoming increasingly popular because of their high sensitivity and superior intensity and stability of the label (De Mey, 1983; Polak and Varndell, 1984). In their most common applications, gold particles are conjugated with the second AB. This can be done according to the method of Faulk and Tailor (1971; cf. De Mey, 1983). However, several types of ABs carrying gold particles of distinct sizes are also commercially available (AuroProbe for LM and EM; Janssen Pharmaceutica).

Gold-labeled structures are not visualized easily under the light microscope without further signal amplification. Only when using dark-field illumination can gold particles be revealed in epoxy resin sections (Varndell et al., 1982). The particles can be enlarged at will by silver intensification procedures, giving rise to a black precipitate (Holgate et al., 1983). The immunogold-silver (IGS) technique has proved very useful both as a sensitive single assay and in double-staining procedures (Anonymous, 1985; Springall et al., 1984). The silver intensification procedure can be carried out in different ways. We used the method of Timm (see Obermayer and Strausfeld, 1980) to detect AKH-like antigens in corpora cardiaca of *Locusta migratoria.* A fine granular black staining results (Fig. 4.2). Immunogold-silver intensification kits are commercially available for immunoblotting applications (IntenSe; Janssen Pharmaceutica). When used for ICC, the sensitivity of these kits is comparable to that of Timm's method, and the better resolution is due to the smaller size of the particles. However, the stain in slides mounted in DePeX is unstable. We achieved good results with goat anti-rabbit antisera (AuroProbe BL) carrying 20-nm gold particles, but 5-nm particles may be preferable for LM (Springall et al., 1984).

AuroProbe EM products (Janssen Pharmaceutica) have been designed for EM. The gold particles come in sizes of 5, 10, 20, and 40 nm. The smaller of these are most suited for high-resolution work, but the larger sizes are easier to detect at low magnifications. Immunolabeling for EM is basically rather simple (see Polak and Varndell, 1984). By using different ABs, each carrying gold particles of a distict size, double-labeling experiments are possible (Varndell et al., 1982). Details on the method used in our laboratory are given in Appendix 8.

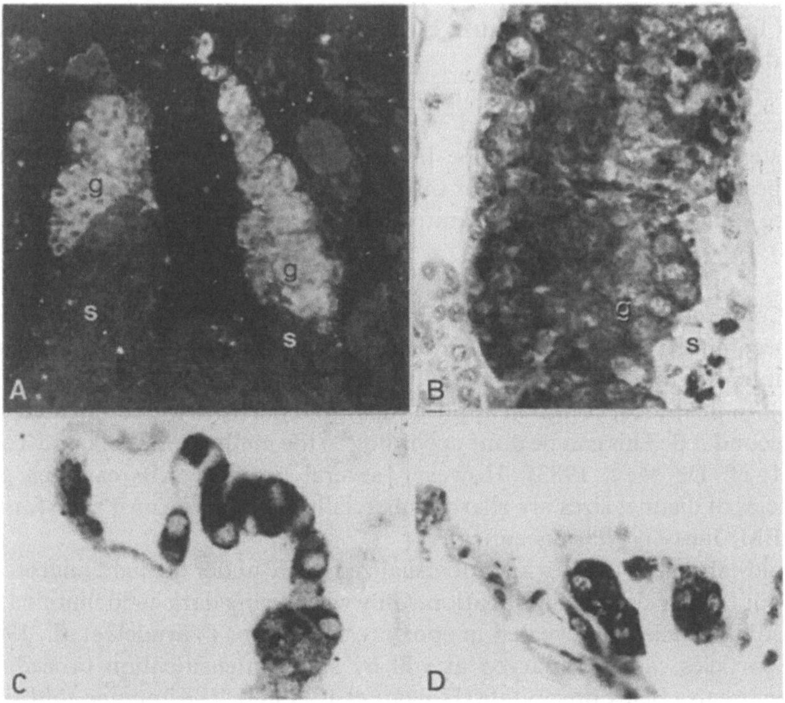

Figure 4.2. Corpora cardiaca (CC) of *Locusta migratoria* immunostained with anti-AKH antiserum (No. 433). The dilutions required depend on the sensitivity of the detection system. *Key*: g, glandular lobe; s, storage lobe of CC. (A) Fluorescence (FITC) staining of total CC. Note fluorescing cells in left and right glandular lobes (g) and the nonreacting storage lobe (s). Antiserum dilution 1:2,000. ×200. (**B**) Peroxidase antiperoxidase (PAP) staining (dark) of glandular lobe (g) associated with nonreacting storage lobe (s). Antiserum dilution 1:16,000. ×500. (C) Immunogold-silver staining of dorsal part of glandular lobe, showing immunoreactive cells (dark) covered with finely granular silver-intensified gold particles. Antiserum dilution 1:32,000. ×500. (D) Streptavidin peroxidase staining of glandular cells similar to those in C. Few cells are stained intensely. Antiserum dilution 1:64,000. ×500.

B. Three-Step Techniques

The unlabeled AB technique that has probably gained greatest popularity is the peroxidase-antiperoxidase (PAP) method of Sternberger (1979). Peroxidase substrates may again be DAB or CN. The advantages of the PAP method are its high sensitivity, flexibility, and reliability. It is our preferred technique when top quality is required, such as for photographic documentation.

PAP complexes can be prepared in the laboratory (Sternberger, 1977, 1979) but are also commercially available from a wide range of suppliers such as

Dakopatts, Jackson ImmunoResearch Lab, Miles Scientific, Nordic Immunochemicals, Sigma, and Sternberger-Meyer. One essential procedural step is the blocking of endogenous peroxidases in the tissue, as these may produce false-positive reactions. These enzymes can be inhibited by pretreatment of the deparaffinized sections, with 3% hydrogen peroxide in methanol or PBS, for 20 min. Our experience, however, is that peroxidases may persist, for instance in hemocytes and additional control experiments comprising the omission of the PAP step are recommended.

Our PAP method is given in Appendix 7. It includes an overnight incubation with primary antiserum in the cold. The concentration of antiserum necessary to give a satisfactory label intensity is 2–3 times lower than with shorter (1–2 h) incubation times at room temperature. Agitation is recommended to obtain even staining over the whole slide. The rinsing steps may be simplified if the quality standards are somewhat lower.

The peroxidase molecule can be substituted by the enzyme alkaline phosphatase, and the resulting alkaline phosphatase-antialkaline phosphatase (APAAP) conjugates can be used in a similar method by making use of phosphatase substrates such as naphthol-AS-MX-phosphate (Van Noorden and Polak, 1983). Once again, the drawback is that the enzyme may be present in the tissue sections, thus possibly causing unacceptable background staining and the necessity of relevant control experiments.

The three-step method is also useful for the labeling of Ags in epoxy resin-embedded tissue for EM. As indicated by Sternberger (1977), the surface of the section must first be "etched" by hydrogen peroxide to expose the reactive sites in the section and to allow AB penetration (see section V). This step is not necessary for the immunogold labeling. The naked nickel grids with sections are next treated essentially analogous to the staining for LM: primary AB, link AB, PAP complex, DAB. Thomsen et al. (1980) thus fixed fat body of the blowfly in glutaraldehyde, embedded in Epon and treated etched sections (5% H_2O_2) with antivitellogenin antiserum. DAB deposits were further amplified in terms of greater electron opacity by a 30-min treatment with 4% unbuffered osmium tetroxide at room temperature. Vitellogenin-containing granules contrasted with the lighter background.

The electron opacity of the immunoreactive cell organelles and the siqual-to-noise ratio are often not quite satisfactory. It would seem that the two-step colloidal gold methods have a higher sensitivity and allow greater flexibility (Hacker et al., 1986).

3. Biotin-Avidin/Streptavidin Techniques

An acceptable alternative to the unlabeled enzyme methods is based on the strong binding forces between biotin, a vitamin of low molecular weight from egg yolk; and avidin, a glycoprotein from egg white; or streptavidin, a protein from *Streptomyces avidinii* (see Buckland, 1986). Avidin and streptavidin can be labeled through biotin bridges (with or without spacer groups) with several labels such as isotopes, fluorophores, enzymes, and colloidal gold (Anonymous, 1984; Bonnard et al., 1984). However, enzymes and in particular peroxidases have so far been preferred (see Boorsma et al., 1986). As Figure 4.1c explains, each biotin moiety attached to the AB molecule gives rise to the binding of up to three peroxidase molecules. As several biotin groups may be attached to the AB without affecting its binding properties, the amount of label increases even more, causing a highly amplified final signal over the section.

The more convenient indirect method implicates the biotinylation of second ABs which can be used for several purposes. The biotinylated reagents can all be made as described by Hsu et al. (1981) and for streptavidin by Bonnard et al. (1984). They can also be purchased from commercial suppliers (E.Y. Laboratories, San Mateo, CA). Complete staining kits are also available—for instance, the ABC kit (Vectastain, Vector Laboratories). We found that this kit gives a stronger signal than the PAP method, provided the components

Table 4.1. Comparison of Sensitivity of Different Staining Methods[a]

Method	Dilution factor 1st antibody	Second antibody	Third step	Label	Staining Intensity
Fluorescence	1:2,000	Swine anti-rabbit FITC[b] 1:50	—	FITC	Good
Immunogold	16,000	Goat anti-rabbit gold[c] 20 nm, 1:30	Silver intensification[d]	Gold-silver	Good
Avidin-biotin	16,000	Goat anti-rabbit biotin[e] 1:200	Avidin peroxidase[e] 1:100	DAB	Good
PAP	8,000 16,000 32,000 64,000	Swine anti-rabbit[b] 1:50	Rabbit PAP[b] 1:200	DAB	Good Weak Very weak Negative
Streptavidin-biotin	8,000 16,000 32,000 64,000	Goat anti-rabbit biotin[e] 1:200	Streptavidin peroxidase[f] 1:100	DAB	Strong Good Acceptable Weak

[a] Paraffin sections of GPA-fixed glandular cells in the corpora cardiaca of *Locusta migratoria* were incubated overnight with different dilutions of anti-AKH antiserum 433.
[b] Nordic Immunological Laboratories, Tilburg, The Netherlands.
[c] Janssen Pharmaceutica, Beerse, Belgium.
[d] According to Timm's intensification method (Obermayer and Strausfeld, 1980).
[e] Vector Laboratories, Burlingame, CA.
[f] Amersham International, Amersham, U.K.

of the reagents are mixed more accurately than recommended (droplet counting).

A possible drawback of avidin is that it has a high isoelectric point (above 10) and it sticks to DNA and cell membranes because it is glycosylated. In contrast, streptavidin has an isoelectric point between 5.5 and 6.5 and shows less unspecific binding and lower background and should therefore be a better candidate for ICC (see Buckland, 1986). Streptavidin can be provided with several labels, including, ^{125}I, enzymes (Amersham International) and colloidal gold (Janssen Pharmaceutica). An important point is that the number of labeled groups per streptavidin molecule is practically unlimited, so that elaborate molecules can be constructed (Figure 4.1c). It remains to be seen if steric hindrance of these large molecules requires that the technique be used only in cases of exceptionally low Ag concentrations. Our first experiences with a streptavidine kit (Amersham) are also encouraging (see section VII.6). It appeared that the concentration of streptavidine conjugate must be increased fourfold with regard to the recommended concentration; a 100-fold dilution was satisfactory. To obtain a staining intensity comparable to that of the PAP or ABC method, the first AB could be diluted much further (Table 4.1; Fig. 4.2D).

4. Protein A Techniques

Protein A from the bacterium *Staphylococcus aureus* shows the ability to bind to the Fc portion of several types of IgG in the tissue as well as to colloidal gold particles. This makes the protein A–colloidal gold complex a suitable marker for antigen-antibody reactions in ICC at the EM level (Roth et al., 1978; Roth, 1984). This simple two-step procedure has found wide applications for post-embedding ICC. The following factors have contributed to that success: (1) one reagent (i.e., protein A complexes) can be used to demonstrate several IgG classes of different animal species; (2) protein A–gold particles can be produced in different nonoverlapping sizes for double-staining procedures (Slot and Geuze, 1984); and (3) protein A may react in a 1 : 1 ratio with the primary antibody, which is useful for quantitative ICC. The complexes are commercially available in sizes from 5 to 15 nm (Janssen Pharmaceutica). A staining procedure used successfully for the demonstration of AKH containing secretion granules in Lowicryl-embedded corpora cardiaca of the migratory locust (Diederen et al., 1987) is given in Appendix 9 (Fig. 4.3).

Protein A–gold has been used for LM ICC (plastic sections) (Roth, 1984), but research reports using this method are still scarce. The protein A can be provided with a variety of other labels including enzymes and fluorophores. Some of these are commercially available (Janssen Pharmaceutica). These conjugates can probably be used in most of the cases where labeled, second-layer antisera are used. However, as protein A is easy to use and recognizes most IgG classes, it is widely used for EM ICC. Moreover, it can be used as an alternative for second antisera in double-labeling experiments.

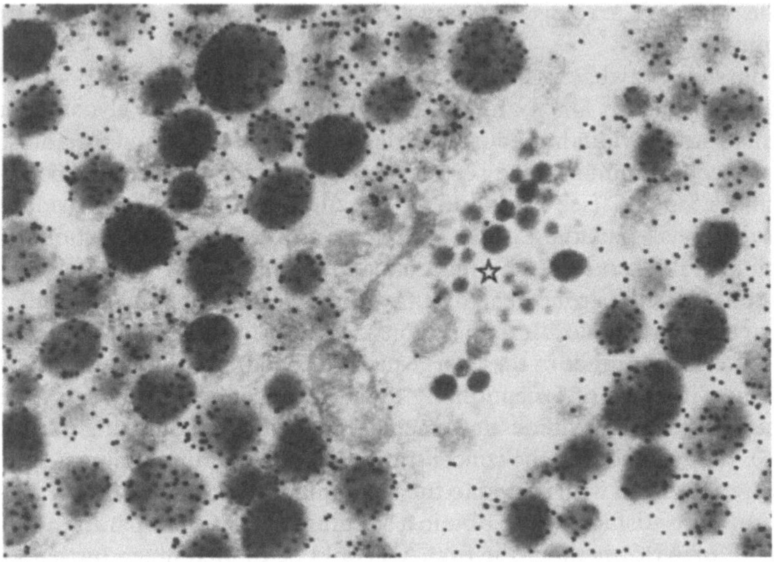

Figure 4.3. Electron micrograph of glandular cell of corpus cardiacum of *Locusta migratoria*. The location of secretory granules containing AKH-like substances revealed by anti-AKH antiserum 433 (diluted 1 : 3,000) is demonstrated by the protein A–immunogold method using 16-nm gold particles. Note that granules in the nerve fiber (asterisk) invading the glandular tissue are left "unstained." (Tissue fixed in 2% glutaraldehyde and 2% paraformaldehyde in buffer and embedded in Lowicryl) (Courtesy Dr. JHB Diederen, State University of Utrecht, The Netherlands). Magnification × 100,000.

5. Controls: Method Specificity

Several control experiments must be conducted to ensure that the ICC reaction is specific and that the observed label is a true reflection of the binding of AB with an Ag. A false-positive reaction may be caused by any of the following factors.

1. Presence in the tissue of substances giving the impression of a positive reaction, such as fluorescent groups (plant debris in gut lumen, trachea) or dark pigments (melanin, remains of larval stemmata in adult holometabolous brain, necrotic tissue within tumors) resembling reaction products of the peroxidase substrate. Such phenomena can be excluded by deparaffinizing the sections and inspecting them in the absence of any staining step.
2. Presence of compounds causing a positive reaction by themselves, such as endogenous peroxidase or other enzymes reacting with the final enzyme substrate. This can be checked by incubating the sections with that particular enzyme substrate alone.

3. Nonspecific binding of the first or second AB to any tissue constituent. This possibility can be checked by performing the whole ICC procedure in the absence of one or both antiserum incubation steps.
4. The presence in one of the ICC solvents of groups recognized by one of the reagents. An example is the presence of IgG in, for instance, normal swine serum (NSS) due to an earlier infection of the donor hog. The addition of 10% NSS to antiserum dilutions is meant to reduce nonspecific binding of reagents by occupying unspecific protein binding sites in the tissue. But this and other IgGs will react with protein A if that is used as the second layer for Ag detection, which will cause a very high background. BSA can be better used in that case than NSS.
5. The presence in the blood of the host species of immunoreactive ABs developed during the host's life prior to immunization.

Having performed these control steps, positive staining may indeed indicate binding of the antiserum. As will be explained in more detail in section IX.A, this does not necessarily mean that a particular Ag is present in the tissue. A common procedure to confirm that an AB-Ag reaction is "specific" is to preabsorb the antiserum with the Ag or immunogen and check if the staining is then completely inhibited. This control experiment is not always easy to perform for a variety of reasons.

One reason may be that the immunogen for monoclonal ABs is not available in sufficient quantities or that it is impure—e.g., if tissue extracts or homogenates are used. Another, frequently encountered reason is that the affinity of the AB for tissue Ags is stronger than for Ags in the preincubation solution. Both types of binding are equilibrium reactions, and prebound Ag may dissociate readily during this control staining of the tissue. This problem can be alleviated by taking the AB-Ag complexes out of the solution prior to ICC. In the case of immunogens having more than one epitope, such as proteins, AB-Ag complexes will be formed that precipitate and can be removed by centrifugation. In the case of smaller Ags, the easiest way is to couple the Ag chemically to a solid-phase carrier, let immune complexes form on this carrier, and finally remove these complexes from the medium through centrifugation. Several carriers are available for these "solid-phase" absorption experiments, but the one most often and easily used for peptide and protein conjugation is the material provided in beads: CNBr-activated Sepharose 4B and activated CH-Sepharose (Pharmacia Fine Chemicals). Antigens can be coupled according to instructions provided by the manufacturer (see Veenstra et al., 1985a). The adsorption and centrifugation routine is carried out three times for complete removal of ABs, each time using an amount of carrier-bound Ag as indicated below for the liquid-phase adsorption experiments. However, if Ag availability is no problem, the procedure may be carried out once, using higher concentrations of carrier-bound Ag.

A final reason for adsorption failure is that ABs are not always formed only to the Ag proper, but also to the chemical moiety linking such (small) Ags to a larger carrier molecule (see below). Preincubation of the antiserum with only the unconjugated Ag is often relatively ineffective—e.g., in the case of biogenic amines conjugated to proteins by glutaraldehyde. To block an antidopamine antiserum, a dopamine concentration of 10^{-3} M was required, compared with 10^{-5} M when the dopamine-glutaraldehyde-BSA conjugates were used (Vieillemaringe et al., 1984). GABA staining in the antennal lobe of the brain of *Manduca sexta* could be blocked by treating the antisera with 60 nM GABA-glutaraldehyde-BSA or 24 nM GABA-glutaraldehyde-keyhole-limpet hemocyanin conjugate (Hoskins et al., 1986). Here and in the case of serotonin (Bishop and O'Shea, 1983) and acetylcholine antisera (Geffard et al., 1985b), preadsorption with the whole conjugate is indicated.

It is a matter of some debate how much Ag should be used to fully adsorb all reactive sites of the antiserum. No guideline can be given. We routinely use a quantity of 10 nM Ag per milliliter diluted antiserum, in overnight shaking incubations at 4°C (Veenstra et al., 1985b) as a starting value. This quantity is perhaps too high for precious Ags or for large proteins. We thus used 2 mg thyroglobulin (MW 650,000) per milliliter antiserum, which was sufficient to eliminate all antithyroglobulin IgG after centrifugation of the resulting precipitate (Schooneveld et al., 1987a). Other investigators use standard peptide doses—e.g., 100 μg peptide per milliliter (El-Salhy et al., 1983). As we showed in a study on the occurrence of AKH-like peptides in corpora cardiaca of *Locusta migratoria* and *Periplaneta americana* (Schooneveld et al., 1985), 10 nM AKH was required to prevent staining of the glandular cells by the antiserum recognizing the N-terminal portion of the molecule, whereas 0.1 nM of AKH residues 1–4 and associated spacer group was sufficient to prevent the staining by the same antiserum. The latter AKH fragment, conjugated to thyroglobulin, was the authentic Ag.

It will be clear that each AB-Ag interaction requires experimentation, preferably with a tissue containing the authentic Ag, to determine the effective dose for immunoadsorption.

6. Comparison of Method Sensitivity

The sensitivity of the available ICC methods is an important factor for choosing the appropriate method for a particular application. If one has a poor antiserum, the most sensitive method is recommended, as this allows relatively low serum concentrations and minimal background staining. We compared a number of current methods and determined the dilution factor necessary to obtain comparable staining intensities for each method. As test tissue we used the glandular lobes of the corpora cardiaca of the migratory locust; these lobes contain cells with widely varying degrees of Ag accumulation.

When using an antiserum (code 433) raised to the amino acids 1–4 at the N-terminal end of locust AKH, most cells stained dark, but others stained intermediate or light. This range of reactions makes it possible to test various concentrations of primary ABs and still observe cells that are properly stained. Results are given in Table 4.1.

Comparing the staining intensities yielding a "good" staining, the fluorescence method appears to be the least sensitive. The PAP method is some four times more sensitive, but the immunogold-silver, the ABC, and the streptavidin methods are each eight times as sensitive, under given conditions. These data must be supplemented by two other considerations. One is that using shorter incubation times, 1 h at room temperature, the ABC method performs relatively well and one-quarter the antiserum concentration gives the same result as the PAP method. The other consideration is that the streptavidine method still performs well, though weakly, at a concentration where the PAP method gives negative results. Even at a 64,000-fold dilution of the anti-AKH serum, some of the cells in the gland are still stained whereas others are not. The signal-to-noise ratio has improved (Fig. 4.2D).

It is difficult to tell how the Ag-to-label relationship evolves in such a diverse tissue. Sternberger and Sternberger (1986) studied the relation between antiserum concentration, Ag, and background staining through measurement of optical densities of label in model tissue sections stained with the PAP and ABC methods. They concluded that the PAP method was always more sensitive than the ABC method. However, with very low antiserum concentrations, the ABC staining intensity dropped less sharply than the PAP staining intensity. (Moreover, the avidin bound nonspecifically to certain cell constituents.) This phenomenon is explained by the fact that the large avidin molecule needs considerable space:

> On a widely dispersed antigen (low antigen concentration) or a widely dispersed first antibody (low antibody concentration), a large deposit of peroxidase is attached to a single immunoreactive site.... With increasing concentrations of antigen or of first antibody, steric hindrance of the complex occurs and staining intensity is not significantly increased. Consequently, it is difficult to differentiate high and low concentrations of antigen ... (Sternberger and Sternberger, 1986).

The same reasoning may be applied to our findings of continued staining of glandular cells with streptavidin (ABC has not been tested).

7. Demonstration of Multiple Antigens

It is often desirable to demonstrate different Ags simultaneously—for instance, when the question is raised whether or not different Ags are colocalized in certain cells. Double- or multiple-staining techniques have been developed that are essentially based on two situations. For details see Sternberger (1979), Varndell *et al.* (1982), Mason (1983), and Vandesande (1983).

1. Immunoreactive structures are located in two or more sections, and sections can be stained separately. The paraffin or plastic sections are distributed alternately over two slides, and each slide is stained with a specific antiserum. By comparing adjacent paraffin sections with two microscopes and superimposing the images, the relative position of IR structures can easily be studied. Veenstra and Schooneveld (1984) thus demonstrated that bovine pancreatic polypeptide and FMRFamide-like immunoreactivities were colocalized in certain neurons of the brain, subesophageal ganglion, and stomatogastric nervous system. Duve and Thorpe (1984) observed that gastrin/CCK and secretin-like immunoreactivities are colocalized in corpus cardiacum glandular cells of the blowfly. The alternate section technique also allows the comparison of subcellular structures if semithin Epon sections are prepared in this way. Peptidergic axon profiles and terminals in the corpora cardiaca of the Colorado potato beetle could be demonstrated to contain bovine pancreatic polypeptide as well as FMRFamide-like immunoreactivities (Veenstra and Schooneveld, 1984). The alternate section technique is the technique of choice whenever an interference of antisera or markers is to be expected on single sections. The same holds true for ultrathin sections. To answer the question whether adipokinetic hormones I and II in corpora cardiaca of *Locusta* and *Schistocerca* are colocalized in the same glandular cell, Diederen et al. (1987) incubated neighboring sections with antibodies specific for either AKH I or AKH II. It was thus demonstrated that both peptides occur in the same cells and even in the same secretory granules.
2. Immunoreactive structures are to be demonstrated in one section. Two approaches are possible, depending on whether different immunoreactivities should be revealed at the same time, or one at a time:
   a. To reveal two or more Ags at the same time, different labels must obviously be used to identify each Ag. Any two contrasting labels may be used—i.e., different enzyme substrates, fluorophores (FITC, rhodamine), enzyme-colloidal gold, enzyme-fluorescent label, etc.—whereas for EM purposes colloidal gold particles of different size can be used (conjugated with either IgG or protein A). An elegant situation arises when the antisera are raised in different species. The antisera can then be applied as a cocktail, and each type of AB can, after binding, be reacted with affinity-purified second ABs carrying different labels (Grimmelikhuijzen, 1983). The result is a two-colored slide, the production of which takes no more time than a single staining procedure, at least when using fluorophore-labeled second ABs.
   b. Both types of Ags are most often revealed successively, in an extended procedure. The main advantage is that antisera can be used that have been raised in the same species. The first normal ICC procedure should result in an insoluble marker, usally DAB. Conditions are chosen so that the second layer fully reacts with the first AB, or a contamination of the later procedure may result. In the procedure of Sternberger and Joseph (1979), the second procedure may simply follow the first without the removal of the first AB. More often, however, the method of Nakane (1968) is used, consisting of the removal of the first AB by

treatment of the sections after DAB staining with 0.1 M glycine-HCl buffer pH 2.2 for 1 h and extensive rinsing. Then the other primary AB is applied, followed by a contrasting label. In the Colorado patato beetle, proctolin proved to be colocalized with a FMRFamide-like substance in the subesophageal ganglion, but neurons in the frontal ganglion contain either one or the other peptide (Veenstra et al., 1985a). Vandesande (1983) took extra precautions to eliminate the first AB by electrophoresis (20 V/cm) in a glycine-HCl buffer pH 2.5. Verhaert et al. (1985) thus demonstrated the relative position of neurons in the brain of the American cockroach containing an AKH or FMRFamide-like peptide.

Double-labeling has not nearly so often been applied in EM ICC. Varndell et al. (1982) and Varndell and Polak (1984) suggested that different Ags in neighboring cells can be stained selectively using colloidal gold particles of different sizes. An interesting suggestion is to use both upper and lower sides of an EM section for independent ICC staining, again with different gold sizes (Anonymous, 1985; Bendayan, 1982). The sections should then be mounted on a gold or nickel grid without further Formvar support; we find this advisable in any event for reducing the formation of precipitate between grid coating and section.

VIII. Lectin Cytochemistry

Lectins are sugar-binding proteins or glycoproteins of plant or animal origin. They are not antibodies. The basis of their use in pathology and cell biology is their ability to recognize complex carbohydrate structures in glycoproteins and glycopeptides, in particular those in cell membranes (Ponder, 1983; Lis and Sharon, 1986). Some of the lectins are commonly used, and FITC and peroxidase-labeled ABs to concanavalin A and other lectins with wide applications are commercially available (BioRad, Dakopatts, E.Y. Laboratories, Miles Scientific, Vector Laboratories). Also, lectins can be directly provided with the usual markers for ICC, but their label density is much reduced as compared with the use of the indirect method (Horisberger, 1984).

Only a few lectin studies have so far been made in insects. Perrone et al. (1986) used a panel of seven biotin-labeled lectins to describe the carbohydrates of the basal lamina in mosquito salivary glands, embedded and sectioned for LM and EM. Some of the lectins associated with specific salivary gland lobes, which indicated the presence of unique surface receptors determining perhaps the specificity of malaria sporozoite invasion. Denburg et al. (1983) used frozen sections to demonstrate chemical differences in the surface of muscles in the cockroach leg. These differences might play a role in the process of cell recognition by the regenerating innervating motor axons.

IX. General Discussion

1. Antiserum Specificity: Possibilities and Impossibilities

The wide applicability of ICC and the high resolution, selectivity, and specificity that can be achieved make it the most powerful and fascinating general histochemical technique available. Although the specificity is very high, it is by no means absolute. Once the method specificity has been established (see section VII.E), one may call the AB-Ag binding "specific." However, this does not necessarily mean that the AB recognizes in the tissue the same molecule to which it was raised. It only means that the AB recognizes a substance that is immunologically indistinguishable from the Ag, at least by this particular AB or antiserum. For this reason, one usually refers to Ag-"like" immunoreactivity. In some cases, the immunoreactive substance in the sections will be identical to the Ag used to raise the antiserum, in which case they are ideal "likes." In most cases, the substances will be similar to the Ag but not identical. Vasopressin and oxytocin, for instance, vertebrate nonapeptides differing in only two amino acid residues, can usually be differentiated by ICC only after special treatment of the antisera. Oxytocin cross-reacting ABs must be removed from the vasopressin antiserum, and vasopressin cross-reacting ABs from the oxytocin antiserum (see Vandesande and Dierickx, 1975; Swaab et al., 1977; Van Leeuwen et al., 1977). Unless the cross-reacting ABs have been removed, both types of antisera will react with both vasopressin and oxytocin producing cells in the hypothalamus, whereas absorption with oxytocin and vasopressin will qualify the immunoreaction as "specific."

Other immunocytochemical "likes" are so different from each other that one may be amazed that they are recognized by the same antiserum. A well-documented example from the vertebrate literature is that antisera to α-MSH also react with neurofilaments, proteins of much larger size than α-MSH. The original ICC observations of α-MSH-like immunoreactive material suggested a much wider distribution of this peptide in the nervous system than later turned out to be the case. The cross-reactivity appears to be due to a resemblance in small fragment of the primary structure of these molecules (see Drager et al., 1983; Shaw et al., 1985). For this reason, one can never be sure of the chemical structure of ICC-localized substances (see Swaab et al., 1977; Van Leeuwen et al., 1982). Other independent assays will always be necessary to reveal the true identity of an ICC-localized molecule.

Yet, these considerations will hardly present a problem where the immunogen in question was first isolated from the test tissue. Thus, insect vitellogenins can be demonstrated and located quite adequately in vitellogenin-containing tissue such as fat body in *Calliphora vicina* (Thomsen et al., 1980), ovary of *Leptinotarsa decemlineata* (Peferoen et al., 1986), and developing eggs of *Blattella germanica* (Tanaka, 1977) with antisera raised against vitellin from the eggs (see Chapters 1, 2). Where the Ag is small, such as in neurotransmitters, and where cross-reactions may easily occur, the useful antisera resulting from the immunization procedure should be carefully selected for their specificity. Moreover, related transmitter substances should be checked for possible cross-reaction (Hoskins et al., 1986). The possibility exists that hitherto unrecognized substances may one day be shown to react with the antiserum.

More problems are usually encountered with peptides, especially in cases where no positive peptide identification has been carried out by analytical chemistry. Conclusions as to the identity of ICC-identified peptides are close to impossible. A good example is the apparent colocalization of pancreatic polypeptide- and FMRFamide-like material in the same neuronal structures in the Colorado potato beetle. Cross-absorption experiments showed that the two antisera recognize the same substance. Based on ICC evidence, this substance seems to be more similar to FMRFamide than to pancreatic polypeptide (Veenstra and Schooneveld, 1984). Such cross-reactivity has also been reported for vertebrates. Thus FMRFamide-like immunoreactive material was originally reported to be present in chicken pancreas (Dockray et al., 1981). Later it was shown that this material was pancreatic polypeptide (Dockray and Williams, 1983). For this reason, it is highly recommended that antisera be carefully characterized by ICC on tissue with known peptides. Although many antisera used in ICC have been characterized as highly specific in RIA, they do not necessarily exhibit the same specificity in ICC (Swaab et al., 1977; Bishop et al., 1981).

2. Potential Uses of ICC

The problems with interpretation just described do not invalidate ICC as a research tool. We feel that ICC will become an increasingly important technique in experimental entomology and wish to suggest only a few of the many exciting possibilities. It seems that the search for proteinaceous substances has profited more from the ICC than other research areas.

One can demonstrate the presence and cellular localization of a wide variety of substances of interest. Limitations are set only by the suitability of the substances as immunogen, possibly after conjugation with larger proteins, and the necessity that the cellular substance be immobilized to the substrate by fixation to prevent it from being washed out. Examples of this approach with polyclonal antisera are numerous: (1) the localization of vitellogenins in fat body, ovary, and embryo (see section III.C); (2) the release of trypsin in the gut of the mosquito after a bloodmeal (Graf et al., 1986); (3) the localization in the CNS of *Periplaneta americana* of proctolin (O'Shea and Schaffer, 1985); (4) the localization in the brain of *Manduca sexta* of the eclosion hormone-producing neurons (Copenhaver and Truman, 1986); and (5) the localization in the CNS, gut, and other organs of neuropeptides recognized by antisera to "vertebrate" peptides (Yui et al., 1980; Hansen et al., 1982; El-Salhy et al., 1983; Duve and Thorpe, 1984; Veenstra et al., 1985; Verhaert and De Loof, 1985; Kramer, 1985). The nature of these immunoreactive materials detected in this way remains uncertain without further chemical identification.

The search for known, partly known, or entirely unknown neuropeptides or other neuronal substances will probably continue. Many more neuron types with ICC-characterized content will be found, to the delight of neuroanatomists. Estimates of the proportion of peptidergic neurons in invertebrates range from 50% to 80% by different investigators (see Scharrer, 1985). But the physiological meaning of these substances remains conjectural.

We have seen (section VI; Chapter 2) that the novel technique of monoclonal ABs has been introduced to find new types of neurons with greater ease. It seems rather simple to obtain large numbers of ABs with a high degree of specificity by injecting mice with rather crude homogenates of, for instance, parts of the CNS. One then finds certain sets of neurons and their processes and acquires new possibilities for identifying neuronal and neurohormonal circuits. Moreover, the extraction of those neurochemical mediators would become feasible using the appropriate immunoaffinity chromatographic procedures. With the progress in methodology for peptide sequencing, chemical identification is within reach. It may be anticipated that ICC methods will play an increasingly important role in neurobiology by directing attention to suitable model circuits.

We have seen that double-labeling experiments make it possible to study the coexistence of peptides and neurotransmitters in certain neurons. It becomes clear from work on both vertebrates and invertebrates that transmitters and peptides are coreleased at synapses and neurohemal sites (see Boer and Van Minnen, 1987). This arrangement indicates that neurons can convey signals of different nature; the effector cells near that area determine which part of the mixture of signals is actually utilized. ICC studies will undoubtedly contribute to the further development of that concept.

Finally, there is the study of homologous neurons. It proved possible to identify neurons and neurosecretory cells in many different insect orders which share certain staining properties, numbers, and positions in the CNS and which are therefore considered to be homologous and to have a common origin (Veenstra, 1984; Veenstra and Yanaihara, 1984; Veenstra et al., 1985b). These observations may further stimulate the development of the concept of the unique identifiable neurons as formulated by Rowell (1977) on the basis of conventional histological staining of neurons and neurosecretory cells. ICC methods now provide better tools. Moreover, our work involving the production and use of antisera with different specificity to AKH reveal AKH-immunoreactive cells in locusts and many other species, suggesting homologies as well as differences in the glandular corpus cardiacum cells (Schooneveld et al., 1983, 1985, 1986, 1987a,b). Also, the cell products proved to be different in the various species, which is in line with recent findings on the nature of AKHs in different species.

It came as a surprise that immunoreactive substances were also present in certain neuron populations in all species investigated. Whereas the localization of the substances in neurons suggests that they play some role as a central neurotransmitter or even as a neurohormone in some cases, the nature of this material is unknown. It was recently confirmed by Moshitzky et al. (1987) that the brain of *Locusta* contains authentic AKH I. It may well be the substance visualized by us in interneurons in this species. Immunoreactive AKH-like substances were present also in the CNS of other invertebrates, including a mollusk (Schooneveld et al., 1987a). We are less confident that this substance is closely related to AKH. But it seems worthwhile to explore the possibility that AKH-related peptides are widely distributed in the animal kingdom, as has been found earlier for certain other neuropeptides (see Scharrer, 1985).

It is clear that ICC comprises powerful techniques for visualizing materials in tissues, but its results can be rendered more profitable when used in conjunction with chemical analyses.

Acknowledgments. We thank Mr. H.M. Smid for carrying out some of the experiments for the completion of the data in this paper, and Dr. J.H.B. Diederen for communicating some of his experimental ICC procedures.

References

Agricola, H., Eckert, M., Ude, J., Birkenbeil, H., and Penzlin, H. (1985). The distribution of proctolin-like immunoreactive material in the terminal ganglion of the cockroach, *Periplaneta americana* L. *Cell Tissue Res.* **239**:203–209.

Anonymous. (1982). Lowicryl K4M. Lowicryl HM20. Instructions for use. Chemische Werke Lowi GmbH, Waldkraiburg, F.R.G.

Anonymous. (1984). The biotin-streptavidin system for ELISA, RIA, immunocytochemical and protein blotting applications. Amersham International, Amersham, U.K.

Anonymous. (1985). Auroprobe EM. Secondary antibody-coated colloidal gold for use in the electron microscopic immunogold staining (IGS) method. Janssen Pharmaceutica, Beerse, Belgium.

Armbruster, B.L., Kellenberger, E., Carlemalm, E., et al. (1984). Lowicryl resins—present and future applications. In: *Proc. 2d Pfefferkorn Conf. Sci. Biol, Spec. Prep. Micr. Microanal.*, 77–81. SEM Inc., AMF O'Hare.

Barlow, D.J., Edwards, M.S., and Thornton, J.M. (1986). Continuous and discontinuous protein antigenic determinants. *Nature* **322**:747–748.

Barnstable, C.J. (1985). Monoclonal antibodies as molecular probes of the nervous system. See: ref. T.A. Springer (1985): pp. 269–289.

Baumiger, S., and Wilchek, M. (1980). The use of carbodiimides in the preparation of immunizing conjugates. *Methods Enzymol.* **70**:151–159.

Bendayan, M. (1982). Double immunocytochemical labeling applying the protein A-gold technique. *J. Histochem. Cytochem.* **30**:81–85.
Birkenbeil, H., and Eckert, M. (1982). Untersuchungen zur Methodik des ultrastrukturellen immuncytochemischen Nachweises von Ecdysteroiden. *Acta Histochem.* **71**:133–143.
Bishop, C., and O'Shea, M. (1982). Neuropeptide proctolin (H-Arg-Tyr-Leu-Pro-Thr-OH): Immunocytochemical mapping of neurons in the central nervous system of the cockroach. *J. Comp. Neurol.* **207**:223–238.
Bishop, C., and O'Shea, M. (1983). Serotonin immunoreactive neurons in the central nervous system of an insect (*Periplaneta americana*). *J. Neurobiol.* **14**:251–269.
Bishop, C., O'Shea, M., and Miller, R.J. (1981). Neuropeptide proctolin (H-Arg-Tyr-Leu-Pro-Thr-OH): Immunological detection and neuronal localization in insect central nervous system. *Proc. Natl. Acad. Sci. USA* **78**:5899–5902.
Boer, H.H., and Van Minnen, J. (1987). Immunocytochemistry of hormonal peptides in molluscs: Optical and electron microscopy and the use of monoclonal antibodies. *Bull. Soc. Zool. Fr.* **111**:14.
Boer, H.H., Schot, L.P.C., Veenstra, J.A., and Reichelt, D. (1980). Immunocytochemical identification of neural elements in the central nervous systems of a snail, some insects, a fish, and a mammal with an antiserum to the molluscan cardioexcitatory tetrapeptide FMRF-amide. *Cell Tissue Res.* **213**:21–27.
Bonnard, C., Papermaster, D.S., and Kraehenbühl, J.-P. (1984). The streptavidine-biotin bridge technique: Application in light and electron microscope immunocytochemistry, In: *Immunolabelling for Electron Microscopy*. J.M. Polak and I. Varndell (eds.), pp. 95–111. Amsterdam: Elseirer.
Boorsma, D.M., Van Bommel, J., and Van den Heuvel, J. (1986). Avidine-HRP conjugates in biotin-avidin immunoenzyme cytochemistry. *Histochemistry* **84**:333–337.
Buckland, R.M. (1986) Strong signals from streptavidin-biotin. *Nature* **320**:557–558.
Carlemalm, E., Villiger, W., Hobot, J.A., Acetarin, J.-D., and Kellenberger, E. (1985). Low temperature embedding with Lowicryl resins: Two new formulations and some applications. *J. Microsc.* **140**:55–63.
Causton, B.E. (1984). The choice of resins for electron immunocytochemistry. In: Polak and Varndell (eds): 29–36. Amsterdam: Elseirer.
Coons, A.H., and Kaplan, M.H. (1950). Localization of antigens in tissue cells. II. Improvement in a method for the detection of antigen by means of fluorescent antibodies. *J. Exp. Med.* **91**:1–13.
Copenhaver, P.F., and Truman, J.W. (1986). Identification of the cerebral neurosecretory cells that contain eclosion hormone in the moth *Manduca sexta*. *J. Neurosci.* **6**:1738–1747.
Cuello, A.C. (1983). Immunohistochemistry. In: *Methods in the Neurosciences*, Vol. 3. Chichester, U.K.: John Wiley.
Davis, N.T. (1985). Serotonin-immunoreactive visceral nerves and neurohemal system in the cockroach *Periplaneta americana*. *Cell Tissue Res.* **240**:593–600.
De Jong, A.S.H., Van Kessel–Van Vark, M., and Raap, A.K. (1985). Sensitivity of several visualization methods for peroxidase and alkaline phosphatase activity in immunoenzyme histochemistry. *Histochem. J.* **17**:1119–1130.
De Mey, J. (1983). Colloidal gold probes in immunocytochemistry. In: Immunocytochemistry. J.M. Polak and S. Van Noorden (eds):82–112. Bristol: Wright PSG.
Denburg, J.L., Eastburn, J.L., and Caldwell, R.T. (1983). Lectin receptors in the

cockroach neuromuscular system. I. Distribution among the set of coxal depressor muscles. *Brain Res.* **289:**241–252.

Denburg, J.L., Caldwell, R.T., and Marner, J.A.M. (1986). Monoclonal antibodies to the cockroach nervous system. *J. Comp. Neurol.* **245:**123–136.

Diederen, J.H.B., Maas, H.A., Pel, Schooneveld, H., Jansen, W.F., and Vullings, H.G.B. (1987). Colocalization of the adipokinetic hormones I and II in the same glandular cells and in same secretory granules of locust corpus cardiacum of *Locusta migratoria* and *Schistocerca gregaria.* An immuno-electron microscopic study. *Cell Tissue Res.* **249:**379–389.

Dockray, G.J., and Williams, R.G. (1983). FMRFamide-like immunoreactivity in rat brain: Development of a radioimmunoassay and its application in studies of distribution and chromatographic properties. *Brain Res.* **266:**295–303.

Dockray, G.J., Vaillant, C., and Williams, R.G. (1981). New vertebrate brain-gut peptide related to a molluscan neuropeptide and an opioid peptide. *Nature* **293:**655–657.

Drager, U.C., Edwards, K.L., and Kleinschmidt, J. (1983). Neurofilaments contain alpha-melanocyte-stimulating hormone (alpha-MSH)-like immunoreactivity. *Proc. Natl. Acad. Sci. USA* **80:**6408–6412.

Duve, H., and Thorpe, A. (1983). Immunocytochemical identification of vertebrate-type brain-gut peptides in insect nerve cells. In: *Functional Neuroanatomy.* N.J. Strausfeld (ed.). Berlin: Springer-Verlag, pp. 250–266.

Duve, H., and Thorpe, A. (1984). Immunocytochemical mapping of gastrin/CCK-like peptides in the neuroendocrine system of the blowfly *Calliphora erythrocephala* (Diptera). *Cell Tissue Res.* **237:**309–320.

Duve, H., Thorpe, A., and Strausfeld, N.J. (1983). Cobalt-immunocytochemical identification of peptidergic neurons in *Calliphora* innervating central and peripheral targets. *J. Neurocytol.* **12:**847–861.

Eckert, M., and Gersch, M. (1978) Immunochemical investigations on the neuroendocrine system of the cockroach, *Periplaneta americana* L. In: *Neurosecretion and Neuroendocrine Activity: Evolution, Structure and Function.* W. Bargmann, A. Oksche, A. Polenov, and B. Scharrer (eds.). Berlin: Springer-Verlag, pp. 365–369.

Eckert, M., and Ude, J. (1983). Immunocytochemical techniques for the identification of peptidergic neurons. In: *Functional Neuroanatomy.* N.J. Strausfeld (ed.). Berlin: Springer-Verlag, pp. 267–301.

Eckert, M., Agricola, H., and Penzlin, H. (1981). Immunocytochemical identification of proctolinlike immunoreactivity in the terminal ganglion and hindgut of the cockroach *Periplaneta americana* (L.). *Cell Tissue Res.* **217:**633–645.

El-Salhy, M., Falkmer, S., Kramer, K.J., and Speirs, R.D. (1983) Immunohistochemical investigations of neuropeptides in the brain, corpora cardiaca, and corpora allata of an adult lepidopteran insect, *Manduca sexta* (L.). *Cell Tissue Res.* **232:**295–317.

Faulk, W.P., and Taylor, G.M. (1971) An immunocolloid method for the electron microscope. *Immunochemistry* **8:**1081–1083.

Fujita, S.C. Zipursky, S.L., Benzer, S., Derrus, A., and Shotwell, S.L. (1982). Monoclonal antibodies against the *Drosophila* nervous system. *Proc. Natl. Acad. Sci. USA* **89:**7929–7933.

Geffard, M., McRae-Degueurce A., Souan M.L. (1985a) Immunocytochemical detection of acetylcholine in the rat central nervous system. Science **229:**77–79.

Geffard, M., Viellemaringe, J., Heinrich-Rock, A.-M., and Duris, P. (1985b) Anti-acetylcholine antibodies and first immunocytochemical application in insect brain. *Neurosci. Lett.* **57**: 1–6.

Giloh, H., and Sedat, J.W. (1982). Fluorescence microscopy: Reduced photobleaching of rhodamin and fluorescein protein conjugates by *n*-propyl gallate. *Science* **217**: 1252–1255.

Graf, R., Raikhel, A.S., Brown, M.R., Lea, A.O., and Briegel, H. (1986). Mosquito trypsin: Immunocytochemical localization in the midgut of blood-fed *Aedes aegypti* (L.). *Cell Tissue Res.* **245**: 19–27.

Grimmelikhuijzen, C.J.P. (1983) Coexistence of neuropeptides in *Hydra*. *Neuroscience* **9**: 837–845.

Hacker, G.W., Springall, D.R., Van Noorden, S., Bishop, A.E., Grimelius, L., and Polak, J.M. (1985). The immunogold-silver staining method. A powerful tool in histopathology. *Virchows Arch. [Pathol. Anat.]* **406**: 449–461.

Hansen, B.L., Hansen, G.N., and Scharrer, B. (1982). Immunoreactive material resembling vertebrate neuropeptides in the corpus cardiacum and corpus allatum of the insect *Leucophaea maderae*. *Cell Tissue Res.* **225**: 319–329.

Hermanns, W., Colbatzky, F., Günther, A., and Steiniger, B. (1986). Ia antigens in plastic embedded tissues: A post-embedding immunohistochemical study. *J. Histochem. Cytochem.* **34**: 827–831.

Holgate, C.S., Jackson, P., Cowen, P.N., and Bird, C.C. (1983). Immunogold-silver staining: New method of immunostaining with enhanced sensitivity. *J. Histochem. Cytochem.* **31**: 938–944.

Horisberger, M. (1984). Lectin cytochemistry. In: *Immunolabelling for Electron Microscopy*. J.M. Polak and I.M. Varndell (eds.), 249–266. Amsterdam: Elsevier.

Hoskins, S.G., Homberg, U., Kingan, T.G., Christensen, T.A., and Hildebrand, J.G. (1986). Immunocytochemistry of GABA in the antennal lobes of the sphinx moth *Manduca sexta*. *Cell Tissue Res.* **244**: 243–252.

Hsu, S.-M., Raine, L., and Fanger, H. (1981) Use of avidine-biotin-peroxidase complex (ABC) in immunoperoxidase techniques. *J. Histochem. Cytochem.* **29**: 577–580.

Hsu, S.-M., and Soban, E. (1982). Color modification of diaminobenzidine (DAB) precipitation by metallic ions and its application for double immunohistochemistry. *J. Histochem. Cytochem.* **30**: 1079–1082.

Huang, S.-N., Minassian, H., and More, J.D. (1976). Application of immunofluorescent staining on paraffin sections improved by trypsin digestion. *Lab. Invest.* **35**: 383–390.

Hudson, L., and Hay, F.C. (1980). *Practical Immunology*, 2d Ed. Oxford: Blackwell.

Hurrell, J.G.R. (1982). *Monoclonal Hybridoma Antibodies: Techniques and Applications*. Boca Raton, FL: CRC Press.

Jaros, P., Dircksen, H., and Keller, R. (1985). Occurrence of immunoreactive enkephalins in a neurohemal organ and other nervous structures in the eyestalk of the shore crab, *Carcinus maenas* L. (Crustacea, Decapoda). *Cell Tissue Res.* **241**: 111–117.

Karnovsky, M.J. (1965). A formaldehyde-glutaraldehyde fixative of high osmolality for use in electron microscopy. *J. Cell Biol.* **24**: 137A–138A.

Klemm, N., Steinbush, H.W.M., and Sundler, F. (1984). Distribution of serotonin-containing neurons and their pathways in the supraoesophageal ganglion of the cockroach *Periplaneta americana* (L.). as revealed by immunocytochemistry. *J. Comp. Neurol.* **225**:387–395.

Kramer, K.J. (1985). Vertebrate hormones in insects. *Comp. Insect Physiol. Pharmacol.* **7**:511–536.

Langone, J.J., and Van Vunakis, H. (Eds.). (1983). Monoclonal antibodies and general immunoassay methods. *Methods in Enzymology* **92**, part E: *Immunocytochemical Techniques*. New York: Academic Press.

Lis, H., and Sharon, N. (1986). Lectins as molecules and as tools. *Annu. Rev. Biochem.* **55**:35–67.

Loeb, M.J. (1986). Ecdysteroids in testes sheaths of *Heliothis virescens* larvae: An immunocytochemical study. *Arch. Insect Biochem. Physiol.* **3**:173–180.

Martin, R., Schäfer, M., and Voigt, K.H. (1982). Enzymatic cleavage prior to antibody incubation as a method for neuropeptide immunocytochemistry. *Histochemistry* **74**:457–467.

Mason, D.Y., Abdulaziz, Z, Falini, B, and Stein, H. (1983). Double immunoenzymatic labelling. In: *Immunocytochemistry*. J.M. Polak and S. van Noorden (eds): 113–128. Bristol: Wright PSG.

Matsuura, S., Sugimoto, T., Nagatsu, I., Yamaguchi, T., and Nagatsu, T. (1983). A specific antiserum against tetrahydrobiopterin: Its preparation and application to immunohistochemistry. In: *Chemistry and Biology of Pteridines*. J.A. Blair (ed.). Berlin: Walter de Gruyter & Co., pp. 827–832.

Miller, C.A. and Benzer, S. (1983) Monoclonal antibody cross-reactions between *Drosophila* and human brain. *Proc. Natl. Acad. Sci. USA* **80**:7641–7645.

Milstein, C., and Cuello, A.C. (1983) Hybrid hybridomas and their use in immunohistochemistry. *Nature* **305**:537–540.

Moshitzky, P., Yamashiro, D., Stuve, L., Ramachandran, J., and Applebaum, S.W. (1987). Determination of locust AKH-I by radioimmunoassay and the identification of an AKH-like factor in the locust brain. *Insect Biochem.* **17**:765–769.

Myers, C.R., Bergtrom, G., Crook, S.M., Boyer, D.R., and Collins, M.L.P. (1986) Postembedding direct immunogold detection of a protein antigen in insect tissue. *J. Histochem. Cytochem.* **34**:221–226.

Nakane, P.K. (1968). Simultaneous localization of multiple tissue antigens using the peroxidase-labeled antibody method: A study in pituitary glands of the rat. *J. Histochem. Cytochem.* **16**:557–560.

Nässel, D.R., and Cantera, R. (1985). Mapping of serotonin-immunoreactive neurons in the larval nervous system of the flies *Calliphora erythrocephala* and *Sarcophaga bullata*. *Cell Tissue Res.* **239**:423–434.

Nässel, D.R., and Elekes, K. (1984). Ultrastructural demonstration of serotonin-immunoreactivity in the nervous system of an insect (*Calliphora erythrocephala*). *Neurosci. Lett.* **48**:203–210.

Newman, G.R., and Jasani, B. (1984). Post-embedding immunoenzyme techniques. In: J.M. Polak and I.M. Varndell (1984): 53–70.

Nishiitsutsuji-Uwo, J., Takeda, M., and Saito, H. (1984). The production of an antiserum to serotonin and serotonin-like immunoreactivity in the cockroach brain-midgut system. *Biomed. Res.* **5**:211–224.

Obermayer, M., and Strausfeld, N.J. (1980). Silver staining cobalt sulfide deposits within neurons of intact ganglia. In: *Neuroanatomical Techniques*. N.J. Strausfeld and T.A. Miller (eds.). Berlin: Springer-Verlag, pp. 407–427.

O'Shea, M. (1982). Peptide neurobiology. An identified neurone approach with special reference to proctolin. *TINS* **5**:69–73.

O'Shea, M., and Schaffer, M. (1985). Neuropeptide function: The invertebrate contribution. *Annu. Rev. Neurosci.* **8**:171–198.

Peferoen, M., Delang, I., and De Loof, A. (1986). Immunocytochemical localisation of vitellogenic and non-vitellogenic yolk proteins in the ovary of *Leptinotarsa decemlineata. Comp. Biochem. Physiol.* **83B**:347–353.

Perrone, J.B., DeMaio, J., and Spielman, A. (1986). Regions of mosquito salivary glands distinguished by surface lectin-binding characteristics. *Insect Biochem.* **16**:313–318.

Pickel, V.M. (1981). Immunocytochemical methods. In: *Neuroanatomical Tract-Tracing Methods.* L. Heimer and M.J. RoBards (eds.). New York: Plenum Press, pp. 483–509.

Polak, J.M., and Van Noorden, S. (1983). *Immunocytochemistry.* Bristol: Wright PSG.

Polak, J.M., and Varndell, I.M. (1984). *Immunolabelling for Electron Microscopy.* Amsterdam: Elsevier.

Ponder, B.A.J. (1983). Lectin histochemistry. In: *Immunocytochemistry.* J.M. Polak and S. Van Noorden (eds): 129–142. Bristol: Wright PSG.

Priestly, J.V. (1984) Pre-embedding ultrastructural immunocytochemistry: immunoenzyme techniques. In: *Immunolabelling for Electron Microscopy.* J.M. Polak and I.M. Varndell (eds): 37–52. Amsterdam: Elsevier.

Reichlin, M. (1980). Use of glutaraldehyde as a coupling agent for proteins and peptides. *Methods Enzymol.* **70**:159–165.

Roitt, I. (1980). *Essential Immunology*, 4th Ed. Oxford: Blackwell.

Roth, J. (1984). The protein A–gold technique for antigen localisation in tissue sections by light and electron microscopy. In: *Immunolabelling for Electron Microscopy.* J.M. Polak and I.M. Varndell (eds): 113–121. Amsterdam: Elsevier.

Roth, J., Bendayan, M., and Orci, L. (1978) Ultrastructural localization of intracellular antigens by the use of protein A-gold complex. *J. Histochem. Cytochem.* **26**:1074–1081.

Rowell, H.F. (1977). The cells of the insect neurosecretory system: Constancy, variability and the concept of the unique identifiable neuron. *Adv. Insect Physiol.* **12**:63–123.

Salvaterra, P.M., Crawford, G.D., Klotz, J.L., and Ikeda, K. (1985). Production and use of monoclonal antibodies to biochemically defined insect neuronal antigens. In: *Functional Neuroanatomy.* N.J. Strausfeld (ed.). Berlin: Springer-Verlag, pp. 223–242.

Scharrer, B. (1985) Invertebrate neurosecretion: A round table discussion. In: *Neurosecretion and the Biology of Neuropeptides.* H. Kobayashi, H.A. Bern, and A. Urano (eds.). Tokyo: Japan Scientific Societies Press, pp. 435–441.

Schooneveld, H. (1986). Localization of AKH/RPCH-related peptides in insects and other invertebrates. In: *Insect Neurochemistry and Neurophysiology.* A.B. Borkovec and D.B. Gelman (eds): 443–446. Clifton: Humana Press.

Schooneveld, H., and Veenstra, J.A. (1985). Insect neuroendocrine cells and neurons containing adipokinetic hormone (AKH)-immunoreactive substances. In: *Neurosecretion and the Biology of Neuropeptides.* H. Kobayashi et al. (Eds.). Tokyo: Japan Sci. Soc. Press, pp. 425–434.

Schooneveld, H., Tesser, G.I., Veenstra, J.A., and Romberg-Privee, H.M. (1983).

Adipokinetic hormone and AKH-like peptide demonstrated in the corpora cardiaca and nervous system of *Locusta migratoria* by immunocytochemistry. *Cell Tissue Res.* **230**:67–76.

Schooneveld, H., Romberg-Privee, H.M., and Veenstra, J.A. (1985) Adipokinetic hormone-immunoreactive peptide in the endocrine and central nervous system of several insect species: A comparative approach. *Gen. Comp. Endocrinol.* **57**:184–194.

Schooneveld, H., Romberg-Privee, H.M., and Veenstra, J.A. (1986). Phylogenetic differentiation of glandular cells in corpora cardiaca as studied immunocytochemically with regio-specific antisera to AKH. *J. Insect Physiol.* **33**:167–176.

Schooneveld, H., Van Herp, F., and Van Minnen, J. (1987a). Demonstration of substances immunologically related to the identified arthropod neuropeptides AKH/RPCH in the CNS of several invertebrate species. *Brain Res.* **406**:224–232.

Schooneveld, H., Romberg-Privee, H.M., and Veenstra, J.A. (1987b). Immunocytochemical differentiation between adipokinetic hormone (AKH)-like peptides in neurons and glandular cells in the corpus cardiacum of *Locusta migratoria* and *Periplaneta americana* with C-terminal and N-terminal specific antisera to AKH. *Cell Tissue Res.* **243**:9–14.

Schürmann, F.W., and Klemm, N. (1984) Serotonin-immunoreactive neurons in the brain of the honeybee. *J. Comp. Neurol.* **225**:570–580.

Shaw, G., Fisher, S., and Weber, K. (1985) Alpha-MSH and neurofilament M-protein share a continuous epitope but not extended sequences. An explanation for neurofibrillary staining with alpha-MSH antibodies. *FEBS Lett.* **181**:343–346.

Siegert, K., Morgan, P., and Mordue, W. (1985) Primary structures of locust adipokinetic hormones II. *Biol. Chem. Hoppe-Seyler* **366**:723–727.

Skowsky, W.R., and Fischer, D.A. (1972). The use of thyroglobulin to induce antigenicity to small molecules. *J. Lab. Clin. Med.* **80**:134–144.

Slot, J.W., and Geuze, H.J. (1981). Sizing of protein A–colloidal gold probes for immunoelectron microscopy. *J. Cell Biol.* **90**:533–536.

Slot, J.W., and Geuze, H.J. (1984). Gold markers for single and double immunolabelling of ultrathin cryosections. In: *Immunolabelling for Electron Microscopy*. J.M. Polak and I.M. Varndell (eds): 129–142. Amsterdam: Elsevier.

Springall, D.R., Hacker, G.W., Grimelius, L., and Polak, J.M. (1984). The potential of the immunogold-silver staining method for paraffin sections. *Histochemistry* **81**:603–608.

Springer, T.A. (1985). *Hybridoma Technology in the Biosciences and Medicine*. New York: Plenum Press.

Sternberger, L.A. (1977). Immunocytochemistry of neuropeptides and their receptors. In: *Peptides in Neurobiology*. H. Gainer (ed.). New York: Plenum Press, pp. 61–97.

Sternberger, L.A. (1979) *Immunocytochemistry*. New York: John Wiley.

Sternberger, L.A. and Joseph, S.A. (1979). The unlabelled antibody method. Contrasting color staining of paired pituitary hormones without antibody removal. *J. Histochem. Cytochem.* **27**:1424–1429.

Sternberger, L.A. and Sternberger, N.H. (1986). The unlabelled antibody method: Comparison of peroxidase-antiperoxidase with avidine-biotin complex by a new method of quantification. *J. Histochem. Cytochem.* **34**:599–605.

Sternberger, L.A. Harwell, L.W., and Sternberger, N.H. (1982). Neurotypy: Regional individuality in rat brain detected by immunocytochemistry with monoclonal antibodies. *Proc. Natl. Acad. Sci. USA* **79:** 1326–1330.

Storm-Mathisen, J., Leknes, A.K., Bore, A.T., et al. (1983). First visualization of glutamate and GABA in neurones by immunocytochemistry. *Nature* **301:** 517–520.

Swaab, D.F., Pool, C.W., and Van Leeuwen, F.W. (1977). Can specificity ever be proved in immunocytochemical staining? *J. Histochem. Cytochem.* **25:** 388–391.

Szelke, M. (1983). Raising antibodies to small peptides. In: *Immunocytochemistry.* J.M. Polak and S. Van Noorden (eds): 53–68. Bristol: Wright PSG.

Tainer, J.A., Getzoff, E.D., Alexander, H., Houghten, R.A., Olson, A.J., and Lerner, R.A. (1984). The reactivity of anti-peptide antibodies is a function of the atomic mobility of sites in a protein. *Nature* **312:** 127–133.

Tanaka, A. (1977). Immunohistochemical studies of vitellogenin during embryogenesis in the cockroach *Blattella germanica. J. Embryol. Exp. Morphol.* **38:** 49–62.

Tesser, G.I., De Hoog-Declerck, R.A.O.M.M., and Westerhuis, L.W. (1975). Some new bifunctional reagents for the conjugation of a protein with an amino compound in water. *Hoppe-Seyler's Z. Physiol. Chem.* **356:** 1625–1629.

Thomsen, E., Hansen, B.L., Hansen, G.N., and Jensen, P.V. (1980). Ultrastructural immunocytochemical localization of vitellogenin in the fat body of the blowfly, *Calliphora vicina* Rob.-Desv. (erythrocephala Meigen.) by use of the unlabeled antibody-enzyme method. *Cell Tissue Res.* **208:** 445–455.

Thorpe, A., and Duve, H. (1984). Immunochemical applications in the study of insect neuropeptides with special emphasis on the peptides of vertebrate type. In: *Insect Neurochemistry and Insect Physiology.* A.B. Borkovec and T.J. Kelly (Eds.). pp. 197–222. New York: Plenum Press.

Tijssen, P. (1985). Production of antibodies. In: *Laboratory Techniques in Biochemistry and Molecular Biology*, Vol. 15: *Practice and Theory of Enzyme Immunoassays.* Amsterdam: Elsevier, pp. 43–78.

Vandesande, F. (1983) Immunohistochemical double staining techniques. In: *Neuroimmunocytochemistry.* E. Cuello (ed.). New York: John Wiley, pp. 257–272.

Vandesande, F., and Dierickx, K. (1975). Identification of the vasopressin producing and of the oxytocin producing neurons in the hypothalamic magnocellular neurosecretory system of the rat. *Cell Tissue Res.* **164:** 153–162.

Van Leeuwen, F. (1982) Specific immunocytochemical localization of neuropeptides: An utopian goal? In: *Techniques in Immunocytochemistry*, Vol. 1. G.R. Bullock and P. Petrusz (eds.). London: Academic Press, pp. 283–299.

Van Noorden, S., and Polak, J.M. (1983). Immunocytochemistry today. In: *Immunocytochemistry.* J.M. Polak and S. Van Noorden (eds.): 11–42. Bristol: Wright PSG.

Varndell, I.M., and Polak J.M. (1984). Double immunostaining procedures: Techniques and applications. In: *Immunolabelling for Electron Microscopy.* J.M. Polak and I.M. Varndell (eds.): 155–177. Amsterdam: Elsevier.

Varndell, I.M., Tapia, F.J., Probert, L., et al. (1982). Immunogold staining procedure for the localization of regulatory peptides. *Peptides* **3:** 259–272.

Veenstra, J.A. (1984). Immunocytochemical demonstration of a homology in peptidergic neurosecretory cells in the suboesophageal ganglion of a beetle and a locust with antisera to bovine pancreatic polypeptide, FMRFamide, vasopressin and alpha-MSH. *Neurosci. Lett.* **48**:185–190.

Veenstra, J.A., and Schooneveld, H. (1984). Immunocytochemical localization of peptidergic neurons in the nervous system of the Colorado potato beetle with antisera against FMRFamide and bovine pancreatic polypeptide. *Cell Tissue Res.* **235**:303–308.

Veenstra, J.A., and Yanaihara, N. (1984). Immunocytochemical localization of gastrin-releasing peptide/bombesin-like immunoreactive neurons in insects. *Histochemistry* **81**:29–34.

Veenstra, J.A. Romberg-Privee, H.M., and Schooneveld, H. (1984). Immunocytochemical localization of peptidergic cells in the neuroendocrine system of the Colorado potato beetle, *Leptinotarsa decemlineata*, with antisera against vasopressin, vasotocin and oxytocin. *Histochemistry* **81**:29–34.

Veenstra, J.A., Romberg-Privee, H.M., and Schooneveld, H. (1985a). A proctolin-like peptide and its immunocytochemical localization in the Colorado potato beetle, *Leptinotarsa decemlineata*. *Cell Tissue Res.* **240**:535–540.

Veenstra, J.A., Romberg-Privee, H.M., Schooneveld, H., and Polak, J.M. (1985b). Immunocytochemical localization of peptidergic neurons and neurosecretory cells in the neuroendocrine system of the Colorado potato beetle with antisera to vertebrate regulatory peptides. *Histochemistry* **82**:9–18.

Verhaert, P., and De Loof, A. (1985). Immunocytochemical localization of a methionin-enkephalin-resembling neuropeptide in the central nervous system of the American cockroach, *Periplaneta americana*. *J. Comp. Neurol.* **239**:54–61.

Verhaert, P., Grimmelikhuijzen, C.J.P., and De Loof, A. (1985). Distinct localization of FMRFamide- and bovine pancreatic polypeptide-like material in the brain, retrocerebral complex and suboesophageal ganglion of the cockroach *Periplaneta americana*. *Brain Res.* **348**:331–338.

Verhaert, P., De Loof, A., Huybrechts, R., et al. (1986). A new alternative for simultaneous immunohistochemical screening of 96 hybridoma clones for tissue specific antibody production selects a monoclonal antibody to insect corpus cardiacum. *J. Neurosci. Methods* **17**:261–268.

Vieillemaringe, J., Duris, P., Geffard, M., et al. (1984). Immunohistochemical localization of dopamine in the brain of the insect *Locusta migratoria migratorioides* in comparison with the catecholamine distribution determined by the histofluorescence technique. *Cell Tissue Res.* **237**:391–394.

Wilson, I.A., Haft, D.A., Getzoff, E.D., Tainer, J.A., Lerner, R.A., and Brenner, S. (1985). Identical short peptide sequences in unrelated proteins can have different conformations: A testing ground for theories of immune recognition. *Proc. Natl. Acad. Sci. USA* **82**:5255–5259.

Wolosewick, J.J. (1984). Cell fine structure and protein antigenicity after polyethylene glycol processing. In: *Proc. 2d Pfefferkorn Conf. Sci. Biol. Specimen Prep. Micr. Microanal:* 83–95. SEM Inc., AMF O'Hare.

Yui, R., Fujita, T., and Ito, S. (1980) Insulin-, gastrin-, pancreatic polypeptide-like immunoreactive neurons in the brain of the silkworm, *Bombyx mori*. *Biomed. Res.* **1**:42–46.

Zola, H., and Brooks, D. (1982). Techniques for the production and characterization of monoclonal hybridoma antibodies. In: *Monoclonal Hybridoma Antibodies: Techniques and Applications*. J.G.R. Hurrell (ed): 1–57. Boca Raton, FL: CRC Press.

Appendices

Like most other laboratories, we use ICC staining methods based on recipes from several sources. These include citations in the books of Sternberger (1979), Polak and Van Noorden (1983), Polak and Varndell (1984); leaflets from manufacturers of reagents; and expertise of collegues. Usually, small modifications were necessary to meet the specific requirements of insect tissue. The following recipes proved to be useful.

Reagents marked with an asterisk may be substituted for antiserum raised in different species if the nature of the first antiserum requires it.

Appendix 1: General Preparatory Procedures

1. Cryostat sections. Mount sections on PLL-coated slides (see page 101). Allow to dry at room temperature for 30–60 min. Postfix as required.
2. Paraffin sections. Optimally fixed and paraffin-embedded tissue is sectioned, and sections are deparaffinized in xylene or other solvent. Bring section to water through graded alcohols. Remove sublimate crystals with iodine (lugol) solution, and bleach in sodium thiosulphate if necessary.
3. Block activity of endogenous peroxidases when peroxidase is used as a marker by immersion for 20 min in 0.05% hydrogen peroxide in PBS or methanol.
4. Rinse twice in 0.15 M PBS (Dulbecco "A"; Oxoid Corporation, U.K.) for 10 min.
5. Block background staining by incubation of horizontally placed slide with 50–200 μl 10% normal swine serum, diluted 1 : 10 in PBS. Use 0.1% BSA instead if protein A–based staining techniques are used. Prevent slides from drying out by keeping them in a moist chamber (Petri dish, closed refigerator box, with fluid on the bottom, etc.). For next step, do not rinse, but drain excess fluid with tissue.
6. Incubate sections with primary AB in the appropriate concentration, diluted in PBS. Some prefer to add 10% normal swine or goat serum or 0.1% BSA. A single section requires some 50 μl fluid, standard full microscope slides (200 μl). Incubations may last from 1 h at room temperature to 16 or 40 h at 4°C. Conditions may be choosen according to desired speed, quality of the antiserum, and signal-to-noise ratio. Agitation of the system by placing the slides on a platform rotating under an oblique angle is highly recommended for an equal density of AB binding in all sections. No cover glasses are used. We saw no advantage to the use of detergents.
7. Rinse twice in PBS (5 and 15 min, respectively), and preincubate again in 10% normal swine serum for 10 min.

Appendix 2: Two-Step Immunofluorescence

8. Second layer: FITC-, TRITC-, or rhodamine-conjugated goat or swine anti-rabbit IgG,* 1 : 50, 1 h.
9. Rinse twice in PBS (5 and 15 min).
10. Mount in warm glycerine jelly (mixture of 10 ml PBS, 100 mg paraphenylene amide, and 90 ml glycerol or glycerol jelly).

11. Examine under fluorescence microscope with epiillumination equipped with the appropriate excitation filters recommended for the type of fluorophore used.

Appendix 3: Two-Step Immunoperoxidase Staining

8. Second layer: peroxidase-labeled swine or goat anti-rabbit* IgG (or peroxidase rabbit anti-mouse IgG for monoclonal ABs), 1 h at room temperature.
9. Rinse in PBS (10 min).
10. Rinse in 0.05 M TRIS-HCl buffer pH 7.6 (5 min).
11. DAB staining. Dissolve 5 mg 3,3′-diaminobenzidinetetrahydrochloride in 10 ml TRIS-HCl buffer; add hydrogen peroxide to a final concentration of 0.01% just before use. Stain for 5–10 min; check development of brown reaction product. DAB is a suspected carcinogen; therefore use gloves and neutralize waste solutions and glassware by immersion in dilute sodium hypochloride. To increase stain intensity, any of the methods given by De Jong et al. (1985) and Hsu and Soban (1982) may be tried. Other, safer peroxidase substrates are 4-chloro-l-naphthol and 3-amino-9-ethylcarbazole, but the color intensity of the oxidized products (blue and red, respectively) is less intense, and sections must then be mounted in a watery medium.
12. Rinse in water.
13. Counterstain, if desired, lightly with Mayer's hemalum.
14. Dehydrate.
15. Mount in DePeX.

Appendix 4: Immunogold-Silver Staining

8. Second layer: goat anti-rabbit* Auroprobe LM (Janssen Pharmaceutica, Beerse, Belgium), 1:35, overnight. Several Auroprobe reagents that happen to be present in the laboratory may be tried to prevent spillage of the precious reagents, but special reagent and 5-nm particles are recommended for light microscopy (Springall et al., 1984). Optimal dilutions must be determined and will depend on concentration of first antiserum as well as the selected silver intensification method.
9. Rinse twice in PBS.
10. Silver intensification procedure based on the method of Timm (Obermayer and Strausfeld, 1980), with minor modifications. In our laboratory, the slides are placed horizontally and subjected to the following treatments.
    a. Incubate for 1 h at 60°C in 10 ml solution of 25% gum arabic in 30% ethanol, 2 ml 10% citric acid, and 2 ml 1% hydrochinon.
    b. Incubate for 30 min at 60°C in the dark in mixture (see a) to which 100 μl 1% silver nitrate is added.
    c. Incubate for 10–15 min at 60°C in the dark in a mixture of 10 ml 25% gum arabic, 2 ml 10% citric acid, 2 ml 5% hydrochinon, and 100 μl 10% silver nitrate.
    d. Stop reaction when visual inspection under red darkroom illumination indicates that sufficient black reaction product has formed (usually 10 min) by immersion for 10 min in 200 ml 0.5% acetic acid in the dark.
    e. Rinse twice in water (in the light).
11. Dehydrate and mount in DePeX.

Appendix 5: Biotin-Avidin ("ABC") Staining Kit

8. Second layer: biotinylated goat anti-rabbit* IgG AB (Vector Laboratories, Burlingame, CA), diluted 1:200 in PBS, 30–60 min at room temperature.
9. Rinse twice in PBS (5 and 10 min).
10. Incubate 30–60 min in Vectastain ABC reagent (prepared by adding 20 μl avidin and 20 μl biotinylated peroxidase complex in 20 ml PBS and allowing the mixture to stand for 30 min at room temperature) (Vector Laboratories).
11. Rinse in PBS (10 min).
12. Rinse in TRIS-HCl buffer pH 7.6 (5 min).
13. DAB staining as in Appendix 3.
14. Rinse in water, dehydrate, and mount in DePeX.

Appendix 6: Biotin-Streptavidin Staining

8. Second layer: biotinylated goat anti-rabbit* IgG AB, diluted 1:200 in PBS, 30–60 min at room temperature.
9. Incubate in preformed streptavidine-peroxidase complex (Amersham International, Amersham, U.K.), diluted 1:100 in PBS, 30–60 min at room temperature.
10. Proceed as in steps 10–14 in Appendix 5.

Appendix 7: Three-Step Peroxidase Antiperoxidase (PAP) Staining

8. Second layer: link AB: swine anti-rabbit* IgG (Fc fragment), diluted 1:50, 60 min at room temperature.
9. Rinse twice in PBS (5 and 10 min).
10. Block unspecific staining by 10% normal swine serum for 10 min.
11. Incubate in rabbit PAP, diluted 1:200, for 60 min at room temperature.
12. Proceed as in steps 9–15 in Appendix 3.

Appendix 8: Immunogold Staining of Epoxy Resin EM Sections

The procedure is based essentially on that of Varndell et al. (1982). Use glutaraldehyde-fixed and Epon-embedded tissue mounted on nickel grids without supporting Formvar film.

1. "Etch" Epon sections, floating section side down, on 10-μl droplets on parafilm support, by 10% hydrogen peroxide for 10 min.
2. Wash on distilled water drops and in water jets, and air-dry.
3. Incubate in normal swine serum 1:30 for 30 min; drain.
4. Incubate in first antiserum, overnight at 4°C. Dilution is to be determined starting with the concentrations used for the PAP staining for LM. Dilutions are made in PBS containing 1% BSA.
5. Rinse in TRIS-HCl buffer pH 7.2 and next in jets of buffer for 20 min; drain the buffer.
6. Incubate in 1% BSA in TRIS buffer pH 7.2.
7. Incubate in second antiserum: goat anti-rabbit* IgG labeled with colloidal gold particles of desired size and dilution for 30–60 min.
8. Rinse with large volumes of TRIS buffer (at least 20 ml per grid).
9. Rinse with distilled water.

10. Postfix the IgG complexes with 2% glutaraldehyde, or counterstain with uranyl acetate and lead citrate as desired. Keep the contrast low to facilitate the location of gold label over dark organelles.
11. Rinse and dry.

Appendix 9: Protein A-Colloidal Gold Staining of Lowicryl-Embedded EM Sections

The following procedures of Diederen et al. (1987) give acceptable preservation of ultrastructure (as osmium tetroxide has to be omitted, poor preservation of membranes results) and give specific labeling of adipokinetic hormone-containing secretion granules in the corpora cardiaca of *Locusta migratoria* and *Schistocerca gregaria* (Fig. 4.3). Fixation and embedding.

1. Fixate corpora cardiaca for 1 h at 0°C in 2% glutaraldehyde and 2% formaldehyde (after Karnovsky, 1965), buffered to pH 7.3 by 0.13 M phosphate buffer.
2. Dehydrate through graded alcohols under progressively lower temperatures as indicated for embedding in Lowicryl K_4M (Chemische Werke Lowi, Waldkraiburg, F.R.G.) (Anonymous, 1982; Carlemalm, 1982). The following steps (a to i) took place under continuous agitation:
   a. 0.13M phosphate buffer pH 7.3, 0°C, 5 + 5 min
   b. 70% ethanol, 4°C, 15 + 15 min.
   c. 80% ethanol, −10°C, 15 + 15 min.
   d. 90% ethanol, −20°C, 15 + 15 min.
   e. 96% ethanol, −30°C, 15 + 15 min.
   f. 96% ethanol/Lowicryl K_4M 1:1 (v/v), −30°C, 30 min.
   g. 96% ethanol/Lowicryl K_4M 1:2 (v/v), −30°C, 30 min.
   h. Lowicryl K_4M, −30°C, 30 min.
   i. Lowicryl K_4M, −30°C, overnight.
   j. Lowicryl K_4M (freshly prepared), 30 min rotation at −30°C, followed by 4 min under vacuum.
   k. Polymerize in transparant gelatin capsules by ultraviolet irradiation for 24h at −30°C.
   l. Capsules in desiccator over $CaCl_2$, overnight, at room temperature.

The Lowicryl K_4M was freshly prepared before use from the following proportions of cross-linker A:monomer B:initiator C = 3.2:17.3:0.1 g.

On-grid staining procedure. Ultrathin sections were stained as follows:

a. Grids were floated, sections down, on a drop of first antiserum, for 22 h at 4°C.
b. Grids were rinsed for 5 sec in each of three changes of 0.1 M phosphate buffer pH 7.3 at room temperature.
c. Grids were floated, sections down, on a drop of protein A–gold complex, at 4°C, for 7 h.
d. Grids were rinsed three times in distilled water for 5 sec.
e. Sections were poststained in a saturated solution of uranyl acetate in 70% methanol, no longer than 1 min, or in 2% osmium tetroxide in 0.1 M phosphate buffer, for 15 min.

The specificity of the anti-AKH staining reactions (Fig. 4.3) was checked by liquid-phase absorption of the antisera with antigens in a concentration of 10^{-5} M for 1 h at 37°C or overnight at 4°C.

Chapter 5

Immunoassay of Pesticides

Bruce W. Hermann

I. Introduction

Immunochemical techniques have become a powerful tool in the trace analysis of organic compounds in a wide variety of biological and environmental matrices. However, until recently their application has been largely limited to the clinical chemistry, endocrinology, and biochemistry laboratory. Acceptance of the techniques outside these areas has been slow, but interest is finally growing in their application to other analytical problems such as pesticide residue analysis.

Pesticide chemists have been reluctant to accept immunochemical techniques largely because of the success of the large number of chromatography based techniques (Hammock et al., 1986). These techniques were tailor made for the analysis of synthetic organic pesticides—mostly small, hydrophobic molecules, easily extracted with organic solvents, and separated by gas liquid chromatography (GLC) and detected with a wide range of element selective detectors. High-pressure liquid chromatographic (HPLC) techniques soon followed for the analysis of more polar or thermally labile compounds, though these are still hampered by a limited repertoire of detectors.

The success of these techniques led to a generation of analytical chemists who turned to chromatographic instrumentation for the analysis of new pesticides much as the preceding generation of chemists used on wet chemical techniques. As analytical complexities increase, detectors have become more sophisticated and now provide increased specificity and sensitivity as demonstrated with the use of selective ion monitoring for routine quantitative analysis. Chromatographic resolution has also improved with the development of capillary chromatography capable of separating even polar compounds of relatively limited volatility with highly inert, bonded phase fused silica capillary columns. The use of supercritical fluid chromatography and microbore HPLC has begun, but neither has been used to date as a routine analytical technique.

Many of the pesticides being developed are pushing these techniques to their current capabilities. Chromatographic powers of separation are continually challenged by the number of hard-to-distinguish "me-too" compounds that soon follow the development of a pesticide with novel activity, eventually leading to an entire class of closely related compounds such as the triazine herbicides or the insect growth regulants represented by diflubenzuron and its lookalike BAY SIR 8514. Both geometric and chiral isomers comprise many pesticides, such as the synthetic pyrethroids. Most if not all of the biological activity may be confined to a single isomer. These isomers may also differ in their metabolic or environmental fate, making isomer discrimination imperative (Hammock et al., 1986). Because many of the newer pesticides are more polar, they require single or multiple derivatizations prior to GLC analysis. Even if these derivatizations are successful, they may require an internal standard to monitor recoveries. The new generation of pesticides such as those derived from fermentation products or protein toxins are difficult to either separate or detect using chromatographic instrumentation. Also, the increasing efficacy of the new pesticides, together with growing concerns over their possible adverse effects on the environment, has meant lowering detection limits to the current ppb or ppt level. These same concerns have generated more stringent regulatory requirements, requiring producers to account for all major metabolites as well as the parent compound. This requirement further compounds the analytical problems previously mentioned.

At the very least, the analytical chemist is faced with an extensive sample preparation prior to chromatographic analysis. To analyze these compounds at biologically relevant levels, sample preparation has necessarily become the most time-consuming and labor-intensive part of any pesticide residue method. This expense often leads to the unwillingness of an investigator to monitor applications adequately with the analysis of background and zero day spray samples, and this in turn has resulted in misleading field efficacy and residue trials. Also, minor crops are often denied a possibly effective pesticide, because the potential market is too small to justify the cost of a residue trial.

Because of the increasing inability of chromatography-based techniques to easily accomplish pesticide trace analysis, scientists have begun to investigate alternate techniques. Among the most promising of these are the immunochemical-based techniques. Unlike other trace analytical methods, the success of these techniques is not dependent on any particular chemical property, allowing them to be widely applied to more chemical compounds than other analytical techniques. The only requirement for a successful immunoassay is that the analyte be of a certain minimum size to allow strong interaction with an antibody (AB). The strength of this interaction is the summation of weak forces, including dispersion forces, electrostatic and hydrophobic interactions, and hydrogen bonding. Because these forces are short-distance interactions, there needs to be a good fit or complementarity between the surfaces of the ligand (antigen, or Ag) and the AB for them to be effective, thus giving the assay its high selectivity. There appears to be no upper size or complexity limit to the technique. Thus immunochemistry will become more important as increasingly complex synthetic molecules, fermentation products, and protein toxins are developed.

Immunochemical assays should not be viewed as a replacement for chromatography-based analytical techniques but rather as a complementary analytical tool that may not be appropriate for every pesticide residue analysis problem. For instance, immunochemistry should probably not be used when the compounds are easily analyzed by GLC or where limited numbers of samples are involved. The specificity of the Ag-AB interaction offers the analyst many advantages over other analytical techniques (Table 5.1), including the ability to perform analysis in the crude sample matrix and bypass the often extensive sample preparation and chromatographic separations associated with trace analysis, allowing a high throughput of samples. In addition, the development of a wide variety of labeling techniques has given the method a flexibility to fit many analytical needs, from rapid and moderately sensitive to extremely precise and sensitive analyses. An unfortunate aspect of this flexibility is that it has given rise to a dizzying array of acronyms to describe the various immunochemical combinations and permutations that have been developed (Edwards, 1985). This language problem is compounded by apparent contradictions in terminology, with different authors using the same terms to convey different meanings. This can be intimidating to an analyst approaching immunochemical analyses without an understanding of the basic concepts and definitions (Hammock et al., 1986).

Table 5.1. Advantages of Immunochemical Technology

Generally applicable	No special chemical property needed other than ability to interact with antibodies
Specific	Lessened dependence on sample cleanup prior to analysis
Sensitive	From ppm to ppb or below (ELISA)
Precise	Low within- and between-run variation
Complementary	Best for compounds difficult to analyze by chromatographic techniques
Simple	Common procedure for all analytes—amenable to automation
Portable	Visible end point (ELISA); requires minimal instrumentation

As mentioned earlier, immunochemistry was first applied extensively to problems in endocrinology (Yalow and Berson, 1960). Much as chromatography has done in the pesticide analysis laboratory, immunochemistry has given rise to a generation of clinical chemists who first turn to the technique even when it may not be appropriate. This widespread application is due to the compounds of interest—i.e. drugs, hormones, proteins—not being amenable to the existing chromatographic analytical techniques. With immunoassay, they found a tool that was able to give them a high degree of sensitivity and precision.

The development of immunochemistry as a pesticide analysis tool has been slowed not only by the success of conventional chromatographic techniques but also by a distorted perception of the technique. Immunochemical determinations are often thought by analytical chemists to be biological assays and therefore incapable of providing precise results. In reality, these techniques are precise, physical assays based on the laws of mass action with the rate of Ag-AB complex formation being proportional to the reactant concentration (Hammock and Mumma, 1980). However, biological systems, either animal (polyclonal) or a cell line through hybridoma technology (monoclonal ABs), produce the extremely selective ABs that give immunoassays their sensitivity (Hurrell, 1985). The extreme sensitivity reported for immunoassays, with assays reported with subnanomole sensitivity (Johannson et al., 1986), can be a concern, though such sensitivities come only at the expense of involved sample preparation and long analysis times. However, in attempting to reach extreme limits of sensitivity, a major advantage of immunoassay, its ability to analyze a large number of samples rapidly at biologically relevant concentrations, is lost. False-positive results due to unexpected cross-reactivity by sample matrix components can be addressed with the present use of controls. Immunoassays have not been as successfully applied to the complex variety of matrices that are found in pesticide residue analysis (Allen, 1986) as the limited matrices found in clinical samples: blood, urine, and cerebral spinal fluid. Finally, though immunochemical analysis is not a new technique, its application to residue analysis still lacks a set of standard procedures, which has often resulted in haphazard and frustrating method development efforts.

A pesticide laboratory may be faced with hundreds to thousands of samples in support of product registration and stewardship. For many of these laboratories, the majority of these samples submitted contain residues too low to be biologically significant and therefore below the limits of detection set in the chromatographic methodology. Thus, since the immunoassay does not require extensive sample preparation, it can be used to eliminate "less-than" samples so that the analyst can concentrate on those samples containing significant residues.

The most attractive immunoassay technique available for the residue laboratory is the enzyme-linked immunosorbent assay (ELISA) (Van Weeman and Schuurs, 1971; Engvall and Perlman, 1971, 1972). This assay

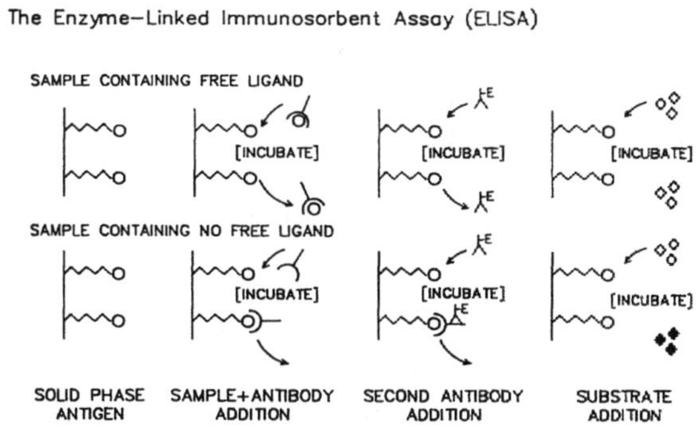

Figure 5.1. The enzyme-linked immunosorbent assay (ELISA). In sample containing free ligand, antibody binds to free ligand in solution and is inhibited from binding to the solid-phase ligand.

takes advantage of the discriminatory power of ABs and the extremely high catalytic power and specificity of enzymes. The method has proven applicability and is rugged and simple, with stable and commercially available reagents; ABs developed by Langone and Van Vunakis in 1975 were still effective 11 years later (Ferguson, 1986). Another enzyme immunoassay gaining acceptance is the enzyme-mediated immunosorbent test (EMIT), a proprietary technique licensed by SYVA.

The ELISA (Fig. 5.1) typically used in trace analysis is a competitive assay with three distinct stages: attachment of immunoreactant to the solid phase, incubation with the test sample, and signal amplification. Each stage can be divided into additional steps to allow many layers of immunoreactants to be adsorbed to the solid phase or to amplify the enzyme signal further. The assay is based on the competition of free and bound ligand for a limited number of AB-binding sites. If there is free ligand in the sample, it will bind with the AB and prevent its binding to the solid support. The amount of AB bound to the support is quantified with an enzyme. This enzyme label is conjugated either to the AB or to an AB-specific AB. The amount of compound in the sample (free ligand) is quantified by the amount it inhibits the AB from binding to the solid-phase ligand as measured by enzyme activity inhibitor.

ELISAs are usually performed in a plastic 96-well microtiter plate, though they can be run in plastic tubes or cuvettes or on a variety of solid adsorbents. For the sake of simplicity, the assays discussed will be performed in the microtiter plate format unless noted otherwise. To perform the assay (Fig. 5.1), samples are dissolved in buffer containing the Ag-specific AB. After reaching equilibrium, the sample is added to wells coated with the solid-phase Ag. After allowing time for the unbound AB to react with the solid-phase Ag, the liquid phase is removed, and another AB, specific for immunoglobulins, is added. This AB typically contains the enzyme label. Again, the liquid phase is removed, and the chromogenic enzyme substrate is added. After the color is allowed to develop for a set period of time, the enzyme-catalyzed reaction is stopped, and the absorbance is read. This color end point is inversely proportional to the degree of inhibition of AB binding which correlates to the unknown compound concentration in the sample.

Several techniques have been developed to amplify further the enzyme signal and lower the ELISA detection limits. Among these are biotin-avidin labeling and enzyme channeling systems (Fig. 5.2). Biotin is a low-molecular-weight enzyme cofactor that is easily conjugated to ABs without affecting the Ag-binding capacity (Guesdon et al., 1983) and has an exceptionally high affinity for avidin (affinity constant = 10^{12} M^{-1}) (Green, 1963). Biotin

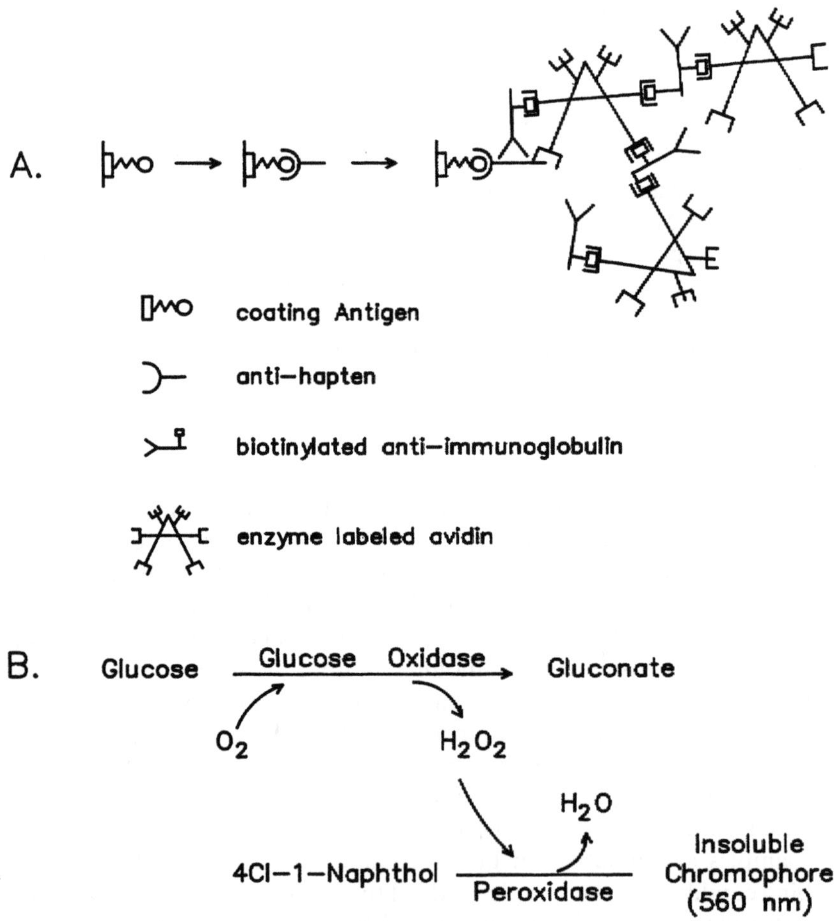

Figure 5.2. Enhanced enzyme amplification systems. (A) Biotin-avidin ELISA. (B) Enzyme channeling ELISA (Johannson et al., 1986).

analogs used for conjugation include biotinyl-*p*-nitrophenyl ester and biotinyl-*N*-hydroxysuccinimide (NHS) ester. The biotin-binding sites are deep within avidin, making the use of a linking group necessary. E-amino caproic acid has been used as a spacer between reactive NHS and biotin (Costello et al., 1979; Hofmann et al., 1980; Leary et al., 1983). Avidin has four binding sites, allowing it to bridge biotinylated reactants. By labeling avidin with enzyme and allowing it to complex with biotinylated antiimmunoglobulin (Fig. 5.2), the number of enzymes labeling each solid-phase IgG can be increased beyond that which can be covalently bound to an AB. Originally used to enhance immunocytochemical staining (Bayer et al., 1976), it has been used to improve ELISA detection limits dramatically (Kendall et al., 1983; Shamsuddin and Harris, 1983). Guides for purifying and labeling avidin and the biotinylation of amino reactants are available (Tijssen, 1985).

The enzyme channeling system employs two sequential catalytic reactions (Johannson et al., 1986). In one system, alkaline phosphatase dephosphorylates nicotinamide adenine dinucleotide phosphate (NADP) to produce nicotinamide adenine dinucleotide (NAD), which catalytically activates a redox cycle involving the enzymes alcohol dehydrogenase and diaphorase. At each cycle, *p*-iodonitrotetrazolium violet is reduced to an intensely colored formazan. The assay can detect 0.01 attamoles of alkaline phosphatase and was used to detect as little as 0.0013 mIU/L of thyroid-stimulating hormone (TSH).

An alternative to antiimmunoglobulin is protein A, which has a high affinity (10^8 M^{-1}) to the Fc portion of many IgGs (Forsgren and Sjoquist, 1966; Kronval et al., 1974). This affinity is higher than that of most antiimmunoglobulins, and its high specificity results in a lower background. Methods for its purification are documented (Sjoquist et al., 1972; Kessler, 1981), and it is commercially available. Because it is specific for the portion of the AB not involved in the Ag binding (Fc portion), protein A does not inhibit the AB-Ag interaction.

Though a multistep procedure, the only manipulations in running the ELISA are the preparation and addition of reagent solutions (many of these solutions can be purchased premixed). Alastair Voller, a pioneer in applying immunoassay to clinical analyses, claims, "Anyone who can make a cup of tea and wash up can do an ELISA" (Allen, 1986). The ELISA is inexpensive, not only in manpower costs (owing to reduced sample handling) but also in reagents. It does not require the quantity of expensive, high-purity solvents, carrier gases, and derivatization agents needed to prepare samples for chromatographic analysis. Though the AB is expensive to develop, only minute amounts are needed (working dilutions (titers) of serum usually range from 1:5,000 to 1:500,000) for each analysis. Instrumentation costs are also low, because the only special apparatus needed is a microtiter plate reader.

The EMIT, in contrast to ELISA, is a homogeneous assay and does not

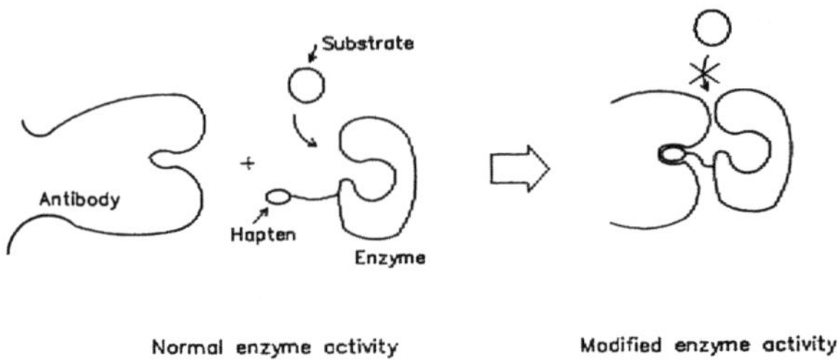

Figure 5.3. The EMIT homogeneous assay (Kabakoff and Greenwood, 1981).

require the separation of free and bound ligand/AB for quantification (Rubenstein et al., 1972; Kabakoff and Greenwood, 1981). Such homogeneous assays are based on a property of a marker or ligand label that is modulated by the ligand-AB interaction (Fig. 5.3). This can be modulation of either a physical property or a biological activity. In EMIT, the AB modulates the activity of a ligand-bound enzyme either by steric exclusion or by a conformational change in the enzyme (Rowley et al., 1975, 1976; Bastiani, 1979). Free ligand in the sample (analyte) will prevent AB binding to the labeled ligand, making the enzyme activity proportional to the amount of ligand in the sample (Belfield et al., 1977; Brunk et al., 1976; Finley et al., 1976, 1977; Kabaoff and Greenwood, 1981). This assay has been adapted to the analysis of macromolecules by attaching the substrate to a high-molecular-weight polymer (Kabakoff and Greenwood, 1981). There are additional variations of label placement such as attaching ligand to substrate (Burd et al., 1977a,b), to enzyme (Wei and Reibe, 1977) or enzyme cofactor (Carrico et al., 1976) to the AB, and an assay based on avidin binding of biotin-containing enzyme being modulated by AB interaction with avidin-labeled ligand blocking inactivation of the enzyme (Bacquet and Twumasi, 1984). The advantages of this technique are its speed (typically on the order of seconds to minutes), sensitivity and precision (typically in the ng/ml range, with $<10\%$ SD) (Plomp et al., 1979), and the elimination of the need to separate free and bound AB. The assay can be used as both a qualitative (Mule et al., 1974; Schneider et al., 1973; Brattin and Sunshine, 1974) and quantitative tool (Gushaw et al., 1977; Cobb et al., 1975; Chang et al., 1975; Pippenger, 1979; Booker and Darcey, 1975). Like ELISA and other immunochemical procedures, EMIT has shown good correlation with GLC and HPLC (Dietzler et al., 1980; Chamberlain et al., 1977; Dellamonica et al., 1978; Johannessen, 1977; Kupferberg, 1970) as well as RIA (Kampa et al., 1978; Wong et al., 1979, 1980; Spiehler et al., 1976; Boeckx et al., 1979). Disadvantages are that it is a proprietary technology and it loses the heterogeneous assay advantages of limiting sample interferences and reducing background noise. This limits its utility to colorless matrices or the need to use kinetic measurements to compensate for matrix absorbance.

II. Assay Development

1. Analyte Selection.

The discussion will be largely limited to the application of ELISA to pesticide analysis, though by analogy, much of the discussion can also be applied to EMIT as well as many of the other immunoassay techniques. This analogy is especially true for the early stages of immunoassay development: Ag selection, conjugation procedures, immunization protocols, and antisera characterization and preparation. The first and probably most important stage in assay development, especially for those applying the technique for the first time, is the selection of the analyte. The primary prerequisite of the compound selected must be its ability to elicit and interact with AB. As in any technique, some compounds are easier to analyze than others. The most successful assays to date have been in the analysis of compounds consisting of relatively polar groups separated by hydrophobic groups. The compound needs to be large enough to complex strongly with an AB with the summation of hydrophobic and electrostatic forces to a total at least 40 kJ/mol (Tijssen, 1985) of hydrophobic and electrostatic forces to produce a sensitive and specific assay. The compound should have some water solubility, though absolute solubility is not usually a problem at the levels at which the assay is usually undertaken. Rather, the problem is more of extraction: partitioning the compound from an oily, hydrophobic matrix into the aqueous phase where it can interact with the AB. The compound does not necessarily have to be the most difficult to analyze by chromatographic techniques, but there should be enough analyses planned to make use of the immunoassay's ability to deal with large numbers of samples.

2. Hapten Synthesis

The ABs needed in the immunoassay must be produced in response to the analyte. To elicit this production, the analyte must be both large (> 10,000) and recognized as foreign. Though this large size is needed to elicit AB production, the actual recognition site is quite small, only 4–6 amino acids. Thus once ABs are elicited, pesticides are able to complex strongly with ABs. For pesticides to elicit this AB production, they must first be conjugated to a larger molecule or carrier, usually a protein, to produce an immunogen conjugate. Such compounds that react with an AB produced against an immunogenic conjugate are referred to as haptens—thus the term hapten immunoassay. Proteins are chosen because of their size and solubility and also because they contain functional moieties for conjugating analogues of the analyte (hapten). Also important is their heterogeneity, necessary for T-cell recognition. The most commonly used proteins for immunization include thyroglobulin, chicken gamma globulin, bovine serum albumin (BSA), keyhole limpet hemocyanin (KLH), and ovalbumin.

To ensure success in eliciting antibodies to the hapten, the hapten must be covalently bound to the protein. This is especially true with most pesticides, which cannot complex strongly enough to the protein to remain complexed after immunization. This covalent attachment has been accomplished in a wide variety of ways, depending on the functional moieties available on the hapten and the protein. A linking group is often used for attachment to extend the hapten away from the protein surface and make it freely available to interact with the AB. Care must be taken that the attachment does not change the chemical properties of the hapten or that the properties of the linking group dominate the hapten-AB interaction so that one ends up with an assay that will recognize the synthesized hapten and not the analyte of interest. To keep the AB's recognition of the linking group from interfering in the assay, different conjugation procedures can be used to synthesize the Ags used in immunization and in coating the solid phase.

If the pesticide has a functionality suitable for conjugation, then it serves as the hapten. However, functionalities such as an amino or carboxyl moiety are not always available, and a reactive analogue of the analyte must be synthesized. Hapten design and conjugation procedure need careful consideration, being the most important variables in determining immunoassay sensitivity and specificity. The AB interacts most strongly with the part of the hapten distal or farthest away from the point of attachment, making this the most critical portion of the molecule in determining AB specificity. By careful selection of the conjugation site and which portion of the molecule will be the distal end, immunoassays can be designed that are specific for either a particular compound or a pesticide class.

The importance of the site of conjugation is demonstrated in the RIA development for S-bioallethrin (1R, 3R, 4′s) (Fig. 5.4). The S-bioallethrin was conjugated to the protein via the hydroxyl functionality of the rethrelone propene side chain (pyrethrin "A") (Wing et al., 1978). The assay showed high specificity for the absolute configuration of the chrysanthremate moiety, and little for the allethrelone. The AB also could not distinguish between allethrin and pyrethrin 1 (Table 5.2).

Once the functionalized analog has been synthesized, the analyst has a variety of methods available to conjugate it to a protein. Several of these procedures also insert a linking group between the hapten and carrier protein. Sometimes this linking group is the conjugation reagent itself, whereas other times a more deliberate insertion of the linking group must be made.

Functionalities of the proteins available for conjugating haptens are practically limited to OH, NH, and COOH, though proteins can also be modified to provide SH functionalities. The procedures to couple the hapten

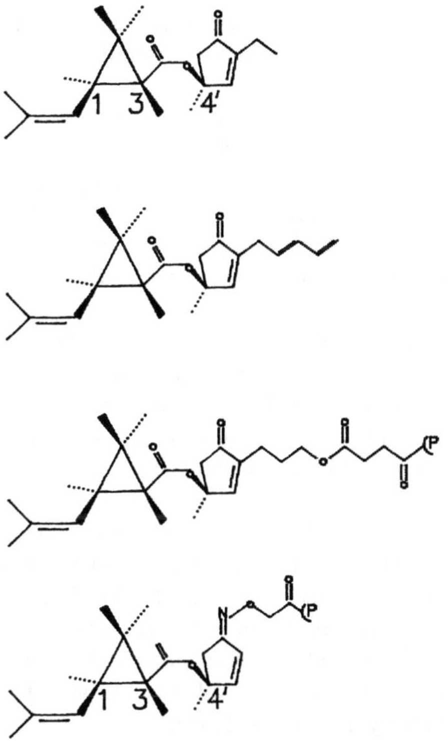

S–Bioallethrin

Pyrethrin I

Pyrethrin "A"

Pyrethrin "B"

Figure 5.4. Pyrethrin I, S-bioallethrin, and two pyrethrin hapten analogues (Wing et al., 1978).

Table 5.2. S-Bioallethrin Antibody Specificity for Allethrin Isomers

Absolute Configuration			Nmoles needed for 50% Inhibition
1	3	4′	
R	R	S	1.2
R	R	R	2.1
S	S	S	24
R	S	R	41
S	R	S	60
S	S	R	160
R	S	S	1,000
S	R	R	1,000

Absolute configuration denotes stereochemistry at the specified carbon (Fig. 5.4). That antibody specificity is highest for the portion of the hapten farthest from the carrier protein is readily illustrated by isomer 2 (1R, 3R, 4′R). Isomer 2 differs from S-bioallethrin only at the 4′ position (closest to the protein) and is the most cross-reactive.

to these groups are extensively covered in the literature and laboratory guides and will only be summarized here, with examples limited to pesticide chemistry (Tijssen, 1985).

There is no consensus on the optimum number of hapten molecules per carrier, but at least two molecules are required if immunoprecipitation tests are desired. Many early studies used a high loading (large hapten-protein ratio), and such heavily loaded immunogens have raised useful AB titers. However, there is a danger that too high a hapten-carrier ratio may lead to tolerance (Tjissen, 1985). Lower loading may raise antibody titers with a higher average specificity for the hapten.

Numerous proteins have been successfully used as carriers. Bovine and human serum albumin (HSA) are very commonly used, because they have numerous free amine groups and are still remarkably soluble when cross-linked or heavily loaded with haptens. Many have found that KLH is highly immunogenic, though solubility problems are sometimes encountered with its use. Other commercially available and exotic proteins have also been used as carriers, but in general, one can be relatively certain that a protein will be immunogenic if it has a molecular weight exceeding 10,000 daltons and if it is immunochemically foreign to the animal receiving the Ag. Because ABs will be raised against the carrier as well as the hapten, one should consider the ultimate use for the AB when choosing the carrier protein. For instance, HSA would be a poor carrier for a hapten if the resulting AB were to be used to monitor human blood samples by immunodiffusion.

As an alternative to the protein carrier, Surflan herbicide was conjugated to Affi-gel 10, a cross-linked agarose gel bead support modified with *N*-hydroxy succimimide ester to aid in coupling (Kuniyuki and McCarthy, 1986). Conjugation was through the sulfonamide at position 1 of Surflan (Fig. 5.5) and accomplished by incubating the two reagents in methanol. This carrier may be useful in conjugating pesticides like Surflan that are only sparingly soluble in water.

Pioneering work on immunochemical assays for pesticides involved the synthesis of haptens for dichloro diphenyl trichloroethane DDT and mala-

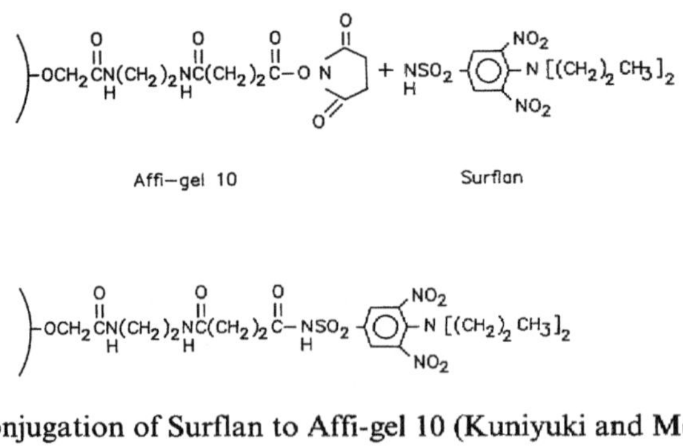

Figure 5.5. Conjugation of Surflan to Affi-gel 10 (Kuniyuki and McCarthy, 1986).

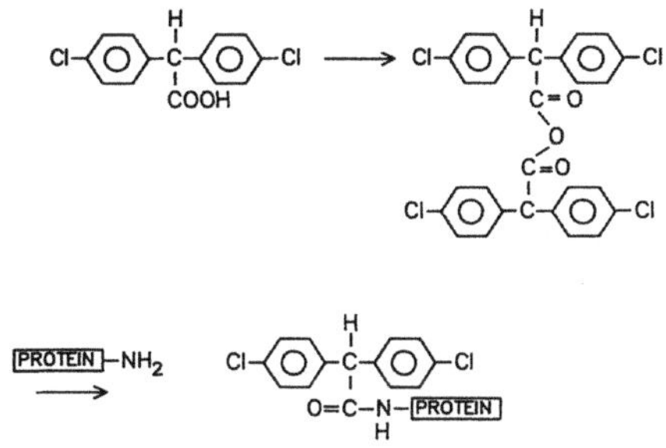

Figure 5.6. Synthesis of the DDT immunogen (Haas and Guardia, 1986).

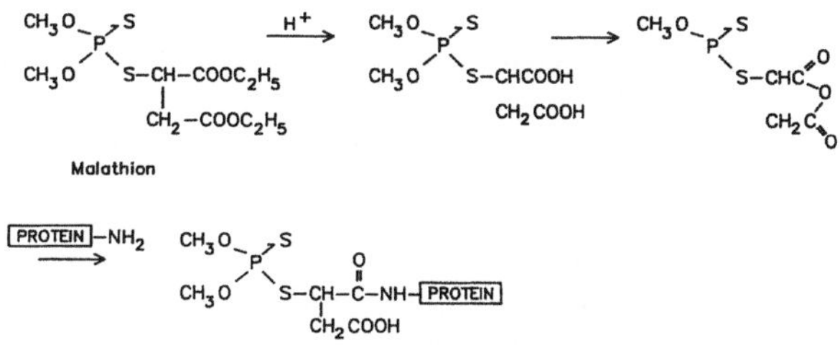

Figure 5.7. Synthesis of the malathion immunogen (Centeno et al., 1970).

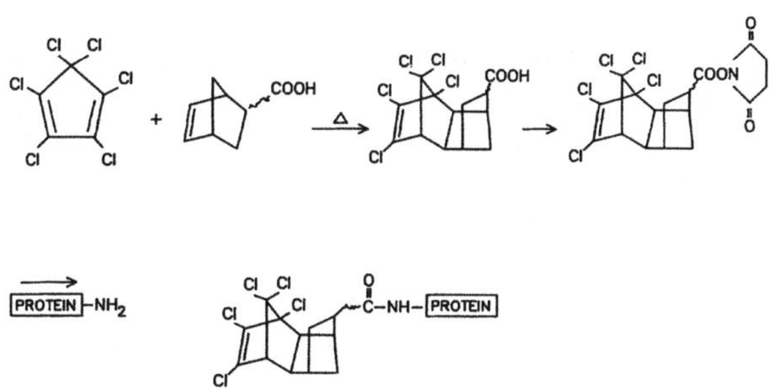

Figure 5.8. Synthesis of the dieldrin immunogen (Langone and Van Vunakis, 1975).

thion. Haas and Guardia (1968) used the acid chlorides of malathion half ester and DDA (2,2-*bis*-[*p*-chlorophenyl]acetic acid) (Fig. 5.6) for conjugation, and Centeno et al. (1970) used the anhydrides of DDA and malathion diacid (*0,0*-dimethyl) *S*-[1,2-*bis*-carboxyethyl]phosphorodithi-oate) (Fig. 5.7).

Langone and Van Vunakis (1975) used an *N*-hydroxysuccinimide active ester of a carboxyl substitute analog of dieldrin for conjugation to HSA (Fig. 5.8). Similar chemistry was used to couple the carboxylic acid derivatives of allethrin and diflubenzuron. The active ester can be purified, being fairly stable under acidic conditions.

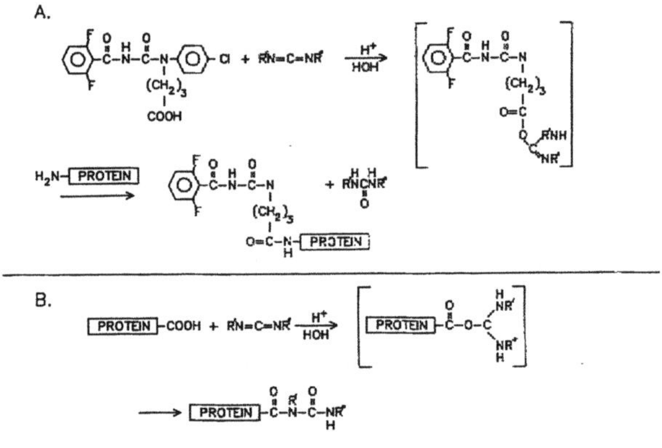

Figure 5.9. Hapten-carrier conjugation using carbodiimide. (A) Synthesis of the diflubenzuron immunogen (Wie and Hammock, 1982b). (B) Diimide reaction with protein to form a guanidine derivative (Hammock and Mumma, 1980).

Water-soluble carbodiimides (CDI) can be used effectively to couple amine and carboxyl moieties and are especially useful for conjugating water-soluble or unstable haptens, since it is unnecessary to isolate the active ester (Fig. 5.9). Reagents available for this include 1-cyclohexyl-3-(2-morpholinyl-4-ethyl) carbodiimide methyl *p*-toluene sulfonate or 1-ethyl-1-3-(3-dimethylaminopropyl) carbodiimide hydrochloride. In this procedure, the carrier protein is dissolved in water, and the solution is titrated to a slightly acidic pH. The CDI is added followed by the carboxylic acid hapten dissolved in dimethylformamide, and the solution is allowed to react for several hours. Some cross-linking of the carrier protein occurs, which may aid antigenicity but will reduce solubility. This diimide linkage was used to conjugate the carboxylic acids of diclofop-methyl to BSA (Schwalbe et al., 1984), triazine herbicides (Sharp et al., 1986), and metalaxyl (Newsome, 1985). A disadvantage of this reaction is that the water-soluble diimide may react directly with the protein, resulting in ABs being raised against the resulting guanidino or acyl derivative (Hammock and Mumma, 1980).

The mixed anhydride procedure *was* used to produce labeled hapten for the diclofop-methyl enzyme and fluorescence immunoassays (Schwalbe et al., 1984) (Fig. 5.3). Here the hapten was dissolved in a solution of dry dimethyl formamide and triethylamine and then treated with isobutylchloroformate. The reaction solution was slowly added to the conjugate, in this case horseradish peroxidase, for the enzyme label and fluorescein amine for the fluorescent label. A similar procedure was used to couple the carboxylic acid hapten for coating Ags used in triazine (Sharp et al., 1986, Fig. 5.10), paraquat (Van Emon et al., 1986), diflubenzuron (Wie and Hammock, 1982), and metalaxyl (Newsome, 1985) ELISAs.

Pesticides can often be hydroxylated or have some hydroxylated me-

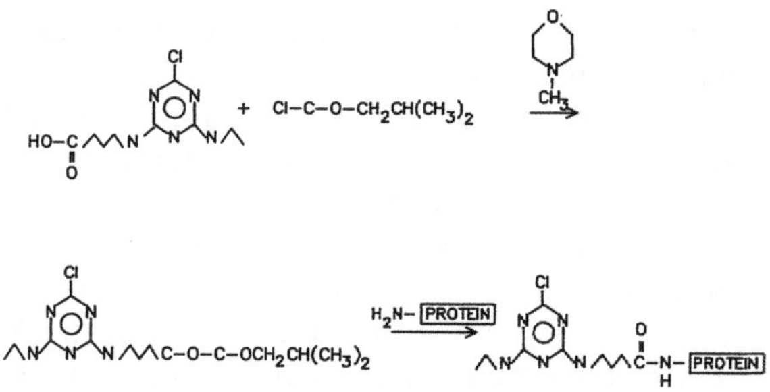

Figure 5.10. Mixed anhydride conjugation of the triazine hapten (Sharp et al., 1986).

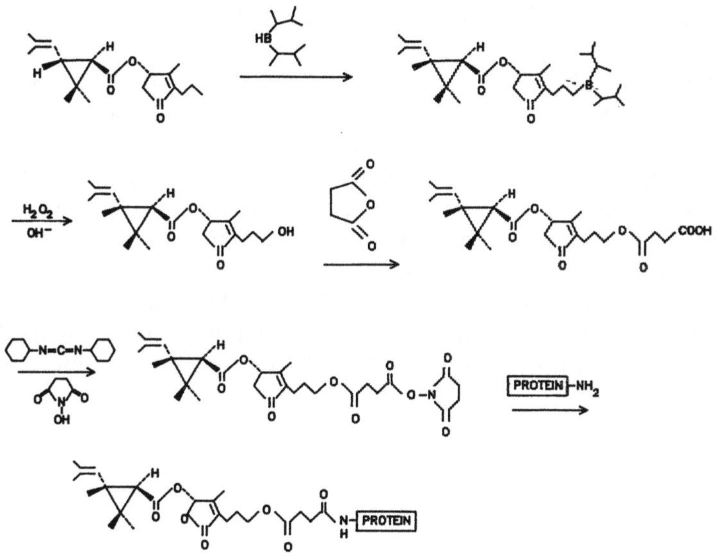

Figure 5.11. Synthesis of the S-bioallethrin-protein conjugate (Wing et al., 1978).

tabolites available for conjugation. Such compounds with an available primary or secondary alcohol can be treated with succinic anhydride in pyridine to lead to a hemisuccinate as shown for allethrin derivatives (Fig. 5.11). Hydroxyl groups can also be exposed to equimolar phosgene resulting in a chlorocarbonate that will react with amino groups of proteins, or alternately reacted with ethyldiazoacetate followed by hydrolysis to give a carboxymethyl ether.

Phenols treated with diazotized *p*-aminobenzoic acid introduce a free carboxyl group. Ethylbromoacetate was used to derivatize phenolic metabolites of the insecticide diflubenzuron (Wie and Hammock, 1982a) under anhydrous conditions. The resulting ethyl ester was hydrolyzed in dilute methanolic base without hydrolyzing diflubenzuron. If desired, longer spacers can be introduced by using bromopropionates and butyrates, but harsher conditions are required for these less reactive bromides. Phenols can also be coupled using other divalent reagents such a tolylene-2,4-diisocyante or cyanuric chloride (Hammock and Mumma, 1980).

The reaction of aldehydes or ketones with carboxymethoxylamine hemihydrochloride (CMA) results in the formation of an oxime (CMO) with a free carboxyl group as shown in Figure 5.4 for the allethrin CMO derivative (pyrethrin "B"). Unfortunately, this particular CMO derivative was found to be quite unstable and could not be used as a hapten. This fact emphasizes the need for rigorous structural proof of hapten structure.

Lukens et al. (1977) attached the benomyl degradation product 2-amino-benzimidazole (2-ABZI) to ovalbumin by treating the amine of 2-ABZI with thiophosgene to produce the isothiocyanate which was then linked to ovalbumin (Fig. 5.12). The C-6 amine of benzo[a]pyrene was treated with phosgene to form the isocyanate for conjugation to BSA. Similar approaches were applied to the development of an immunoassay for tetrachlorodibenzo-p-dioxin using 1-amino-2,3,7,8-tetrachlorodibenzo-p-dioxin as a hapten.

Aromatic amines may be converted to their diazonium salts with nitrous acid (Fig. 5.13). By mixing the protein and diazonium salt under basic conditions, the hapten salt may then be bound via an azo linkage to the tyrosine, histidine, lysine, and possibly arginine and tryptophane residues of the carrier protein. Pauling et al. (1942) and Landsteiner (1945) used this method in their classic immunochemical studies. More recently amino-parathion

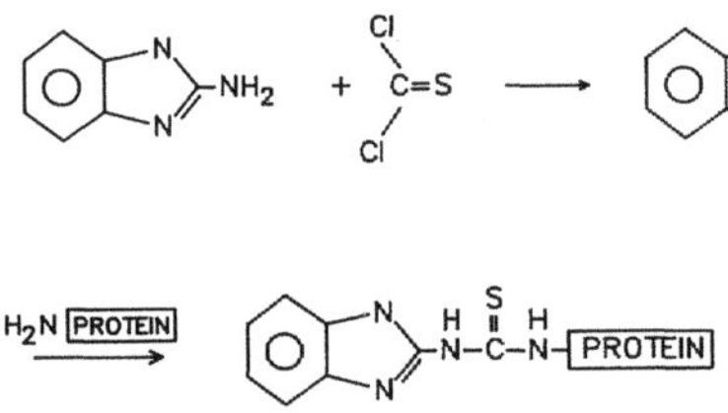

Figure 5.12. Synthesis of the benomyl immunogen (Lukens et al., 1977).

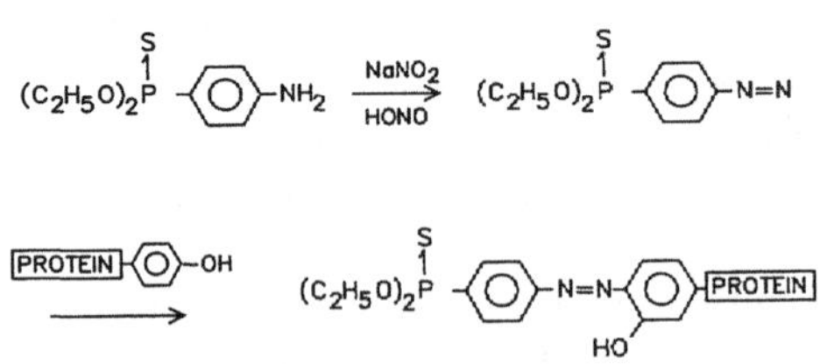

Figure 5.13. Synthesis of amino parathion and its conjugation to BSA (Ercegovich et al., 1981).

Table 5.3. Cross-Reactivity of Triazine Herbicide Immunoassays Developed with Three Different Haptens (Sharp et al., 1986; Huber and Hock, 1985)

	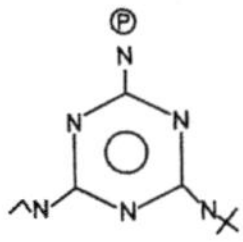	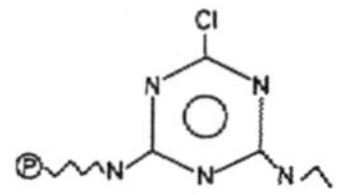	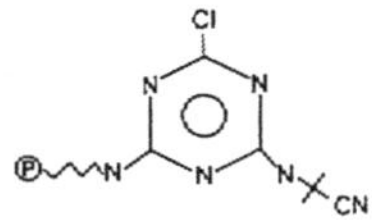
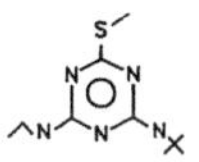	100%	50%	0.04%
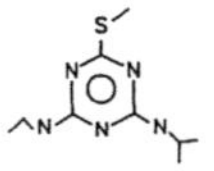	50	—	—
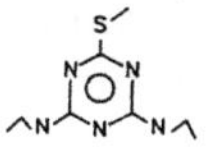	<10	—	—
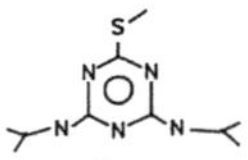	<10	33	—
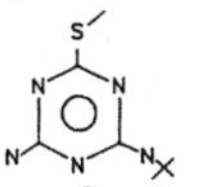	64	—	—
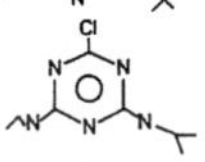	55	100	ND
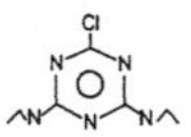	12	50	ND
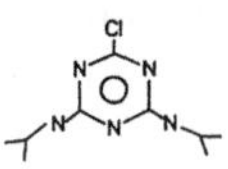	9	25	0.04
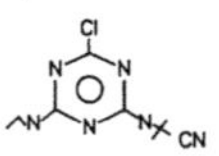	—	8.3	100
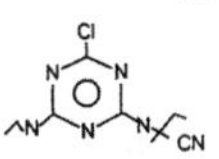	—	—	2.1
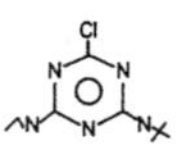	—	—	2.4
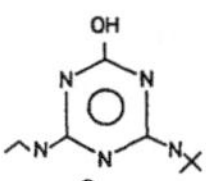	78	—	—
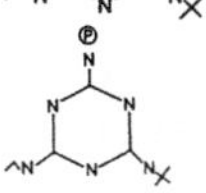	66	—	—

(Ercegovich et al., 1981), amino-paraoxon (Hunter and Lenz, 1982), and amino-chlorsulfuron (Kelly et al., 1985) were coupled by this procedure. The S-triazine, terbutryne, was oxidized to the sulfoxide with *m*-chloroperbenzoic acid in chloroform (Hammock et al., 1981). After TLC purification, the derivative was dissolved in ethanol and then added to BSA in carbonate buffer (Huber and Hock, 1984) to produce the hapten conjugate (Table 5.3).

III. Immunogen Purification and Characterization

Dialysis is the most obvious method of purification, but some lipophilic molecules cannot easily pass through a dialysis membrane. Gel filtration provides an alternative method, or, if the protein is not too badly denatured, repeated precipitation with an organic solvent such as ethanol can remove lipophilic impurities.

The most quantitative means of determining the moles of hapten bound per mole of carrier protein is to use radiolabeled hapten. Alternatively, a change in the absorbance of the hapten carrier conjugate could be monitored if the hapten absorbs in a spectral region where the protein itself does not absorb. An elemental analysis such as the phosphorus determination used to monitor parathion conjugation is yet another strategy. If means of monitoring the hapten directly cannot be employed, the proteins' reactive groups (such as the free amine) can be monitored both before and after conjugation. However, careful controls are necessary with these procedures, because self-conjugation or denaturation may decrease or even increase the apparent functionalities available for binding (Hammock and Mumma, 1980).

1. Immunization

Once the immunogen is synthesized, the conjugate is introduced into an animal to trigger the immune system and elicit antibody production. There is no standardized protocol for immunization, with most approaches being largely empirical. Most protocols, though usually overdone, still produce useful antisera. A wide variety of animals have been used, ranging from mice to horses. However, since a little antiserum goes such a long way, rabbits or guinea pigs are usually the choice for polyclonal antiserum production, especially when the assay is limited to in-house use. Once a good producer rabbit is identified, it can, over a lifetime, produce enough antiserum for hundreds of thousands or even millions of assays (Murphy and Lehotay, 1977). Some investigators have used chickens (Jensenius, 1981; Rose et al., 1974; Bar-Joseph and Malkinson, 1980), though the variety of prepared labeled antibody aganist chicken IgY is limited when compared to that available for the IgG of rabbits, guinea pigs, or goats.

To challenge with the immunogen, the compound is injected into the animal, usually after being emulsified in a combination of saline solution and complete Freund's adjuvant (CFA) (Freund and McDermott, 1942). CFA is a stabilized heavy mineral oil emulsion that will form a stable emulsion with buffer to form depots that serve both to protect the immunogen and to slow its introduction into the animal, providing a low-level continuous stimulation of the immune system (Herbert, 1973). In addition, CFA contains killed mycobacteria, which help stimulate the immune system. Other adjuvants used include soluble and insoluble aluminum salts, bentonite, methylated BSA (Sueoka and Cheng, 1962), and heat-inactivated *Bordetella pertussis*. Routes for introduction have included intradermal, perilymphnodal, subcutaneous, intramuscular, and even footpads. Most routes have resulted in the production of usable antisera, demonstrating that the painful (footpad) or highly involved (multisite intradermal) protocols are unnecessary.

To illustrate, three routes of immunization were tried with the triazine immunogens: a subcutaneous challenge and boosters, intradermal challenge followed by intramuscular boosters, and a perilymphnodal challenge followed by intramuscular boosters (Sharp et al., 1986). All these protocols raised ABs of similar sensitivity and specificity, with higher titers than earlier protocols where animals were immunized at each interval with 2–5 mg. Thus, the quality of the immunogen and the amount and timing of its introduction are more important than the route of introduction.

The initial challenge is followed by a series of immunization boosters. A good general rule is to introduce the immunogen at a rate of 15 μg/kg body weight of the animal (Tijssen, 1985). The boosters usually consist of the immunogen emulsified in adjuvant without the mycobacteria (incomplete Freund's adjuvant) to prevent severe allergic reactions. Intervals between boosters vary, but good antisera have been produced with a series of monthly boosters begun after the declination of the initial immune response. Bleeds are usually taken 10 days after the booster. A prebleed should be taken before immunizations are begun to check for presence of preexisting ABs.

Since the ELISA is a competitive assay, it is usually not necessary to do any purification beyond serum isolation. Purification can be helpful to maximize AB concentration on the solid phase (in a direct ELISA, where labeled and unlabeled ligand compete for solid-phase AB binding site) or where nonspecific interactions have proved to be a problem. Non-IgG proteins can be precipitated with organic compounds such as ethodine, rivanol (Horejse and Smetana, 1956; Gordon et al., 1962), or caprylic acid (CA) (Steinbuch and Audran, 1969). Salting out can be effective in selectively precipitating IgGs (Tijssen, 1985), since the conditions that cause precipitation vary with the protein type. Other methods of purification include weak anion exchange, strong anion exchange, weak cation exchange chromatography (Peterson, 1970; Baumstark et al., 1964; Perper et al., 1967; Reif, 1969; Levy and Sober, 1960), hydrophobic chromatography (Doellgast and Plaut, 1976), zone electrophoresis (Jaton et al., 1979), isoelectric focusing (Williamson et al.,

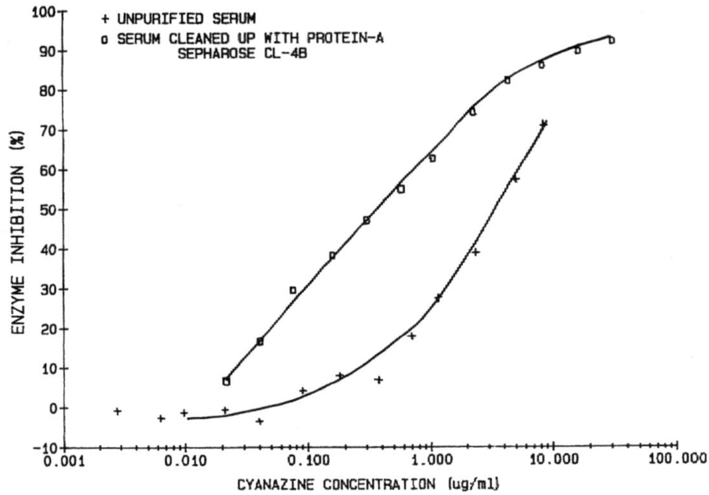

Figure 5.14. Cyanazine ELISA enzyme inhibition curve generated with anticyanazine antisera before (+) and after (□) protein A–Sepharose CL-4B chromatography cleanup.

1973; Braun et al., 1979; Radola, 1974; Schalch and Braun, 1979), and density gradient ultracentrifugation (McCall and Potter, 1973). Protein A Sepharose affinity chromatography was successfully used in our hands to yield useful ABs after we had failed to produce a useful assay with the crude antisera (Fig. 5.14). Haptens can be immobilized on a solid support and used to capture hapten-specific IgG for later isolation. Small haptens such as pesticides are immobilized with the use of a spacer to prevent steric interference between the solid support and the AB (Ferrua et al., 1979). The best results have been with low ligand concentrations and 0.5% Tween 80 in the buffer to eliminate nonspecific interactions (Eveleigh and Levy, 1977). The captured ABs are eluted with a reduction in pH or a pH reduction in combination with a reducing reagent, dimethylformamide (Fig. 5.15). Unfortunately, denaturation often results from the extreme conditions necessary to elute these high-affinity ABs.

2. Antisera Characterization

The only characterization needed to develop an immunoassay is the AB's affinity for the desired ligand and the potential for interference from similar compounds or the sample matrix in the assay quantification. Further characterization such as isotype is not necessary unless one is developing monoclonal cell lines.

A variety of techniques are available to test for antiserum activity against

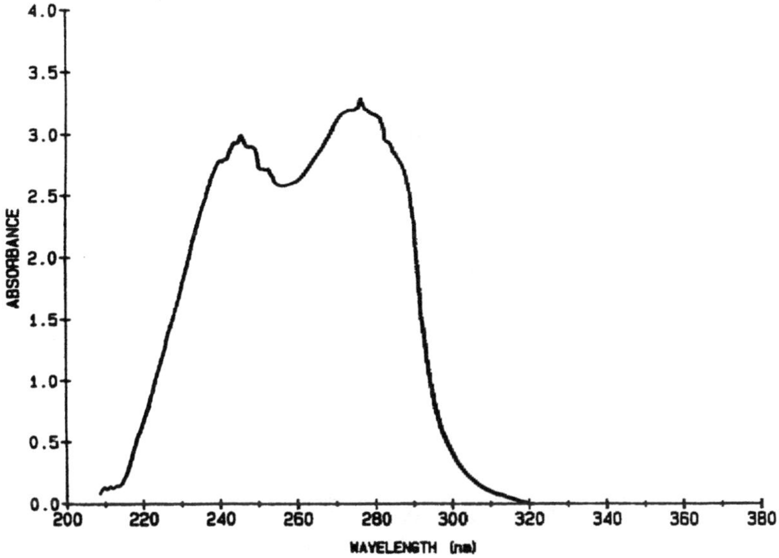

A

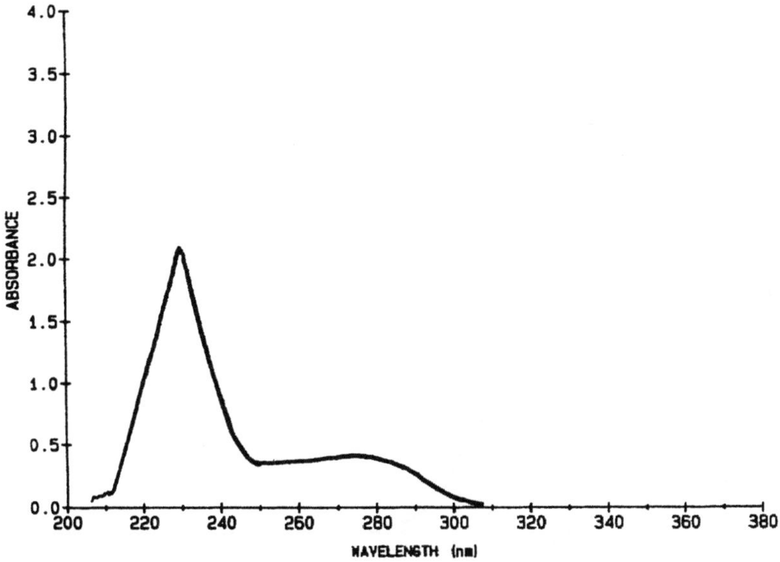

B

Figure 5.15. UV spectra of serum fractions eluted from protein A–Sepharose CL-4B. The absorption maximum at 270 nm is due to protein. (A) Unretained proteins. (B–D) Fractions taken after addition of 0.1 M acetic acid to column. Fraction C contained greater than 90% of the anticyanazine activity.

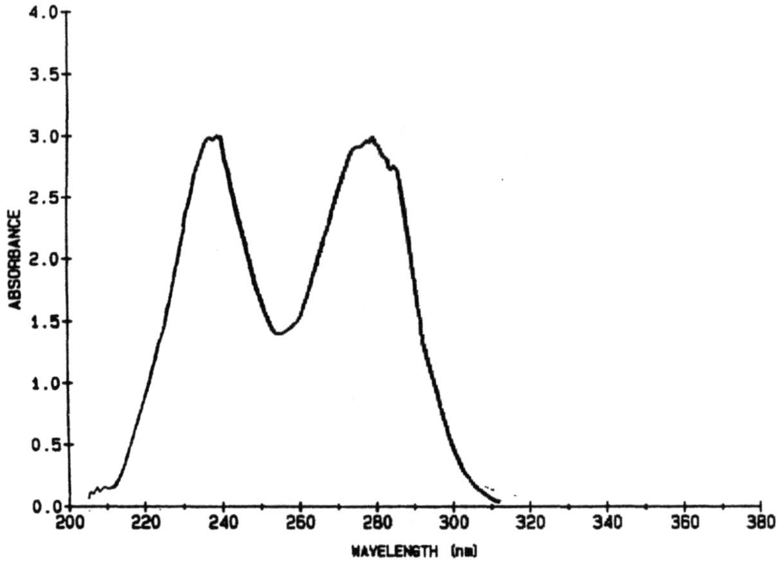

C

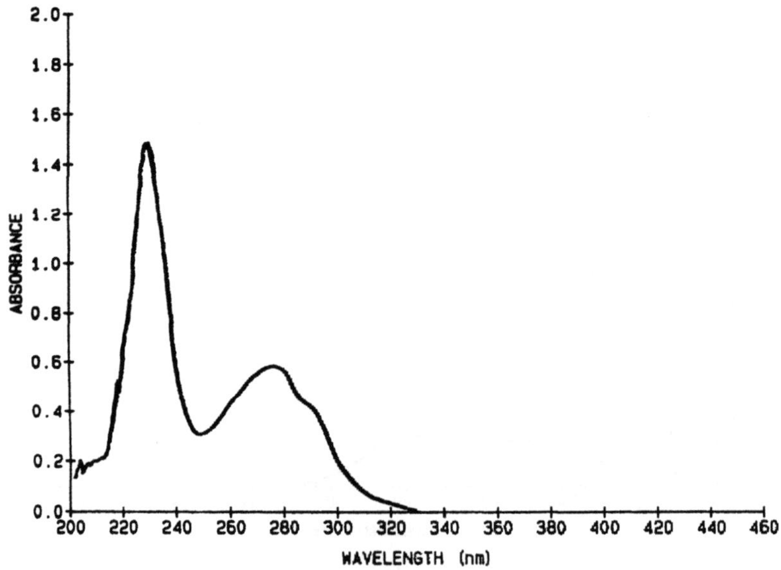

D

Figure 5.15 (*continued*)

the Ag as well as the linking group and the carrier. These include Ouchterlony double diffusion (Ouchterlony, 1958), immunoelectrophoresis, immunoprecipitation, and immunoprecipitation inhibition (Al-Rubae, 1978). These procedures are also not needed to develop an effective ELISA. One needs only to characterize the AB by how well it will perform in the assay using the established criteria of AB affinity and cross-reactivity.

The AB affinity is determined by the intrinsic affinity constant, K_0, which is the affinity when half the AB binding sites are filled, otherwise stated as the reciprocal of the free hapten concentration at equilibrium when half the binding sites of a divalent AB are filled. This can best be determined with a Scatchard plot. Data reduction packages are available to ease this determination. Other methods of determining AB affinities include methods based on Michaelis-Menton kinetics (Van Heyningen et al., 1983), equilibrium dialysis (Ercegovich et al., 1981), quantitative immunoelectrophoresis (Birkmeyer et al., 1981), the Farr test (Farr and Nakane, 1958; Griswold and Nelson, 1984), and Sips plots (Sips, 1948; Nisonoff et al., 1975).

The working titer of the antisera and the concentration of coating Ag used to coat the sample well are most easily determined with a checkerboard assay. The ELISA is performed with each of the six inside rows of the microtiter plate coated with a decreasing concentration of coating Ag, usually ranging from 10 μg to 100 ng/ml or less of Ag. The 10 inside columns are filled from left to right with serial dilutions of the antisera. As with the competitive assay, the outside wells of the plate are usually left empty because of their low precision (Sharp et al., 1986). A guideline for selecting optimum concentrations is to choose a coating level where a further increase in Ag concentration is no longer beneficial and an antiserum dilution that produces an absorbance of 1.0 in 20–30 min. The checkerboard assay can also give an indication of the affinity constant (K) of the AB, with a steep drop-off in activity indicative of a high K value. To develop a sensitive assay, it is essential to use as dilute an AB concentration as possible, since at a dilute concentration only those ABs with a high K value will be reacting, and weak AB, which tend more toward nonspecific binding, will remain dissociated.

3. Enzyme-AB Conjugate Syntheses

Enzyme conjugates can be obtained from commercial supply houses. Alternatively they can be synthesized in house at much lower cost. In-house conjugation gives the analyst increased flexibility in the identity and placement of the label, which in turn gives more control in defining assay characteristics (working range and sensitivity), tailoring it to the desired application. Several enzymes have been used, the most common being horseradish peroxidase, alkaline phosphatase, and B-galactosidase. Conjugation is usually done chemically, originally with glutaraldehyde and more recently with more specific divalent reagents that better preserve the enzyme activity (Tijssen, 1985).

IV. Applications

The examples for this discussion are of the application of immunochemistry to the analysis of insecticides and herbicides, since similar problems are encountered in both analyses (Table 5.3). The challenge facing analytical chemists in the application of immunoassay to pesticide analysis, especially trace level analysis, is in overcoming the myriad of matrix effects that may interfere with the Ag-AB interaction. The prevention of AB nonspecific binding and the analysis of lipophilic compounds found in hydrophobic environments are only two of the many problems that have been solved with only limited success. Thus, most of the early examples of environmental pesticide analysis have been limited to water or soil extracts.

The ELISA is performed as described earlier. The steps in the procedure remain relatively unchanged for all analytes, though certain variables can be changed to increase assay precision and sensitivity. The variables studied included the makeup of the wash and buffer solutions, microtiter plate pretreatment, and coating procedures enzyme label, as well as the time and temperature of the various incubation steps.

An advantage of the hapten immunoassay is that, unlike in many macromolecular assays, the chemical nature of the Ag is well known, and conclusions can be made of the importance of the different weak forces involved in its interaction with the AB (Freedmen et al., 1968; Nisonoff et al., 1975). This can be used to design the AB-Ag incubation buffers that maximize the strength as well as specificity of the Ag-AB interaction. pH (Eisen, 1980), temperature (Weiner, 1957; Tijssen, Su, and Kurstak 1982), buffer strength, and the presence of organic solvents (Chan-Shu and Blair, 1979) can all have important effects, depending on the nature of the Ag-AB interaction.

The Ag-AB incubation buffer is also where any sample matrix effects that may contribute to both specific and nonspecific interactions will be seen. Historically, the nonionic surfactant Tween 20 was first added to limit nonspecific interactions (Voller), but it has proved ineffective unless the AB has high affinity for the analyte. Another strategy includes adding blocking agents, usually other proteins, in high concentrations relative to the antiserum concentration to displace the AB from absorption sites on sample surfaces. Blocking agents can be used alone or in combination with other proteins or surfactants such as Tween 20 or Triton X. Proteins used include BSA and gelatin (Sharp et al., 1986), casein (Kenna et al., 1985), or IgG (Voller and Bidwell, 1986). Protein mixtures such as casein or powdered nonfat milk are effective as blocking agents, perhaps because of protein variety of size, degree of glycosylation, sulfation, and surface charge. An antibacterial such as thimerosal has also been shown to reduce nonspecific binding by controlling bacteria that adsorb on the solid phase and may have ABs against them preexisting in the sera. The effectiveness of this antibacterial illustrates the importance of keeping all immunochemical apparati clean to minimize contamination and controlling bacterial growth in all protein and buffer solutions.

The function of the wash steps is to prevent carryover of the immunoreactants. Historically, PBS–Tween 20 was prescribed as the wash solution, though distilled water may be just as effective. Others just hold the plate under running tap water (Ashorn and Krohn, 1986) or use multiple washes with buffer solutions containing blocking agents and/or antibacterials for very sensitive assays (Kenna et al., 1986). In addition to removing previously added reagents, multiple washings of the sample solution from the plate can be one means of eliminating cross-reactivity caused by weakly binding, less specific ABs due to their short complex half-life (Tijssen, 1985).

Pretreatment of microtiter plates with UV light prior to coating has been reported to increase assay precision (Zouali and Stollar, 1986; Sharp et al., 1986). Washing the plates before coating, with either water or dilute acid, has also been beneficial. Once used, plates are usually discarded, though they can be reused after soaking overnight in 6 *N* hydrochloric acid.

1. Organochlorine Compounds

The hapten utilized for the RIA developed by Langone and Van Vunakis (1975) elicited an AB that, though selective for dieldrin and aldrin, had significant cross-reactivity with heptachlor and chlordane. To prepare the hapten, a mixture of hexachlorocyclopentadiene and bicyclo(2.2.1)-5-heptene-2-carboxylic acid was heated to produce 6,7-dihydro-6-carboxyaldrin which was purified by recrystallization (Fig. 5.8). The demonstrated cross-reactivity shows that the AB is specific to the hexachlorobicyclic ring. The extremely low cross-reactivity to lindane (1/325) may be due to activity toward particular isomers with similar conformation to the hexachloro ring. To give adequate precision, the assay had to be carried out in a buffer containing 10% horse serum, perhaps needed to help solubilize the dieldrin. The problem of 2,3,7,8-tetrachlorodioxin (TCDD) solubility was overcome by suspending the compound in Triton X-305 or dimethyl sulfoxide (DMSO) (Vanderlaan et al., 1986). This ELISA has a limit of detection of 1 ng.

2. Organophosphorus Compounds

Parathion immunoassays were first successfully developed with an azo linkage as described in the hapten synthesis section (Ercegovich et al., 1981). The ABs displayed strong affinity for parathion (1.5×10^7 M^{-1}) and discriminated against methyl parathion, paraoxon, and reduced parathion. The RIA could detect as low as 10–20 ng in fortified human plasma and crude extract of various plant tissues. The same hapten was used to develop an ELISA (Al-Rubae, 1978). In this procedure, the crops were analyzed by macerating with methanol, taken to dryness, then picked up into DMSO to a crop concentration of 2 g/ml for analysis. The ABs had a binding constant of 1.5×10^7 M^{-1} and produced assays capable of detecting from 0.025 ppm in tomatoes to 0.05 ppm in biological fluids such as cow rumen, human serum, and urine. Some nonspecific inhibition was caused by the sample matrix, though no cleanup was necessary for any of the analyses. Little cross-reactivity was seen with other pesticides, including methyl parathion.

Attempts were made to increase the selectivity of the assay by connecting a linking group to the 2- or 3-phenyl position, preserving both the nitro and the phosphoroester functionalities (Vallejo et al., 1982). The Ags elicited ABs that were highly selective for the synthesized haptens but did not recognize parathion. Even when reduced parathion was used with longer bridging groups, glutarimine or diazobenzoic acid, the ABs did not recognize unaltered parathion. Thus the parathion hapten originally used, with no new functionalities or determinant groups introduced and with the altered function well masked, seems to be the optimum design.

A paraoxon hapten was similarly linked to KLH to raise ABs used to develop a sensitive ELISA (28 pg in buffer and 280 pg in serum) (Hunter and Lenz, 1982). The ion exchange chromatography purified IgG fraction (DE-52) was further purified on Sepharose-4B coupled with paraoxon-BSA. An RIA was developed for paraoxon with a sensitivity of 0.2 ng (Rong et al., 1986). The antiserum was specific with an affinity constant of 2.13×108 M^{-1} and could be used directly in untreated blood samples. Monoclonal ABs were developed against paraoxon in another attempt to improve assay selectivity (Brimfield et al., 1985). The ABs exhibited an affinity of 1×10^5 M^{-1} with significant cross-reactivity with parathion and amino paraoxon. The ELISA had a minimum detectable concentration of 1 ppm, significantly higher than the polyclonal assay, due largely to the decreased AB affinity.

3. Pyrethroids

Antisera raised against the hemisuccinate of an alcohol derivative of the allethrin propene side chain were used to develop an RIA (Wing et al., 1978) and later an ELISA (Van Emon et al., 1986a). In both assays, the AB was highly selective for the conformation of the chrysanthemate moiety, distal to the site of conjugation. The analytical range of the ELISA assay was 0.0016–1.6 nmole; for the RIA it was 0.5–10 nmole. While the assays could be used directly in milk, there was a 10-fold loss in sensitivity compared to the assay performed in buffer.

An assay for 2-(chlorophenyl)isovaleric acid, a urinary metabolite of fenvalerate, was developed to monitor human exposure (Onishi et al., 1986). Rabbits were immunized with 2-(chlorophenyl)isovalericaminobutyric acid conjugated to BSA. The antisera exhibited low cross-reactivity with other fenvalerate metabolites. The RIA (^3H label) was performed by incubating a

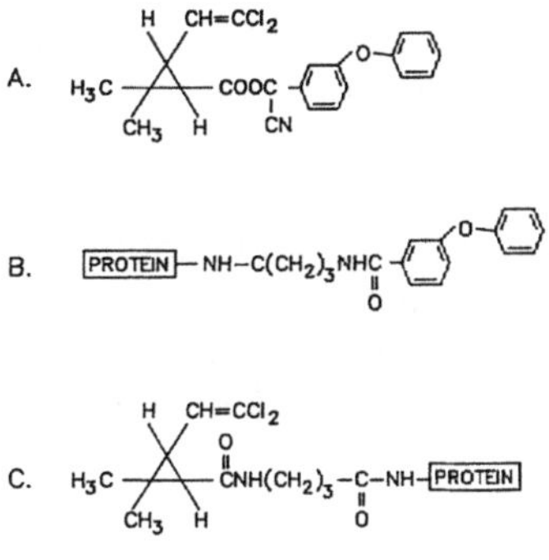

Figure 5.16. The immunogens used in developing a cypermethrin immunoassay. (A) Cypermethrin. (B) 3-Phenoxybenzoic acid hapten. (C) Dichlorovinyl cyclopropane-carboxylic acid hapten (Wraith et al., 1986).

mixture of urine, diluted antiserum, and [^{3}H]CPIA for 18 h at 4°C, then separating the bound and free fractions by the dextran-coated charcoal method. The detection limit was 2 ng per sample.

In the development of a cypermethrin ELISA, ABs were raised both against 3-phenoxybenzoic acid (PBA) and the dichlorovinylcyclopropane–carboxylic acid (CPA) in addition to being raised against a cypermethrin conjugate (Wraith et al., 1986) (Fig. 5.16). The PBA ABs were sensitive to PBA and to a wide class of synthetic pyrethroids, whereas the CPA ABs were only responsive to cypermethrin and permethrin. There have been problems in applying the assays owing to sample matrix interferences, even after extensive sample cleanup.

4. Insect Growth Regulators

ELISAs were developed for the benzoylphenylurea insect growth regulators diflubenzuron, BAY SIR 8514, and some of their analogues. ELISAs were based on ABs raised against an *N*-carboxypropyl hapten of diflubenzuron (Wie and Hammock, 1982a) (Fig. 5.9). A diflubenzuron phenylacetate derivative coupled to a different carrier from the immunizing Ag was used as the coating Ag. Despite being challenged with the same immunogen, one rabbit produced antisera sensitive for diflubenzuron, BAY SIR 8514, and closely related IGRs, but a second rabbit produced ABs selective only for diflubenzuron (Wie and Hammock, 1982). Neither of the assays demonstrated significant cross-reactivity with the benzamide, urea, phenylurea, or aniline components of diflubenzuron or BAY SIR 8514. Each of the assays was shown to be as sensitive as the recommended HPLC methodology for the analysis of diflubenzuron in water. Diflubenzuron could also be detected in a solution of up to 50% milk at a level of 2 ppb without any sample extraction procedure, demonstrating the AB's ability to extract diflubenzuron out of the milk micelles. The within-run and between-run variations were 3% and 5%, respectively, over a period of several months. Sample cleanup led to still more sensitive assays, though in the crude sample matrix provided precision and sensitivity adequate for residue analysis. In another study, these same authors also demonstrated that by using different coating Ags with the AB, the selectivity and sensitivity of the assay could be changed (Wie and Hammock, 1984) to suit the particular analytical requirements.

5. Bacillus thuringiensis

Bacillus thuringiensis bacteria produce protein protoxins during sporulation. This toxin is of commercial interest not only as a pesticide formulation, but because of the potential for manipulating cloned protoxin genes (Aronson et al., 1986). *B. t.* var. Kurstaki has been expressed in plants and in an alternate bacterial species. ELISAs are invaluable in monitoring commercial production, supplanting a long and tedious bioassay (Smith and Ulrich, 1983) and serving as a residue procedure. They also serve to complement gene probes, providing a measure of genetic expression. Monoclonal ABs as described elsewhere in this volume have helped characterize the active site of the toxin (Huber-Lukac et al., 1982, 1983).

6. Triazine Herbicides

The terbutryn immunoassay developed by Huber and Hock (1985) showed cross-reactivity with dialky-substituted S-triazine and chlorotriazine herbicides (Table 5.4). Aqueous solutions could be analyzed directly with a linear measuring range of 4.8–289.7 ng terbutryn per assay. There was high cross-reactivity with the metabolite hydroxyterbutryn, which would allow for the simultaneous detection of the herbicide and its hydroxy metabolite. If desired, selective detection could be performed with this AB by employing an on- or off-line chromatographic separation.

Some of the cross-reactivity seen in the terbutryn ELISA could be due to the lack of a linking group, with some masking of the compound's antigenic determinants by the protein carrier, though an immunoassay against an atrazine hapten provided with a linking group showed similar cross-reactivity with S-triazine herbicides with similar N-alkyl substituents (Sharp et al., 1986). However, this assay proved to be both more sensitive and more specific than that developed for terbutryn, probably because of the use of a linking group in the immunogen conjugation. A cyanazine ELISA developed by these same authors showed virtually no cross-reactivity with any of the tested triazines and had a lower limit of detection of 0.5 ppb for cyanazine in water and 0.01 ppm for cyanazine in soil. To perform the assay, the soil samples were sonicated with methanol, and an aliquot of the solvent was taken to dryness

Table 5.4. Examples of Pesticide Immunoassays

Chemical	Hapten	Detection Limit	Assay type	Reference
Aldrin		700 pg	RIA(^{125}I)	Langone and Van Vunakis, 1975
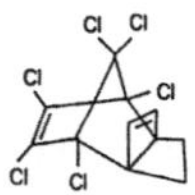	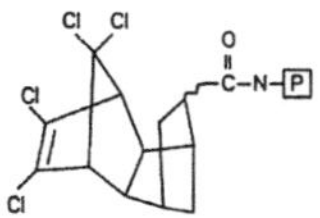			
Atrazine		1 pmole 0.01 ppm (soil)	ELISA	Huber and Hock, 1985
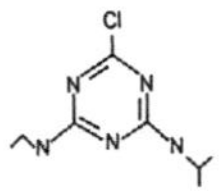	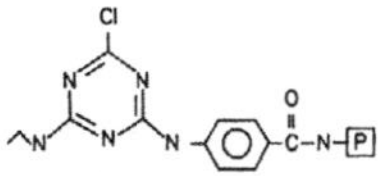			
Bacillus thuringiensis	*B. t. K*	105 ng	ELISA	Wie et al., 1984
BAY SIR 8514		9.6 ng	ELISA	Wie and Hammock, 1984
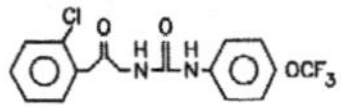	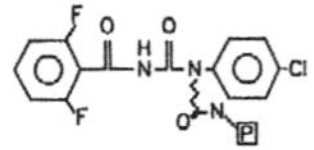			
Benomyl		0.1 ng 1.25 ng	FIA RIA(^{14}C)	Lukens et al., 1977 Newsome and Shields, 1981
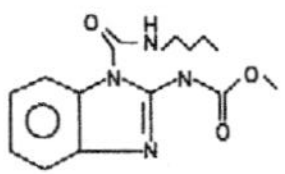 as 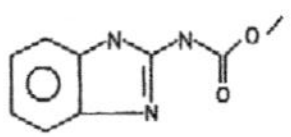	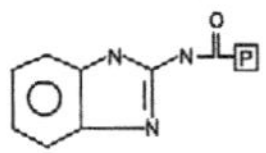			

(*continued*)

Table 5.4. (*continued*)

Chemical	Hapten	Detection Limit	Assay type	Reference
Chlorsulfuron 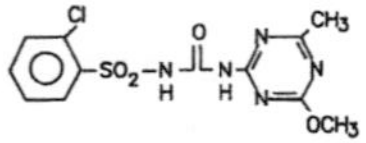	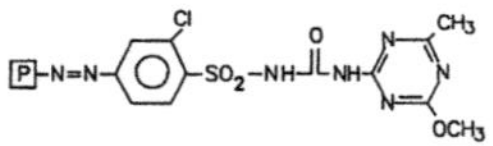	0.4 ppb (buffer) 1.2 ppb (soil)	ELISA	Kelley et al., 1985
2-(4-Chlorophenyl)isovaleric acid 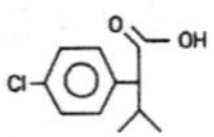	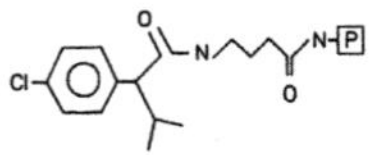	2 ng	RIA(^{3}H)	Onishi et al., 1986
Cyanazine 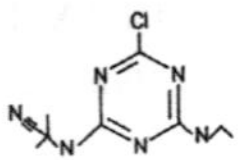	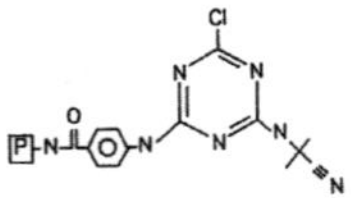	0.5 ppb (water) 10 ppb (soil)	ELISA	Sharp et al., 1986
DDT 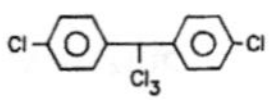	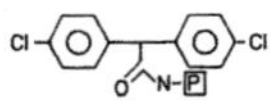	NA	RIA	Centeno et al., 1970
Diclofop-methyl 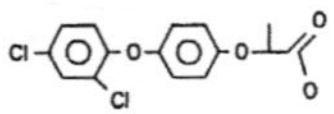	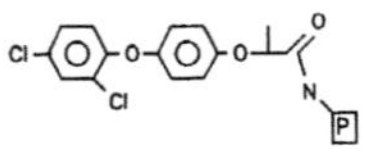	45 ng/ml 23 ng/ml	FIA ELISA	Schwalbe et al., 1984
Dieldrin 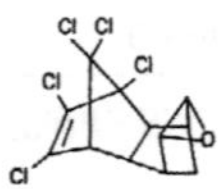	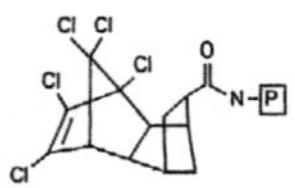	150 pg 10 pg	RIA(^{125}I) EMIT	Langone and Van Vunakis, 1975 Ferguson, 1986

(*continued*)

Malathion		—	RIA	Centeno et al., 1970
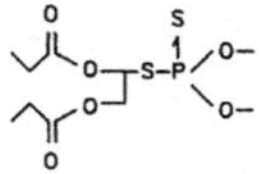	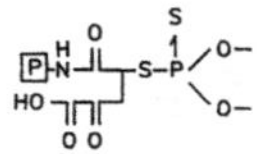			
Metalaxyl		0.1 ppm	ELISA	Newsome, 1985
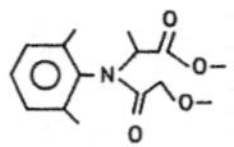	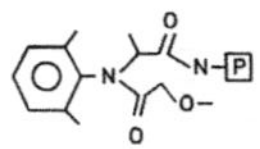			
Paraoxon		28 ppb (buffer)	ELISA	Hunter and Lenz, 1982
		280 ppb (serum)		
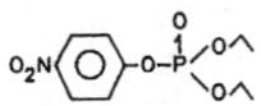	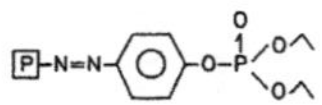	1 ppm	ELISA	Brimfield et al., 1985
Paraquat		10 ppb	RIA(^{3}H)	Fatori and Hunter, 1980
		0.25 ppb	RIA(^{131}I)	
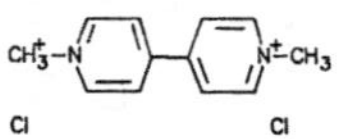	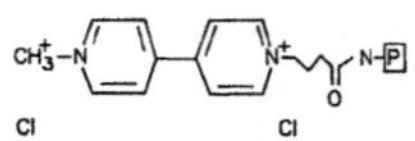	0.1 ppb (buffer)	ELISA	Van Emon et al., 1986
		1 ppb (serum)		
Parathion		4 ng	RIA	Ercegovich et al., 1981
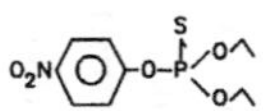	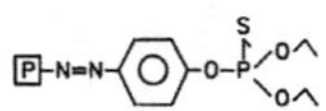	7.8 ng	ELISA	Al-Rubae, 1978
		2.5 ng	ELISA	Environmental Diagnostics
Terbutryn		25 ppb	ELISA	Huber and Hock, 1985
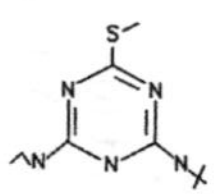	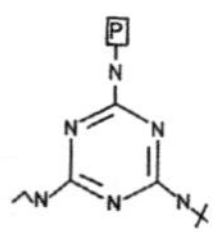			

A.

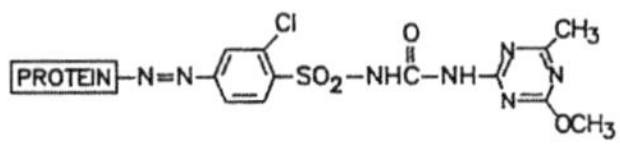

B.

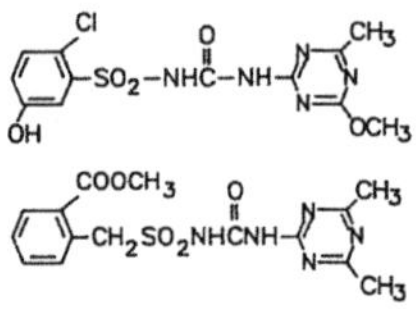

C.

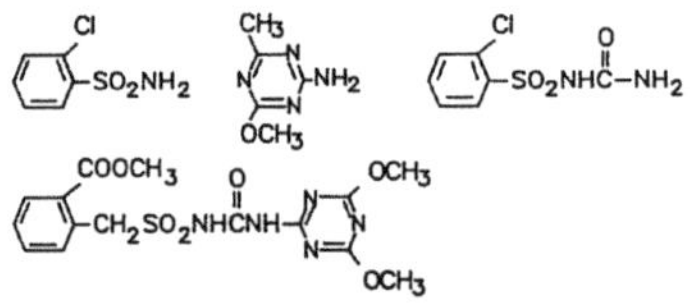

Figure 5.17. Chemicals tested for cross-reactivity in the chlorsulfuron immunoassay. (A) Chlorsulfuron immunogen. (B) Compounds showing significant cross-reactivity. (C) Compound with minimal cross-reactivity (Kelley et al., 1985).

and reconstituted in PBS-Tween with 0.5% BSA and 0.1% gelatin. A 9-ml water sample was mixed with 1 ml of a 10 × PBS-Tween buffer with 5% BSA and 1% gelatin.

7. Sulfonylurea Herbicides

An ELISA was developed for the analysis of chlorsulfuron in crude soil extracts (Kelley et al., 1985). The chlorsulfuron diazonium salt was prepared from *N*-[4-methoxy-6-methyl-1,3,5-triazin-2-yl-amino]carbonyl-4-amino-2-chlorobenzensulfonamide and then conjugated through an azo linkage to KLH (immunogen) or BSA (coating Ag). In addition to chlorsulfuron, the assay detected some sulfonylurea herbicides that are chemically similar to chlorsulfuron, but it did not detect certain other sulfonylurea herbicides or the primary degradation products of chlorsulfuron (Fig. 5.17). Crude, unfiltered soil extracts did cause some inhibition, though chlorsulfuron could still be detected in the soil at the ppb level.

8. Paraquat

A paraquat assay was developed for the assessment of human exposure (Van Emon et al., 1986). The hapten was synthesized from 4,4′-bipyridine and iodomethane in dry chloroform to form N-methyl-4-(4-pyridyl)pyridinium iodide (Fig. 5.18), then treated the product with 5-bromovalerate in anhydrons dimethylformamide. The resulting ester was hydrolyzed to give the desired acid, *N*-(4-carboxybut-1-yl)-*N*′methylbipyridilium. This was coupled

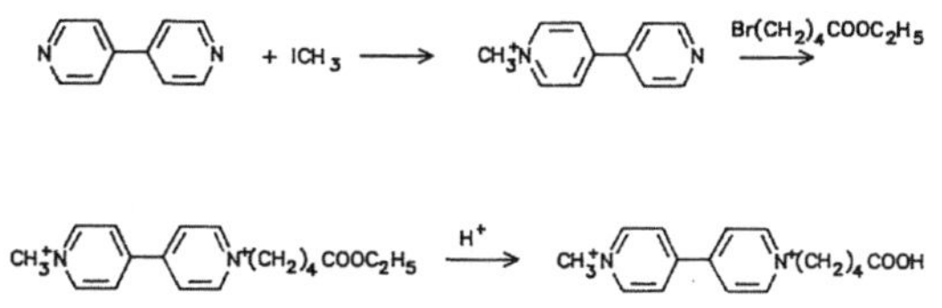

Figure 5.18. Synthesis of the paraquat hapten (Van Emon et al., 1986a).

to conalbumin for immunization and KLH for coating the microtiter plates. With the resulting AB at a dilution of 1/5,000, concentrations of paraquat cation in the range 0.1–27 ng/ml could be measured. The limit of detection ranged from 0.1 to 1.0 ng/ml, depending on the sample matrix, with between-run and within-run variations of 5% and 4%, respectively. The assay showed minimal cross-reactivity with ethylparaquat, diquat, paraquat metabolites, and related compounds. In comparison with a gas chromatographic method, the ELISA was not only faster but gave higher recoveries and was more sensitive. The ELISA was applied to high-volume glass fiber filters, personal air monitors, worker clothing patches, and hand washes collected during aerial spraying of cotton.

9. Other Pesticides

Up to 50% field water could be directly assayed in an molinate ELISA (Hammock, 1986) with no significant effect on the standard curve. Since water can be analyzed directly, the assay is able to analyze the large number of samples necessary to evaluate the dynamics of molinate in rice culture or to evaluate water in the field before release to eliminate potential fish kills. For increased sensitivity, the water is solvent-extracted and concentrated, and the residue is exchanged into propylene glycol. Up to 5% propylene glycol could be added to the ELISA with no effect on AB binding.

A Surflan assay was performed in wells coated with Surflan dissolved in DMSO with monoclonal AB raised against the Affi-gel conjugate described earlier (Kuniyuki and McCarthy, 1986). This coating procedure may be appropriate in developing assays for compounds only sparingly soluble in water. A nicotine immunoassay (Langone and Van Vunakis, 1982, Fig. 5.19) has shown subnanogram sensitivity in various body fluids (Bjercke et al., 1986).

V. Validation

Like any new analytical method, immunoassay should be validated by comparison against an alternate accepted analytical technique. This is especially important for the methodology to gain acceptance in pesticide analysis.

The triazine immunoassay has been compared to the gas chromatographic (capillary gas chromatography with nitrogen-phosphorus detection) determination for water and soil samples with residues ranging from 1.0 to 30 ppb for water and 0.1 to 3 ppm for soil (Sharp et al., 1986). In this comparison, the natural logarithm of the determined value was used to make the variation in the analytical results more nearly constant across the wide level of cyanazine in the sample. In addition, when considering the ratio of the ELISA and GC results, the logarithmic transformation serves to make the ratios average appropriately (e.g., the average of 2 ratios of 0.5 and 2.0 is 1.0 if logarithms are used but 1.25 if not). Percent recovery of the ELISA and GC methods on the fortified sample of each media type was assessed by calculating the percent recovery of each sample by each method and performing a paired *t*-test for each fortification level and pooled over all levels.

Recoveries for the two methods were not significantly different. To illustrate how the two methods correlated, the values determined by ELISA were plotted against those determined by GLC. This correlation was further assessed using the ELISA and GLC results. The mean ratios, standard deviations, and standard errors were calculated and compared to a ratio of 1.0 for perfect correlation (Table 5.5). No differences in average ratios were noted between fortified and natural samples, nor did the ratio vary with the amount found. For water samples, the two methods gave comparable values; for soil samples, the ELISA results averaged 15% higher than the GLC results.

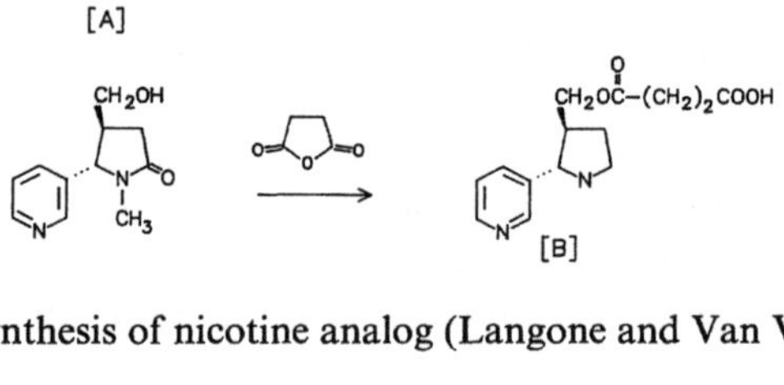

Figure 5.19. Synthesis of nicotine analog (Langone and Van Vunakis, 1982).

Table 5.5. Ratio of ELISA to GLC/NPD Results

Media type	Number of samples	Mean ratio	Standard deviation	Standard error
Water	39	1.03	0.24	0.04
Soil	42	1.15	0.24	0.04

VI. Field Portable Assays

Some of the most exciting applications of immunoassay are in the area of rapid, on-site testing. These are best applied where there is a need for rapid qualitative analysis. Such tests could be used to monitor pesticide residues in the field for farm worker safety and possibly adverse environmental effects by migration away from the site of application. Assays for water-soluble pesticides such as the thiocarbamates could be valuable in monitoring surface water (Hammock et al., 1986), and assays for triazines and acetanalides would be useful in groundwater monitoring. Assays for insecticides could be utilized by wholesalers and importers to ensure that crop residues are within tolerance and by pesticide companies in their stewardship programs. Assays in kit form have been designed that require minimal instrumentation, calibration, and operator expertise. Enzyme immunoassays are particularly well suited for these on-site applications, and their chromogenic end points make them useful for qualitative analysis without the need of any instrumentation.

An enzyme channeling immunoassay is the basis for a portable test strip immunoassay (Fig. 5.2). These assays use a pair of enzymes that participate in sequential reactions such that the product of the first enzyme serves as substrate for the second assay. The catalytic efficiency of the coupled reaction is highest when the two enzymes are held in close proximity. The enzyme pair utilized in the work described below is made up of glucose oxidase and horseradish peroxidase, where glucose oxidase generates hydrogen peroxide to catalyze the oxidation of 4-chloronaphthol. AB binding is the event that brings the two enzymes into close proximity. In the assay, ligand AB and one enzyme of the pair are covalently bound to a solid support (cellulose paper). Free ligand and ligand labeled with the complementary enzyme compete for the solid-phase AB binding sites.

Another laboratory assay based on the EMIT is being developed and marketed by ImmunoSystems, Inc. (Ferguson, 1986). In this assay, ligands have been linked to glucose-6-phosphate dehydrogenase, and the enzyme activity measured by monitoring the conversion of NADP to NADPH spectrometrically at 340 nm. In the laboratory version, AB, sample, enzyme-labeled ligand and the substrate are mixed in TRIS buffer and aspirated into thermostated cuvette. Polyclonal AB-based assays have been developed for aldicarb, chlorpyifos, and chlordane, with assay limits of detection in the low-nanogram range. The assays can be made field-portable with the use of a battery-powered photometer.

An assay designed from its inception to be portable is a credit card-size test kit marketed by Environmental Diagnostics, Inc. (Granite Division, Environmental Diagnostics, Inc., Burlington, NC). These tests are qualitative assays, with the lower limit of detection in the ppm to ppb range. The card has two ports (control and sample) containing AB on a filterlike material sandwiched between plastic sheets. Sample is applied to the sample port, and any compound in the sample is bound by the AB as the sample is wicked through. This bound ligand inhibits the binding of the ligand-enzyme conjugate which is added next, followed by the enzyme substrate. Each reagent displaces the previous. Color development is inhibited when ligand concentration in the sample is at a higher level than the sensitivity specified on the card. Color development in the control spot is used to ensure the integrity of the reagents. Parathion and paraquat test kits are among those currently being marketed.

VII. ELISA Automation

As mentioned earlier, an advantage of the immunoassay is the routine, repetitive nature of the steps in the procedure, making it amenable to automation. The 96-well microtiter plate format simplifies the sample handling and reagent addition, allowing the use of multichannel pipettes and wash manifolds. The spectrophotometric measurements are also facilitated, with spectrophotometers available that read the entire place in under a minute.

To increase sample throughput further, microassay systems with various levels of automation are now available, giving the laboratory the option of automating all or portions of the ELISA. Multichannel pipetters speed sample addition, place washers perform multiple washes in seconds, and diluter/dispensers speed reagent additions and perform serial dilutions to speed checkerboard assays. Inexpensive readers are available that can take absorbance readings of an entire microtiter plate in seconds. These can be fitted with devices to feed plates from a stack, and they can be interfaced to a computer for data processing.

One of the first systems to completely automate the ELISA, including plate coating, washing, sample, and reagent addition, as well as reading the plates, was developed in our laboratory together with Zymark, Inc. (Zymark Corp., Hopkinton, MN). The system was able to process up to 10 microtiter plates unattended. Based on experience gained with this and other systems, Zymark has developed and is marketing an improved version. Another automated system, the Microassay System developed by HP Genenchem (South San Francisco, CA) combines a precision robotic arm for manipulating plates and syringe pumps and manifold for performing reagent additions with sophisticated ELISA data handling software.

VIII. Conclusion

The immunochemical methods described will not replace the current methods of trace organic analysis, but they offer advantages that complement the classical techniques. Compounds most difficult to analyze by GC because of polar functional groups and low volatility lend themselves most readily to immunoassay. The isomer specificity offered by immunochemical techniques may make it the regulatory method of choice in analyzing the new chiral pesticides, the assay being made specific for the single most biologically active isomer.

In addition to potentially serving as a regulatory or enforcement analytical method, immunoassay may open up new areas to the pesticide chemist that are now unavailable owing to the time and cost constraints of the current trace residue techniques. If the promise of a rapid and inexpensive assay can be fulfilled, then cost-effective analytical data can be generated in support of field testing to confirm the rate and persistence of applications and to identify potential drift problems. More field residue samples can be taken to determine a compound's field stability and environmental fate more accurately.

Thus, instead of substituting for current analytical methods, immunoassays will supplement them. Since immunochemical assays such as ELISA can be run without the need for time-consuming cleanup procedures, these assays can rapidly screen many samples prior to a more involved analytical technique. When results indicate pesticide residues, the sample can be further analyzed using GLC or HPLC. Conversely, immunoassay can confirm residues found by other techniques. Antisera can also be used to concentrate pesticide residues or increase cleanup effectiveness to allow GLC or HPLC to be performed at greater sensitivity. Immunoassays, especially ELISA and EMIT, may also aid in determining reentry schedules or estimating human exposure by providing simple and inexpensive analyses.

Immunochemical methodology is still not fully accepted as a legitimate pesticide residue technique, making it important that early applications of the technique be successfully applied to demonstrate its utility further. Only appropriate analytes should be chosen for a laboratory's initial attempts at developing the technology. Standardization of the assay is a key, particularly in defining its analytical parameters—sensitivity, precision, and detection limits. Commercial enzyme-labeled antiimmunoglobulins vary widely in both their AB affinity and their enzyme activity. Suppliers should at least supply these parameters as well as suggested working titers to minimize any between-run variation caused by changes in reagents. A large boost to immunochemical applications would be given by EPA's acceptance of this technology in fulfilling new pesticide regulatory requirements. This lack of acceptance is largely due to the hesitation of chemical companies to submit an immunochemical technique, even as a secondary registration method. EPA has, however, accepted ELISA-generated data from a groundwater monitoring program.

References

Allen, J.C. (1986). Immunoassays in food analysis. *BNF Nutr. Bull.* **11**(1):46–54.

Al-Rubae, A.Y. (1978). Thesis, Department of Entomology, Pennsylvania State University.

Aronson, A.I., Beckman, W., and Dunn, P. (1986). *Bacillus thuringiensis* and related insect pathogens. *Microbiol. Rev.* 1–24.

Ashorn, P., and Krohn, K. (1986). Washing of ELISA plates with running tap water. *J. Immunol. Methods* **88**:141–142.

Bacquet, C., and Twumasi, D. (1984). A homogeneous enzyme immunoassay with avidin-ligand conjugate as the enzyme modulator. *Anal. Biochem.* **136**:487–490.

Bar-Joseph, M., and Malkinson, M. (1980). Hen egg yolks as a source of antiviral antibodies in the enzyme-linked immunosorbent assay: A comparison of two plant viruses. *J. Virol. Methods* **1**:179–183.

Bastiani, R.J. (1979). The EMIT system: A commercially successful innovation. *Antibiot. Chemother.* **26**:89–97.

Baumstark, J.S., Laffin, R.J., and Bardaw, W.A. (1964). A preparative method for the separation of 7S gamma globulin from human serum. *Arch. Biochem. Biophys.* **108**:514–522.

Bayer, E.A., Wilchek, M., and Skultelsky, E. (1976). Affinity cytochemistry—localization of lectin and antibody receptors on erythrocytes via avidin-biotin complex. *FEBS Lett.* **68**:240–244.

Belfield, A., Duncan, A.M., and Reavey, P.C. (1977). Enzyme multiplied immunoassay of phenobarbital by LKB reaction rate analyer: A comparison with analysis by gas-liquid chromatography. *Ann. Clin. Biochem.* **14**:218–222.

Birkmeyer, R.C., Keyes, L.L., and Tan-Wilson, A.L. (1981). Determination of relative antigen-antibody affinities by quantitative immunoelectrophoresis. *J. Immunol. Methods* **44**:271–284.

Bjercke, R.J., Cook, G., Rychlik, N., Gjika, H.B., Van Vunakis, H., and Langone, J.J. (1986). Stereospecific monoclonal antibodies to nicotine and cotinine and their use in enzyme-linked immunosorbent assays. *J. Immunol. Methods* **90**:203–213.

Boeckx, R.L., Frith, E.M., and Simons, F.E.R. (1979). Comparison of a high performance liquid chromatographic and an enzyme immunoassay technique for quantitation of theophylline in serum. *Ther. Drug Monitoring* **1**:65–73.

Booker, H.E., and Darcey, B.A. (1975). Enzymatic immunoassay vs. gas-liquid-chromatography for determination of phenobarbital and diphenylhydantoin in serum. *Clin. Chem.* **21**:1766–1768.

Brattin, W.J., and Sunshine, I. (1974). Comparison of available immunoassays for drugs of abuse in urine. In: *Immunoassay for Drugs Subject to Abuse*. pp. 107–116. S.J. Mule, I. Sunshine, M. Braude, and R.E. Willette (Eds.). CRC Press, Cleveland.

Braun, D.G., Hild, K., and Ziegler, A. (1979). Resolution of Immunoglobulin Patterns by Analytical Isoelectric focusing. In: *Immunological Methods*, Vol. 1, pp. 107–121. I. Lefkovits and B. Pernis (Eds.). Academic Press, New York.

Brimfield, A.A., Lenz, D.E., Graham, C. and Hunter, K.W. Jr. (1985). Mouse monoclonal antibodies against paraoxon: Potential reagents for immunoassays with constant immunochemical characteristics. *J. Agric. Food Chem.* **33**:1237–1242.

Brunk, S.D., Hadjiioannou, T.P., and Hadjiioannou, S.I. (1976). Adaptation of the EMIT technique for serum phenobarbital and diphenylhydantoin assays to a miniature centrifugal analyzer. *Clin. Chem.* **22**:905–907.

Burd, J.F., Carrico, R.J., Fetter, M.C., Bucker, R.T., Johnson, R.D., Boguslaski, R.C., and Christner, J.E. (1977a). Specific protein-binding reactions monitored by enzymatic hydrolysis of ligand-fluorescent dye conjugates. *Anal. Biochem.* **77**:56–57.

Burd, J.F., Wong, R.C., Feeney, J.E., Carrico, R.J., and Boguslaski, R.C. (1977b). Homogeneous reactant-labeled fluorescence immunoassay for therapeutic drugs exemplified by gentamicin determination in human serum. *Clin. Chem.* **23:** 1402–1408.

Carrico, R.J., Christner, J.E. Boguslaski, R.C., and Yeung, K.K. (1976). A method for monitoring specific binding reactions with cofactor labeled ligands. *Anal. Biochem.* **72:** 271–282.

Centeno, E.R., Johnson, W.J., and Sehon, A.H. (1970). Antibodies to two common pesticides, DDT and malathion. *Appl. Immunol.* **37:** 1–13.

Chamberlain, R.T., Stafford, D.T., and Maijub, A.G. (1977). High-pressure liquid-chromatography and enzyme immunoassay compared with gas-chromatography for determining phenytoin. *Clin. Chem.* **23:** 1764–1766.

Chang, J.J., Crowl, C.P., and Schneider, R.S. (1975). Homogeneous enzyme immunoassay for digoxin. *Clin. Chem.* **21:** 967.

Chan-Shu, S.A., and Blair, O. (1979). New method of antibody elution from red blood cells. *Transfusion* **19:** 182–185.

Claus, R. (1982). Attempts to develop a milk progesterone-radioimmunoassay on paper strips. *Br. Vet. J.* **138:** 430–435.

Cobb, M.E., Buckley, H., Hu, M.W., Miller, J.B., Singh, P., and Schneider, R.S. (1977). Homogeneous enzyme immunoassay for lidocaine in serum. *Clin. Chem.* **23:** 1161.

Costello, S.M., Felix, R.T., and Giese, R.W. (1979). Enhancement of immune cellular agglutination by use of an avidin-biotin system. *Clin. Chem.* **25:** 1572–1580.

Dellamonica, C., Bory, C., and Baltassat, P. (1978). Comparison of EMIT test with high-performance liquid-chromatography for determination of phenobarbital in serum. *Clin. Chem. Acta* **86:** 1–5.

Dietzler, D.N., Hoelting, C.R., Leckie, M.P., Smith, C.H., and Tieber, V.L. (1980). EMIT assays for 5 major anticonvulsant drugs—an evaluation of adaptations to 2 discrete kinetic analyzers. *Am. J. Clin. Pathol.* **74:** 41–50.

Doellgast, G.J., and Plaut, A.G. (1976). Purification of human IgA by salt-mediated hydrophobic chromatography. *Immunochemistry* **13:** 135–139.

Edwards, R. (1985). *Immunoassay: An Introduction*, William Heineman Medical Books, London, 192 pp.

Eisen, H.N. (1980). *Immunology: An Introduction to Molecular and Cellular Principles of the Immune Responses*, 2d Ed. Harper and Row, Hagerstown, MD, 258 pp.

Engvall, E., and Perlman, P. (1971a). Enzyme linked immunosorbent assay, ELISA. III. Quantitation of specific antibodies by enzyme-labeled anti-immunoglobulin in antigen coated tubes. *J. Immunol.* **109:** 129–135.

Engvall, E., and Perlman, P. (1971b). Enzyme linked immunosorbent assay (ELISA). Quantitative assay of immunoglobulin G. *Immunochemistry* **8:** 871–874.

Ercegovich, C.E., Valejo, R.P., Gettig, R.R., Woods, L., Bogus, E.R., and Mumma, R.O. (1981). Development of a radioimmunoassay for parathion. *J. Agric. Food Chem.* **29:** 559–563.

Eveleigh, J.W., and Levy, D.E. (1977). Immunochemical characterization and preparative application of agarose-based immunosorbents. *J. Solid Phase Biochem.* **2:** 44,45–78.

Farr, A.G., and Nakane, P.K. (1981). Immunostochemistry with enzyme labeled antibodies—a brief review. *J. Immunol. Methods* **47:** 129–144.

Fatori, D., and Hunter, W.M. (1980). Radioimmunoassay for serum paraquat. *Clin. Chim. Acta* **100**: 81–90.

Ferguson, B.S. (1986). Rapid detection and measurement of pesticide residues by homogeneous enzyme immunoassay at part per billion concentrations. Sixth International Congress of Pesticide Chemistry, Poster Session, Ottawa, Canada, August 11–15.

Ferrus, B., Milano, G., Ly, B., Duennec, J.Y., and Masseyeff, R. (1983). An enzyme immunoassay design using labeled antibodies for the determination of haptens—application to methotrexate assay. *J. Immunol. Methods* **60**: 257–268.

Finley, P.R., Williams, R.J., and Byers, J.M. (1976). Assay of phenytoin adaption of EMIT to the centrifugal analyzer. *Clin. Chem.* **22**: 911–914.

Finley, P.R., Williams, R.J., Lichti, D.R., and Byers, J.M. (1977). Assay of phenobarbital—adaption of "EMIT" to the centrifugal analyzer. *Clin. Chem.* **23**: 738–740.

Forsgren, A., and Sjoquist, J. (1966). Protein A from *Staphylococcus aureus*. I. Pseudo-immune reaction with human gamma globulin. *J. Immunol.* **97**: 822–827.

Freedman, M.H., Grossberg, A.L., and Pressman, D. (1968). Evidence for ammonium and guanidinum groups in the combining sites of anti-p-azobenzenearsonate antibodies—separation of two different populations of antibody molecules. *J. Biol. Chem.* **243**: 6186–6195.

Freund, J., and McDermott, K. (1942). Sensitization to horse serum by means of adjuvants. *Proc. Soc. Exp. Biol.* **49**: 548.

Gordon, M.A., Edwards, M.R., and Tompkins, V.N. (1962). Refinement of fluorescent antibody by gel filtration. *Proc. Soc. Exp. Biol. Med.* **109**: 96–99.

Green, N.M. (1963). The use of biotin-^{14}C for kinetic studies and for assay. *Biochem. J.* **89**: 585–591.

Griswold, W.R., and Nelson, D.P. (1984). Theoretical analysis of the Farr antibody-assay with a computer model—importance of antigen concentration and antibody affinity. *J. Immunoassay* **5**: 71–86.

Guesdon, J.-L., Jouanne, C., and Avrameas, S. (1983). An amplification system using BSA-antibody conjugate for a sensitive enzyme immunoassay. *J. Immunol. Methods* **58**: 133–142.

Gushaw, J.B., Hu, M.W., Singh, P., Miller, J.G., and Schneider, R.S. (1977). Homogeneous enzyme immunoassay for theophylline in serum. *Clin. Chem.* **23**: 1144.

Haas, G.J., and Guardia, E.J. (1986). Production of antibodies against insecticide-protein conjugates. *Proc. Soc. Exp. Biol. Med.* **129**: 546–551.

Hamboeck, H., Fischer, R.W., Di Iorio, E.E., and Winterhalter, K.H. (1981). The binding of S-triazine metabolites to rodent hemoglobins appears irrelevant to other species. *Mol. Pharmacol.* **20**: 579–584.

Hammock, B.D., and Mumma, R.D. (1980). Potential of immunochemical technology for pesticide analysis. In: *Pesticide Analytical Methodology*, ACS Symposium Series 136, Jr., J. Harvey and G. Zweig (Eds.). Washington D.C.: American Chemical Society.

Hammock, B.D., Bee, S.J. Cheung, P.Y.K., Miyamoto, T., Goodrow, M.H., and Seiber, J.N. (1986). Utility of immunoassay in pesticide trace analysis. Presented at the Sixth International Congress of Pesticide Chemistry, Ottawa, Canada, August 11–15.

Herbert, W.J. (1973). Mineral oil adjuvants and the immunization of laboratory animals. In: *Handbook of Experimental Immunology*. D.M. Wier (Ed.). Blackwell Scientific, Oxford, U.K., 2d Ed., App. 2 and 3, pp. A2.1–A2.14.

Hofmann, K., Wood, S.W., Brinton, C.C., Montibeller, J.A., and Finn, F.M. (1980). Imino-biotin affinity columns and their application to the retrieval of streptavidin. *Proc. Natl. Acad. Sci. USA* **77**:4666–4668.

Horejse, J., and Smetana, R. (1956). The effect of rivanol on plasma proteins—isolation of gamma globulin from blood plasma by rivanol. *Acta Med. Scand.* **155**:65–70.

Huber, S.J., and Hock, B. (1985). A solid-phase enzyme immunoassay for quantitative determination of the herbicide terbutryne. *Z. Pflanzenkrankheiten Pflanenschutz* **92**(2):147–156.

Huber-Lukac, M., Herbst, H., and Braun, D.G. (1982). Specificities of monoclonal antibodies against the activated delta-endotoxin of *Bacillus thuringiensis* var. *thuringiensis*. *Experientia* **38**:1103–1105.

Huber-Lukac, M., Luthy, P., and Braun, D.S. (1983). Monoclonal antibodies against functionally distinct sites on the delta-endotoxin of *Bacillus thuringiensis* var. *thuringiensis*. *Infect. Immun.* **183**:608–612.

Hunter, K.W. Jr., and Lenz, D.E. (1982). Detection and quantification of the organophosphate insecticide paraoxon by competitive inhibition enzyme immunoassay. *Life Sci.* **30**:355–361.

Hurrell, J.G.R. (Ed.) (1985). *Monoclonal Hydridoma Antibodies: Techniques and Application.* CRC Press, Boca Raton, FL, 231 pp.

Husby, S., Holmskov-Nielson, U., Jensenius, J.C., and Erb, K. (1985). Increased non-specific binding of heat-treated proteins to plastic surfaces analyzed by ELISA and HPLC-fractionation. *J. Immunoassay* **6**:95–110.

Jaton, J.-C., Brandt, D.C., and Vassali, P. (1979). The isolation and characterization of immunoglobulins, antibodies, and their constituent polypeptide chains. In: *Immunological Methods.* I. Lefkovits and B. Perniss (Eds.). Academic Press, New York, Vol. 1, pp. 43–67.

Jensenius, J.C., Anderson, I., Hau, J., Crone, M., and Koch, C. (1981). Eggs—conveniently packaged antibodies—methods for purification of yolk IgG. *J. Immunol. Methods* **46**:63–68.

Johannessen, S.I. (1977). In: *A Comparison of Methods, Antiepileptic Drug Monitoring.* C. Gardner-Thorpe, D. Janz, and H. Mernardi (Eds.). Pitman Medical. Marshfield, MA.

Johannson, A., Ellis, D.H., Bates, D.L., Plumb, A.M., and Stanley, C.J. (1986). Enzyme amplification for immunoassay. Detection limit of one hundredth attamole. *J. Immunol. Methods* **87**:7–11.

Jolley, M.E., Stroupe, S.D., Schwenzer, K.S., et al. (1981). Fluoroescence polarization immunoassay. II. An automated system for therapeutic drug determination. *Clin. Chem.* **27**:1575–1579.

Kabakoff, D.S., and Greenwood, H.M. (1981). Recent developments in homogeneous enzyme immunoassay. In: *Recent Advances in Clinical Chemistry*, Vol. 2. K.G.M.M. Alberti and C.P. Price (Eds.). Churchill Livingstone, Edinburgh, pp. 1–30.

Kampa, I.S., Jarzabek, J., and Hundertmark, J.M. (1978). Comparison of EMIT assay with 2 iodinated radioimmunoassays for diphenylhydantoin. *Clin. Biochem.* **11**:167–168.

Kelley, M.M., Zahnow, E.W., Petersen, W.C., and Toy, S.T. (1985). Chlorsulfuron determination in soil extracts by enzyme immunoassay. *J. Agric. Food Chem.* **33**:962–965.

Kendall, C., Konescu-Matiu, I., and Dreesman, G.R. (1983). Utilization of the biotin/avidin system to amplify the sensitivity of the enzyme-linked immunosorbent assay (ELISA). *J. Immunol. Methods* **56**:329–339.

Kenna, J.G., Major, G.N., and Williams, R.S. (1985). Methods for reducing non-specific antibody binding in enzyme-linked immunosorbent assays. *J. Immunol. Methods* **85**:409–419.

Kessler, S.W. (1981). Use of protein A–bearing staphylococci for the immunoprecipitation and isolation of antigens from cells. *Methods Enzymol.* **73**:442–459.

Kronvall, G., Seal, U.S., Svensson, S., and Williams, R.C. (1974). Phylogenetic aspects of stephylococcal protein A–reactive serum globulins in birds and mammals. *Acta Pathol. Microbiol. Scand.* **B82**:12–18.

Kuniyuki, A.H., and McCarthy, S. (1986). A novel method for the development of monoclonal antibodies to Surflan. Sixth International Congress of Pesticide Chemistry, Poster Session, Ottawa, August 11–15.

Kupferberg, H.J. (1970). Quantitative estimation of diphenylhydantoin, primidone, and phenobarbital in plasma by gas-liquid chromatography. *Clin. Chim. Acta* **29**:283–288.

Landsteiner, K. (1945). *The Specificity of Serological Reactions*. Harvard University Press, Cambridge, MA (Dover Publications, New York, 1962), 330 pp.

Langone, J.J., and Van Vunakis, H. (1975). Radioimmunoassay for dieldrin and aldrin. *Res. Commun. Chem. Pathol. Pharmacol.* **10**:163–171.

Langone, J.C., and Van Vunakis, H. (1982). Radioimmunoassay of nicotine, cotinine, and gamma-(3-pyridyl)-gamma-oxo-N-methylbutyramide. *Methods Enzymol.*, **84**:628–640.

Leary, J.J., Brigati, D., and Ward, D.C. (1983). Rapid and sensitive colorimetric method for visualizing biotin labeled DNA probes hybridized to DNA or RNA immobilized on nitrocellulose. *Proc. Natl. Acad. Sci. USA* **80**:4045–4049.

Levitt, T. (1979) Determination of paraquat in clinical practice using radioimmunoassay. *Proc. Anal. Div. Chem. Soc.* **16**:72–76.

Levy, H.B., and Sober, H.A. (1960). Preparation of gamma globulin. *Proc. Soc. Exp. Med. Biol.* **103**:250–252.

Lukens, H.R., Williams, C.B., Levison, S.A., Dandliker, W.B., and Murayama, D. (1977). Fluorescence immunoassay technique for detecting organic environmental contaminants. *Environ. Sci. Technol.* **11**:292–297.

McCall, J.S., and Potter, B.J. (1973). *Ultracentrifugation*. Balliere Tindall, London, 126 pp.

Mule, S.J., Bastos, M.L., and Jukafsky, D. (1974). Evaluation of immunoassay methods for detection in urine of drugs subject to abuse. *Clin. Chem.* **20**:243–248.

Mumma, R.O., and Brady, J. (1986). Immunological assays for agrochemicals. Sixth International Congress of Pesticide Chemistry, Poster Session, Ottawa, Canada, August 11–15.

Murphy, B.E.P., and Lehotay, D.C. (1977). Competitive protein binding assays (including radioimmunoassays). In: *Radiotracer Techniques and Applications*, Vol. 2. E.A. Evans and M. Muramatsu (Eds.). Marcel Dekker, New York, pp. 969–1003.

Newsome, W.H. (1985). An enzyme-linked immunosorbent assay for metalaxyl in foods. *J. Agric. Food Chem.* **33**:528–530.

Newsome, W.H., and Shields, J.B. (1981). A radioimmunoassay for benomyl and methyl 2-benzimidazolecarbamate on food crops. *J. Agric. Food Chem.* **29**:220–222.

Nisonoff, A., Hopper, J.E., and Spring, S.R. (1975). *The Antibody Molecule*. Academic Press, New York, 142 pp.

Onishi, J., Hishikawa, Y., Suzuki, Y., Matsuda, T., and Miyamoto, J. (1986). Application of immunoassay for a major urinary metabolite of fenvalerate to detect human exposure. Sixth International Congress of Pesticide Chemistry, Poster Session, Ottawa, Canada, August 11–15.

Ouchterlony, O. (1959) Diffusion-in-gel methods for immunological analysis. *Prog. Aller.* **5**: 1–78.

Parker, C.W. (1965). Immunochemical mechanisms in penicillin allergy. *Fed. Proc.* **24**:51–54.

Parker, C.W. (1976). *Radioimmunoassay of Biologically Active Compounds*. Prentice-Hall, Englewood Cliffs, NJ, 239 pp.

Pauling, L., Pressman, D., Campbell, D.H., Ikeda, C., and Ikawa, M. (1942). Serological properties of simple substances. I. Precipitation reactions between antibodies and substances containing two or more haptenic groups. *J. Am. Chem. Soc.* **64**:2994–3003.

Perper, R.J., Okimoto, J.T., Cochrum, K.C., Ramsey, H., and Najarian, J.S. (1967). Purification of large quantities of antilymphocytic serum. *Proc. Soc. Exp. Biol. Med.* **125**:575–580.

Peterson, E.A. (1970). Cellulosic ion exchangers. In: *Laboratory Techniques in Biochemistry and Molecular Biology*, Vol. 2, Part 2. T.S. Work and E. Work (Eds.). Elseveir/North-Holland, Amsterdam.

Pippenger, C.E. (1979). Reference sources for therapeutic drug monitoring. *Ther. Drug Monitoring* **1**:3–10.

Plomp, T.A., Drost, R.H., Thyssen, J.H.H., Der Kinderen, P.J., and Maes, R.A.A. (1979). Evaluation of the manual enzyme immunoassay (EMIT) procedure for determination of serum thyroxine. *J. Clin. Chem. Clin. Biochem.* **17**:315–320.

Radola, B.J. (1975). Isoelectric focusing in layers of granulated gels. II. Preparative isoelectric focusing. *Biochim. Biophys. Acta* **386**:181–195.

Reif, A.E. (1960). Batch preparation of rabbit gamma G globulin with DEAE-cellulose. *Immunochemistry* **6**:723–731.

Rinder, D.F., and Flecker, J.R. (1981). A radioimmunoassay to screen for 2,4-dichlorophenoxyacetic acid and 2,4,5-trichlorophenoxyacetic acid in surface water. *Bull. Environ. Contam. Toxicol.* **26**:375–380.

Rong, K.-T., Xu, Q.-H., Shao, R.-X., and Gong, X.-Q. (1986). Radioimmunoassay for paraoxon. *Acta Pharmacol. Sin.* **1**:93–96.

Rose, M.E., Orlans, E., and Buttress, H. (1974). Immunoglobulin classes in hens egg—their segregation in yolk and white. *Eur. J. Immunol.* **4**:521–523.

Rowley, G.L., Rubenstein, K.E., Huisjen, J., and Ullman, E.F. (1975). Mechanism by which antibodies inhibit hapten-malate dehydrogenase conjugates. *J. Biol. Chem.* **250**:3759–3766.

Rowley, G.L., Rubenstein, K.E., Weber, S.T., and Ullman, E.F. (1976). Mechanism by which antibodies inhibit hapten-glucose-6-phosphate dehydrogenase. Presented at the 172nd ACS National Meeting, Abstract No. 151.

Rubenstein, K.E., Schneider, R.S., and Ullman, E.F. (1972). "Homogeneous" enzyme immunoassay. New immunochemical technique. *Biochem. Biophys. Res. Commun.* **47**:846–851.

Schalch, W., and Braun, D.G. (1979). *Immunol. Methods* Isolation of Monoclonal

Antibody by Preparative Isoelectric Focusing in Horizontal Layers of Sephadex G-75 **1:**123.

Schneider, R.S., Lindquist, P., Wong, E.T., Rubenstein, K.E., and Ullman, E.F. (1973). Homogeneous enzyme immunoassay for opiates in urine. *Clin. Chem.* **19:**821–825.

Schwalbe, M., Dorn, E., and Beyerman, K. (1984) Enzyme immunoassay and fluorimmunoassay for the herbicide diclofop-methyl. *J. Agric. Food Chem.* **32:**734–741.

Shamsuddin, A.M., and Harris, C.C. (1983). Improved enzyme immunoassays using biotin-avidin enzyme complex. *Arch. Pathol. Lab. Med.* **107:**514–516.

Sharp, J.K., Robotti, K.M., Ehrmann, P.R., Brown, L.J., and Hermann, B.W. (1986). Quantitative Analysis of Cyanazine in Groundwater and soil by enzyme-linked immunoassay. *J. Agric. Food Chem.* (in press) (E.J. du Pont de Nemours & Co.).

Sips, R. (1948). On the structure of the catalyst surface. *J. Chem. Phys.* **16:**490–495.

Sjoquist, J., Meloun, B., and Hjelm, H. (1972). Protein A isolated from *Stephylococcus aureus* after digestion with lysostaphin. *Eur. J. Biochem.* **29:**572–578.

Smith, R.A., and Ulrich, J.T. (1983). Enzyme-linked immunosorbent assay for quantitative detection of *Bacillus thuringiensis* crystal protein. *Appl. Environ. Microbiol.* **45:**586–590.

Spiehler, V., Sun, L., and Miyada, D.S. (1976). Radioimmunoassay, enzyme immunoassay, spectrophotometry, and gas-liquid chromatography compared for determination of phenobarbital and diphenylhydantoin. *Clin. Chem.* **22:**749–753.

Steinbuch, M., and Audran, R. (1969). Isolation of IgG from mammalian sera with the aid of caprylic acid. *Arch. Biochem. Biophys.* **134:**279–284.

Sueoka, N., and Cheng, T.-T. (1962). Fractionation of nucleic acids with the methylated albumin column. *J. Mol. Biol.* **4:**161–172.

Tijssen, P. (1985). *Laboratory Techniques in Biochemistry and Molecular Biology*, Vol. 15. Elsevier North Holland, Amsterdam, 549 pp.

Tijssen, P., Su, D.-M., and Kurstak, E. (1982). Rapid and sensitive heterologous enzyme immunoassays for densonucleosis virus (Parvoviridae). *Arch. Virol.* **74:**277–291.

Vallejo, R.P., Bogus, E.R., and Mumma, R.O. (1982). Effects of hapten structure and bridging groups on antisera specificity in parathion immunoassay development. *J. Agric. Food Chem.* **30:**572–580.

Vanderlaan, M., Watkins, B., Devivar, R., and Stanker, L. (1986). Enzyme immunoassay for dioxins using monoclonal antibodies. Sixth International Congress of Pesticide Chemistry, Poster Session, Ottawa, Canada, August 11–15.

Van Emon, J.M., Hammock, B.D., and Seiber, J.N. (1986a). Enzyme-linked immunosorbent assay for paraquat and its application to exposure analysis. *Anal. Chem.* **58:**1866–1873.

Van Emon, J.M., Seiber, J.N., and Hammock, B.D. (1986b). Application and selectivity of an enzyme-linked immunosorbent assay (ELISA) for the pyrethroid S-bioallethrin. Sixth International Congress of Pesticide Chemistry, Poster Session, Ottawa, Canada, August 11–15.

Van Heyningen, V., Brock, D.J.H., and Van Heyningen, S. (1983). A simple method for ranking the affinities of monoclonal antibodies. *J. Immunol. Methods* **62:**147–154.

Van Weeman, B.K., and Schuurs, A.H.W.M. (1971a). Immunoassay using hapten-enzyme conjugates. *FEBS Lett.* **24:**77–81.

Van Weeman, B.K., and Schuurs, A.H.W.M. (1971b). Immunoassay using antigen-enzyme conjugates. *FEBS Lett.* **15**:232–236.

Voller, A., and Bidwell, D.E. (1986). Developments in enzyme-immunoassay techniques. In: *Developments in Enzyme-Immunoassay Techniques in Veterinary Viral Diseases: Their Significance in South-East Asia and the Western Pacific.* Academic Press, Sydney Australia, pp. 61–70.

Voller, A., and Bidwell, D.E. Huldt, G., and Engvall, E. (1974). A microplate method of enzyme-linked immunosorbent assay and its application to malaria. *Bull WHO* **51**:209–211.

Wei, R., and Reibe, S. (1977). Preparation of a phospholipase C–antihuman IgG conjugate and inhibition of its enzymatic activity by human IgG. *Clin. Chem.* **23**:1386–1388.

Weiner, W. (1957). Eluting red-cell antibodies: A method and its application. *Br. J. Haematol.* **3**:276–283.

Wie, S.I., and Hammock, B.D. (1982a). Development of enzyme-linked immunosorbent assays for residue analysis of diflubenzuron and BAY SIR 8514. *J. Agric. Food Chem.* **30**:949–957.

Wie, S.I., and Hammock, B.D. (1982b). The use of enzyme-linked immunosorbent assays (ELISA) for the determination of Triton X nonionic detergents. *Anal. Biochem.* **125**:168–176.

Wie, S.I., and Hammock, B.D. (1984). Comparison of coating and immunizing antigen structure on the sensitivity and specificity of immunoassays for benzoylphenylurea insecticides. *J. Agric. Food Chem.* **32**:1294–1301.

Wie, S.I., Sylwester, A.P., Wing, K.D., and Hammock, B.D. (1982). Synthesis of haptens and potential radioligands and development of antibodies to insect growth regulators diflubenzuron and BAY SIR 8514. *J. Agric. Food Chem.* **30**:943–948.

Williamson, A.R., Salaman, M.R., and Kreth, H.R. (1973). *Ann. N. Y. Acad. Sci.* **209**:210–224.

Wing, K.D., Hammock, B.D., and Wustner, D.A. (1978). Development of an S-bioallethrin specific antibody. *J. Agric. Food Chem.* **26**(6):1328–1333.

Wong, R.C., Burd, J.F., Carrico, R.J., Buckler, R.T., Thomas, J., and Boguslaski, R.C. (1979). Substrate labeled fluorescent immunoassay for phenytoin in human serum. *Clin. Chem.* **25**:686–691.

Wong, R.C., George, R., Yeung, R., and Burd, J.F. (1980). Comparison of serum phenytoin determination by the substrate-labeled fluorescent immunoassay with gas chromatography, liquid chromatography, radioimmunoassay and EMIT. *Clin. Chem. Acta* **100**:65–69.

Wraith, M.J., Hitchings, E.J., Woodbridge, A.P., Cole, E.R., and Roberts, T.R. (1986). Development of Immunoassay methods for pyrethroid Insecticides. Sixth International Congress of Pesticide Chemistry, Poster Session, Ottawa, Canada, August 11–15.

Yallow, R.S., and Berson, S.A. (1960). Immunoassay of endogenous plasma insulin in man. *J. Clin. Invest.* **39**:1157–1163.

Zouali, M., and Stollar, B.D. (1986). A rapid ELISA for measurement of antibodies to nucleic acid antigens using UV-treated polystyrene microplates. *J. Immunol. Methods* **90**:105–110.

Chapter 6

Radioimmunoassay: Ecdysteroids

James T. Warren and Lawrence I. Gilbert

I. Introduction

The presence of the phylum characteristic exoskeleton requires that insects molt periodically to grow, and this is elicited by a family of polyhydroxylated steroids—the ecdysteroids. The control of the hemolymph ecdysteroid titer is an extremely complicated process, being a result of both synthetic and degradative events, with regulation at the level of the prothoracic glands being primary. In response to developmental or environmental events, a neurohormone, prothoracicotropic hormone (PTTH), is released from the brain-retrocerebral complex and stimulates the synthesis and then release of ecdysone and 3-dehydroecdysone from the prothoracic glands (Warren et al., 1988). 3-Dehydroecdysone is subsequently converted to ecdysone and the latter to 20-hydroxyecdysone, the principal molting hormone, and a variety of other ecdysteroids (see Rees, 1985; Horn and Bergamasco, 1985).

The 20-hydroxyecdysone, and perhaps other ecdysteroids, then equilibrates with target tissues and initiates the complex events leading to molting and metamorphosis. The active hormone(s) are then reversibly or irreversibly inactivated by subsequent metabolism or are sequestered in the gut and secreted, except in the pupal stage. To characterize this chain of events fully, it is necessary to determine the identity of all the ecdysteroids involved, their hemolymph and tissue compartment concentrations, and the pharmacokinetic processes associated with their biosynthesis, tissue equilibrations, and enzymatic interconversions. Control of these processes determines the rise and fall of hormone titer(s) in the various target tissues, events associated with ecdysteroid action. The radioimmunoassay (RIA) of ecdysteroids, first introduced by Borst and O'Connor (1972, 1974), was the first RIA developed for an insect hormone and has been used to explore these questions. Indeed, without the RIA, progress in ecdysteroid endocrinology would be a result of studies in those few laboratories that can routinely process large numbers of samples via gas-liquid chromatography (GLC)-mass spectrometry (MS).

Prior to 1972, the method available for measuring ecdysteroids other than by physical chemical analysis was the bioassay, in essence an equilibrium binding assay similar to RIA. In RIA, the competition between labeled and nonlabeled ecdysteroid for associations with an artificial binder, the ecdysteroid antibody (AB), is used to measure ecdysteroid levels. In bioassay, a morphological response resulting from the binding of an ecdysteroid to an endogenous ecdysteroid-specific receptor is used to determine the concentration of molting hormone.

In some respects, bioassay is more appropriate than RIA for investigating ecdysteroid physiology, because by definition bioassay ultimately measures the active hormone(s) of interest; i.e., it does not measure morphogenetically inactive ecdysteroids. In tissues thought to be incapable of metabolically altering ecdysteroids, responses such as the puffing of specific gene loci in the giant chromosomes of *Drosophila melanogaster* salivary glands (Ashburner, 1971, 1973; Richards, 1978), the shortening of interganglionic ventral nerve cords (Robertson and Pipa, 1973), and the evagination of imaginal disks (Milner and Sang, 1974), etc. have been used to measure ecdysteroid titers and the relative biological activity (structure-activity) of various ecdysteroids. In some cases, more than one ecdysteroid appears to act differentially on target tissues. It is apparent that additional tissue-specific and ecdysteroid-specific bioassays are required, since one must always extrapolate data from in vitro studies back to the organism. However, metabolic alteration of injected or topically applied ecdysteroids is likely in the more complex bioassay systems, and therefore the active principle eliciting the physiological effects remains unidentified. In fact, short of measuring the competition between labeled and nonlabeled ecdysteroids for specific binding to purified cellular receptors (radioreceptor assay), the metabolism of ecdysteroids in bioassays cannot be excluded (Horn and Bergamasco, 1985). The potential for synthesis of endogenous ecdysteroids by some of these assay systems may further confuse the interpretation of the results (Smith, 1985).

Bioassay procedures are lengthy and involved, requiring relatively large amounts of ecdysteroids and the manipulations of many insects. At the level of the individual insect or tissue preparation, however, bioassays are quite sensitive (Table 6.1). When employed as a means of detection following some chromatographic purification step (thin layer chromatography, TLC; high performance liquid chromatography, HPLC), they have been very useful in assaying individual ecdysteroids. However, only ecdysteroids that are active or that can be converted to active compounds will be detected, and then only if they are not degraded before initiating an adequate response.

Table 6.1. Approximate Detection Limits for 20-Hydroxyecdysone by Various Methods

Method	Limit
A. Biological assay	
Musca domestica	5×10^{-9} g per insect or
	1.5×10^{-7} g per test
Calliphora vicina	7.5×10^{-9} g per insect
Calliphora stygia	5×10^{-8} g per insect
	3×10^{-5} M
Aedes atropalpus	3×10^{-6} M
Chromosome puff	1.0×10^{-8} M
Imaginal disk	4.2×10^{-9} M
Kĉ cell lines	1.0×10^{-8} M
Radioimmunoassay	1.0×10^{-9} g
	7.5×10^{-11} to 5.0×10^{-10} g
	2.0×10^{-12} g
B. Instrumental analysis	
TLC	5.0×10^{-7} g
HPLC	5.0×10^{-9} g
Fluorescence	2.0×10^{-10} g
GLC-ECD	1.0×10^{-15} to 1.0×10^{-12} g
GLC-MS	1.0×10^{-15} to 1.0×10^{-12} g

From Horn and Bergamasco (1985).

Instrumental means of ecdysteroid analysis are relatively nonspecific and require some purification of the ecdysteroid components prior to their detection and quantification. This is particularly true for radiotracer analysis, which can be quite sensitive (0.5 pg at 60 Ci/mmol), but all labeled ecdysteroids give identical responses. For TLC and HPLC, detection of acid-induced ecdysteroid fluorescence (Koolman, 1980) can be far more sensitive than UV absorption, but maximum sensitivity is realized when electron-capture detectors are utilized in GLC and especially when mass spectrometers are interfaced with GLC (Gilbert et al., 1977; Morgan and Wilson, 1980). However, sample analysis efficiency is reduced significantly if an additional chromatographic step is necessary prior to the already time-consuming ecdysteroid derivatization procedure required for GLC. Both qualitative and quantitative ecdysteroid analysis will be greatly simplified when routine HPLC-MS (chemical ionization and fast atom bombardment) becomes available. Until that time, when used in conjunction with HPLC, RIA, because of its great simplicity, sensitivity, and characteristic specificity, is the method of choice for most biologists and biochemists.

Bioassay and RIA have proved quite useful for the direct determination of approximate hemolymph or whole-body ecdysteroid titers during arthropod development, but neither assay is specific for any ecdysteroid. It is unlikely that this situation would improve even if sufficient quantities of purified ecdysteroid receptor or high-affinity binding proteins were available, for they too are not completely specific (Cherbas et al., 1980; Yund and Osterbur, 1985). Mammalian steroid RIAs and radioreceptor assays also exhibit this lack of specificity between structurally similar hormones (Kellie, 1975).

II. Radioimmunoassay of Ecdysteroids

There are six basic techniques involved in ecdysteroid RIA: synthesis of the hapten and immunogen, production of specific ABs (the binder), synthesis of a radioligand (the tracer), separation of AB-bound antigen (Ag) from free Ag, sample preparation, and data analysis. In addition, as with any sensitive assay, problems can be encountered with data interpretation. These techniques will be discussed individually along with specific examples of the applications of the ecdysteroid RIA.

1. Synthesis of Ecdysteroid Haptens

Free ecdysteroids (<500 daltons) are not immunogenic as such but must first be linked covalently to a large protein, usually foreign, before the production of ecydsteroid-specific ABs can be stimulated in mammals (see Chapter 1). Two considerations are important in the design of an immunogen of high efficiency. First, in order to use the variety of available bifunctional hapten-protein-linking (bridging) reagents, a reactive substituent (—OH, —NH, —SH, —C=O, —HC=O, or —COOH) must either be a constituent of the hapten or introduced later using chemical or biochemical means. Second, to achieve the desired AB specificity, the routes of significant ecdysteroid metabolism must be determined in the host organism, and the immunogens must be synthesized accordingly.

Ecdysone has three reactive secondary hydroxyl groups at C-2, C-22, and C-3 in decreasing order of susceptibility to chemical esterification. The tertiary hydroxyls at C-14, C-25, and C-20 are much less reactive (Horn, 1971). Esterification with succinyl chloride or succinic anhydride introduces a four-carbon spacer (1,4-butanedioic acid) ending in a free acid function which can be linked via an amide bond to the susceptible free amino groups of exposed lysine residues in the carrier protein (Fig. 6.1A,B). The formation of these peptide bonds is promoted by dehydrating reagents or by the prior formation of activated esters or mixed anhydride derivatives of the hapten-linker molecule.

In addition, a side chain C-26 primary hydroxyl function can be introduced biochemically as in the ecdysone metabolite 26-hydroxyecdysone (40 μg/g), present as the 26-phosphate conjugate in newly deposited *Manduca sexta*

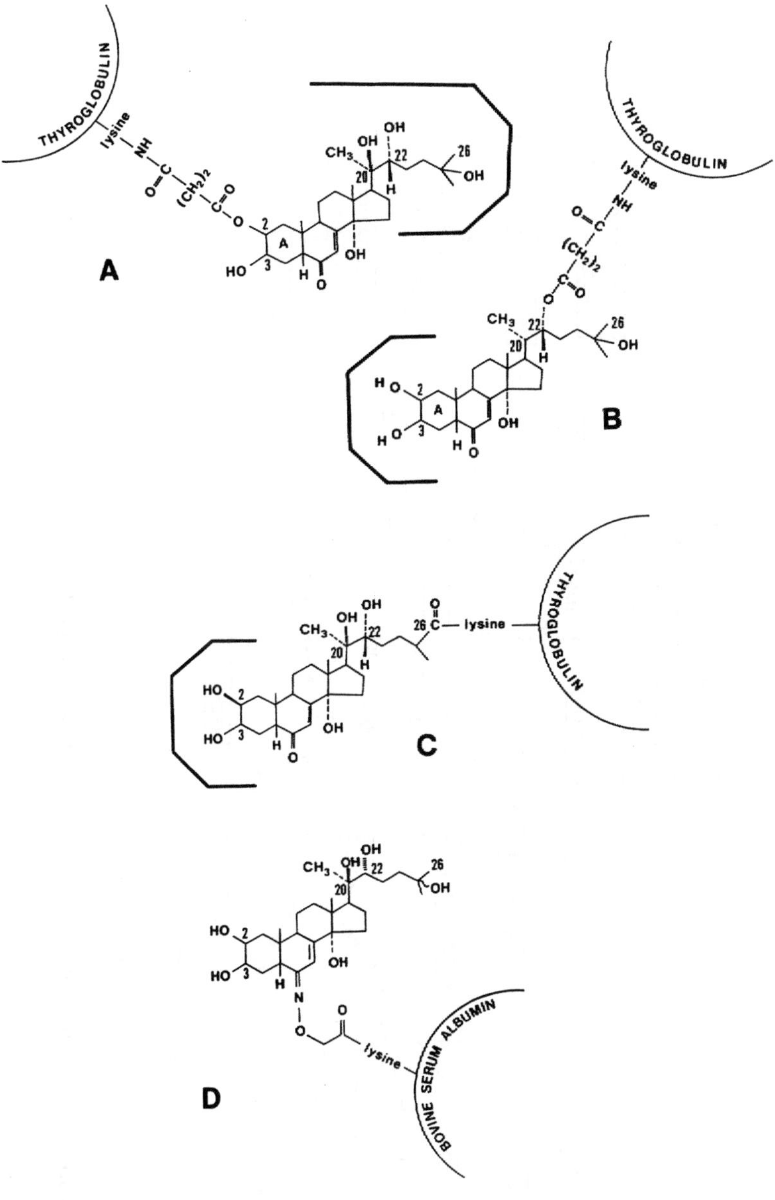

Figure 6.1. Structures of various ecdysteroid immunogens.

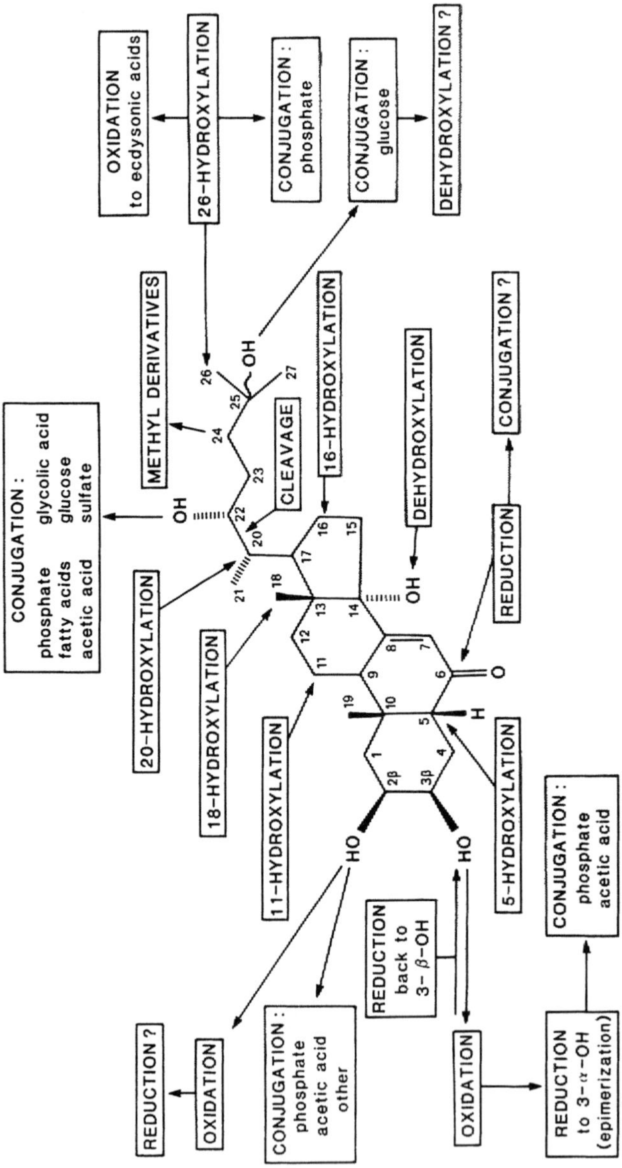

Figure 6.2. Biotransformation reactions of ecdysone (adapted from Lafont et al., 1986; Lafont, 1987).

eggs (Thompson et al., 1985b) or in inokosterone (20,26-dihydroxy-25-deoxyecdysone). Either steroid may then be oxidized chemically or biochemically to the terminal side chain acid which can then be conjugated directly to protein. Thus, inokosterone was oxidized with platinum and oxygen to inokosteronic acid which was then coupled to thyroglobulin (Fig. 6.1C) (Spindler et al., 1976, 1978). Alternatively, the 26-hydroxyl group in inokosterone can be selectively succinylated following the protection of the 20,22-*cis*-diol as the diphenylborate (or acetonide) and the 2,3-*cis*-diol as the acetonide derivative. Both blocking groups are then easily hydrolyzed by dilute acid (Guédin-Vuong et al., 1985), and the free acid is reacted with carrier protein.

For low-molecular-weight hapten-protein immunogens, ABs are produced that recognize (bind to) determinants associated with the molecular shape, polarity, and charge of the hapten far removed from the surface of the molecule that is bonded to and sterically hindered by the immunogenic protein. This reflects the hydrophobic (water exclusion), hydrophilic (hydrogen bonding), and ionic interactions that are the basis of AB-Ag affinity. A response is not generally directed against those determinants that occur at or near the bridging group, so ABs will usually bind to structural analogues containing different alkyl substituents or polar functional groups in this linker region. Heterogeneous antisera will also contain ABs directed against not only the carrier protein but also the bridging group in the hapten conjugate, and these may cause problems in some assays (see section II.E below).

Because the same four hydroxyl groups in ecdysteroids available for conjugation to protein are also involved in most of the phase I (oxidation) or phase II (conjugation) biotransformation reactions of ecdysone, ABs made against ecdysteroid immunogens linked through hydroxyl groups can never be completely hapten-specific, since they will likely bind (cross-react) strongly with metabolites that mimic the structure of the hapten in the immunogen—i.e., have these same oxidized or esterified hydroxyl groups (Fig. 6.2).

ABs generated in the preceding manner can be very useful in both quantifying these known metabolites and detecting new metabolites altered in these positions (Redfern, 1984). Ecdysone biosynthesis and metabolism in *Manduca sexta* have been probed (Warren and Gilbert, 1986; Warren et al., 1986, 1988) using ABs that were elicited by immunogens in which the carrier protein was linked via succinate to either the A-ring hydroxyl at C-2 (H-2 antisera) (Soumoff et al., 1981) (Fig. 6.1A) or to the side chain hydroxyl at the opposite and of the molecule at C-22 (H-22 antisera) (Horn et al., 1976) (Fig. 6.1B). These two regions constitute the major centers of metabolism in ecdysteroids (Lafont and Koolman, 1984; Koolman and Karlson, 1985). The H-22 ABs directed against ecdysone-22-succinyl thyroglobulin amide recognize ecdysone and its 20-hydroxy and/or 26-hydroxy metabolites (Fig. 6.3; Table 6.2). In addition, they cross-react strongly with the many different conjugates formed from these substrates by esterification of the 22- and 26-hydroxyls with various polar and nonpolar groups such as phosphate, sulfate, acetate,

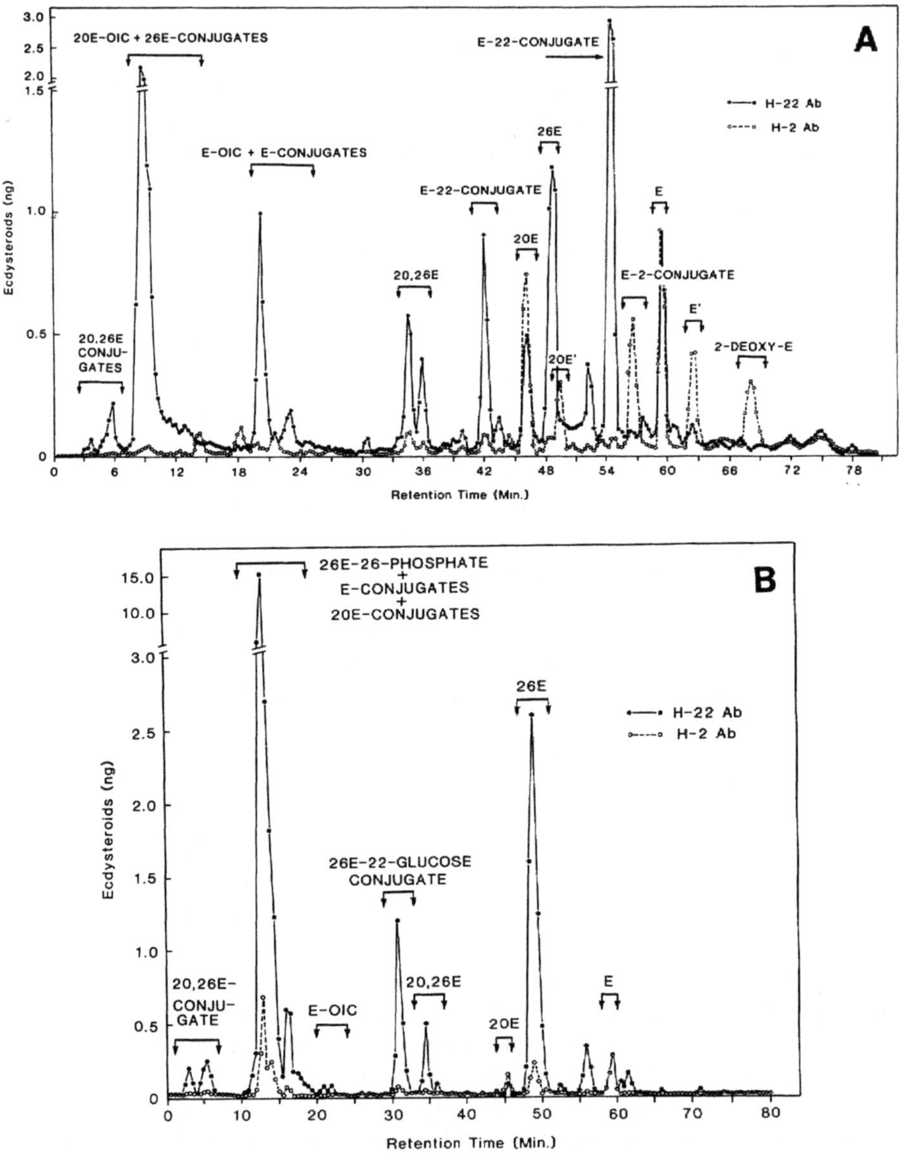

Figure 6.3. Differential ecdysteroid RIA/reverse-phase HPLC analysis of *Manduca sexta* extracts. (A) Pupal day 14 female gut contents (from Warren et al., 1986); (B) 48-h developing eggs (including embryos) (from Warren et al., 1986); (C) media following in vitro incubation of larval prothoracic glands. RP-HPLC conditions: 5-μm C_{18}-resolve column (Waters) with linear 20 mM TRIS-perchlorate buffer (pH 7.4)/ acetonitrile (ACN) gradient, 5% ACN changing to 20% ACN over 60 min. E = ecdysone; E' = 3(α)-epiecdysone; x = a mixture of 2- and 3-dehydroecdysone; 20E = 20-hydroxyecdysone, etc.

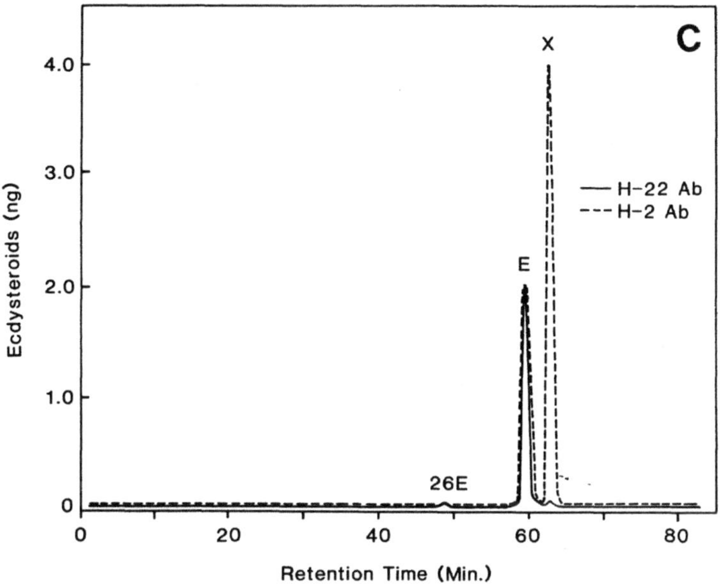

Figure 6.3 (*continued*)

glucose, and palmitate. The oxidation of the primary alcohol function of 26-hydroxyecdysone and 20,26-dihydroxyecdysone to yield the respective ecdysonic acids does not seriously affect binding, nor does the presence of the 24-methyl group of makisterone. However, alteration of the stereochemistry of the *cis*-diol at C-2 and C-3 of the A ring (epimerization to the *trans*-diol), esterification of these A-ring hydroxyls with various inorganic and organic acids, or their oxidation to ketones will abolish AB recognition.

In a complementary fashion, the H-2 ABs against 20-hydroxyecdysone-2-succinyl thyroglobulin amide do not recognize side chain modified ecdysteroids other than ecdysone and 20-hydroxyecdysone. They do, however, detect the metabolites resulting from A-ring hydroxyl oxidation, conjugation, and/or epimerization at *either* the C-2 or C-3 position in addition to those ecdysteroids lacking a hydroxyl function at C-2. The close proximity of the A-ring *cis*-diols or a facile acyl migration of the linker from C-2 to C-3 prior to the synthesis of the immunogen may account for the lack of A-ring specificity exhibited by these ABs (Horn and Bergamasco, 1985). Neither AB preparation is capable of significant binding to metabolites conjugated in *both* the A ring and side chain, nor can they be used to measure certain A-ring epimerized (or oxidized) *and* side chain hydroxylated substances—e.g., 3α(epi)-26-hydroxyecdysone, 3α-(epi)-20,26-dihydroxyecdysone, or the epimeric ecdysonic acids, prominent metabolites of ecdysone in embryos, and guts of some arthropods (Kaplanis et al., 1980; Thompson et al., 1985a). An immunogen

Table 6.2. Cross-Reactivities[a] of Various Ecdysteroids in Radioimmunoassays Employing Three Different Antisera

Ecdysteroids	Horn-22 AB	Horn-2 (DHS1-15)AB	Reum DBL-1 AB
2-Deoxyecdysone	68	3	4.0
20-Hydroxy-2-deoxyecdysone	>100	10	30
3-α-Epiecdysone	>100	4	>100
Ecdysone	1.0	1.0	1.0
Low-polarity ecdysone conjugate (H-2AB)	>100	1.0	>100
Makisterone A	5.0	25	15
Low-polarity ecdysone conjugate (H-22AB)	1.0	>100	>100
3-α-Epi-26-hydroxyecdysone	>100	65	>100
26-Hydroxyecdysone	1.0	15	20
3-α-20-Hydroxyecdysone	>100	12.0	>100
20-Hydroxyecdysone	4.5	3.0	3.5
Midpolar ecdysone conjugate (H-22AB)	4.0	>100	>100
Midpolar 26-hydroxyecdysone conjugate (H-22AB)	4.0	>100	>100
3-α-Epi-20,26-dihydroxyecdysone	>100	>100	>100
20,26-Dihydroxyecdysone	5.0	60	30
3-α-Epiecdysonic acid	>100	>100	>100
Ecdysonic acid	1.0	25	15
Highly polar ecdysone conjugate (H-22AB)	1.0	>100	>100

Highly polar ecdysone conjugate (H-2AB)	>100	1.0	>100
3-α-20-Hydroxyecdysonic acid	>100	>100	>100
20-Hydroxyecdysonic acid	5.0	80	40
Highly polar 26-hydroxyecdysone conjugate (H-22AB)	1.0	>100	>100
Highly polar 20,26-dihydroxyecdysone conjugate (H-22AB)	5.0	>100	>100
3-Dehydroecdysone	30	2	1.5
3-Dehydro-20-hydroxyecdysone	>100	5	5
Ecdysone-22-laurate	1.0	>100	>100
Ecdysone-2-laurate	>100	1.0	>100
Ecdysone-22-acetate	1.0	>100	100
Ecdysone-2-acetate	>100	1.0	60
Ecdysone-2,3-diacetate	>100	3	>100
Ecdysone-2,22-diacetate	>100	>100	>100
20-Hydroxyecdysone-22-acetate	1.0	>100	>100
20-Hydroxyecdysone-2-acetate	>100	1.0	40
20-Hydroxyecdysone-2,3-diacetate	>100	10	>100
20-Hydroxyecdysone-2,22-diacetate	>100	>100	100

[a] $\text{Cross-reactivity}_{50} = \dfrac{\text{Inhibition}_{50\%}\ (\text{ecdysteroid})}{\text{Inhibition}_{50\%}\ (\text{ecdysone})}$

From Warren and Gilbert (1986).

consisting of a 3α-(epi)inokosteronic acid hapten covalently linked to protein would likely produce ABs of suitable specificity that would prove useful in studies investigating the kinetics and control of these important enzymatic inactivation reactions.

The ketone function at C-6 in ecdysone can be reacted with the bifunctional amino acid spacer carboxymethoxyamine to form the oxime acid derivative which is then linked to protein (Fig. 6.1D). This was the first method used to derivatize an ecdysteroid for the synthesis of an immunogen (Borst and O'Connor, 1972, 1974). Since metabolism does not normally occur at this position, it might be expected that the immunogens so derived would elicit ABs highly specific for the hapten, be it 20-hydroxyecdysone or ecdysone. Occasionally, considerable AB preference for binding to a specific ecdysteroid hapten has been realized (Briers and DeLoof, 1981; Reum et al., 1981), but usually, little specificity is observed between ecdysone and 20-hydroxyecdysone. This seems also to be generally true for the ABs derived from hydroxyl group derivatized immunogens (Reum and Koolman, 1979; Chang and O'Connor, 1979; Hirn and Delaage, 1980; Richards, 1981), although exceptions do occur (Lauer et al., 1974). ABs directed against ketone-derivatized ecdysteroid immunogens do not normally recognize any of the 2-, 3-, 22-, or 26-hydroxyl group ester metabolites (conjugates) found in arthropods. In addition, they exhibit low cross-reactivity with side chain-oxidized (C-26) or A-ring-epimerized metabolites, both believed to be inactive ecdysteroid metabolites (Briers et al., 1982; Warren and Gilbert, 1986). Thus, at first glance these ABs appear to be ideal for measuring active hormone levels. However, the absence of a C-2 hydroxyl, oxidation of ecdysteroid A-ring hydroxyls to form 2- and 3-dehydroecdysteroids, hydroxylation at C-5, or reduction of the C-6-ketone to an alcohol does not seriously interfere with the AB-Ag reaction of ketone-derived ABs, and these ecdysteroids may be important intermediates in ecdysone biosynthesis and metabolism in arthropods, other invertebrates, and mammals (Fujimoto et al., 1985; Lafont et al., 1986; Lafont, 1987; Warren et al., 1988).

ABs of different and perhaps useful specificity might be directed against an immunogen formed by the attack of long-chain alkylating agents (i.e., 5-carbon allylic bromoesters) on the C-7 vinyl position in ecdysteroids containing the 6-keto-7-ene system. Following hydrolysis of the resultant ester, these derivatives could be conjugated to protein. Similar immunogens can likely be made with convenient photolysis techniques, since when irradiated at 315 nm, ecdysteroids will bond covalently to proteins (Reum et al., 1982). Alternatively, ecdysteroids may be photolinked to a suitably reactive bifunctional spacer molecule which is then conjugated to protein. In such ecdysteroid-protein conjugates, the attached hapten molecule can still bind to anti-ecdysteroid AB (ketone derived) to form a ternary complex. Thus, either procedure could be used to covalently attach ecdysteroids to enzymes for use as tracers in enzyme immunoassays.

Curiously, the side chain (C-22)-linked immunogen derived ABs do not distinguish between ecdysteroids contaning zero, one, or two hydroxyls at the terminal end of the side chain as though the terminal end of the side chain were not part of the antigenic determinant. In contrast, ketone linker-derived ABs will not bind to ecdysteroids lacking a terminal side chain hydroxyl (ponasterone A) or to those with two such hydroxyls at C-25 and C-26 (26-hydroxyecdysteroids). They cannot, however, distinguish steroids with a single hydroxyl group at either of these positions (inokosterone or 20-hydroxyecdysone). Apparently in this case, a portion of the side chain is a determinant exhibiting a specific degree of end polarity (Borst and O'Connor, 1974).

A. Synthesis of Ecdysteroid Carboxymethyloxime

Excess carboxymethoxyamine (aminooxyacetic acid, 4%) in pyridine is added at 40°C to ecdysone, 20-hydroxyecdysone or other ketone containing ecdysteroids (Borst and O'Connor, 1974). Various reaction times have been used, with overnight being most convenient. Longer reaction times should be avoided, because ecdysteroid derivatives may form under these basic dehydrating conditions owing to the susceptibility of the 14-hydroxyl group to elimination or the 5-β proton to epimerization (Porcheron et al., 1976). If these minor side products are carried through during the production of the immunogen, the resulting AB specificity may be affected. The derivatization process can be followed by TLC and the oxime product(s) purified by preparative TLC. Yields of 60% are average.

The 22-hydroxyl of ecdysone can be selectively succinylated after the 2,3-*cis*-diol function is blocked by the prior formation of the acetonide. Conversely, taking advantage of the different reactivities of the secondary hydroxyl groups to esterification, 2-succinyl-20-hydroxyecdysone is the major product formed from 20-hydroxyecdysone and succinic anhydride in pyridine.

B. 20-Hydroxyecdysone-2-Hemisuccinate

"20-Hydroxyecdysone (300 mg) and succinic anhydride (75 mg) are dissolved in dry tetrahydrofuran (30 ml) under argon in a 100-ml flask fitted with a serum cap. Dry triethylamine (0.75 ml) is added, and the mixture is allowed to react at 20°C for 24 h. The reaction mixture is then evaporated to dryness under vacuum, and the residue is chromatographed on silica gel (10% H_2O, 100 g) using chloroformethanol (70:30, v/v). The main fractions (180 mg) upon recrystallization from ethyl acetate-methanol yield 20-hydroxyecdysone-2-hemisuccinate, m.p. 185–193°C. To characterize the acid, a portion can be methylated with diazomethane. Crystallization from ethyl acetate/methanol yields the methyl ester, m.p. 252–255°C" (Soumoff et al., 1981; see also Delaage et al., 1982).

C. Synthesis of Ecdysone-22-Hemisuccinate

"Phosphomolybdic acid (14.7 mg) is added to ecdysone (225 mg) dissolved in reagent grade acetone (50 ml), and the mixture is stirred at room temperature for 30 min. The solution is then concentrated on a rotary evaporator to 2 ml, and water (50 ml) and potassium hydrogen carbonate (2 mg) are added. This mixture is extracted with ethanol-chloroform (1 : 1). The chloroform layer is washed with water and evaporated to dryness, and the residue is chromatographed on silica gel. Recrystallization of the main fraction (152 mg) from ethyl acetate yields pure ecdysone acetonide, m.p. 251–253°C.

"Dry tetrahydrofuran (100 ml) and pyridine (600 μl) are added to the ecdysone acetonide (152 mg) contained under argon in a 250-ml flask fitted with a serum cap and containing a magnetic stirrer. The mixture is cooled to O°C, succinyl chloride (400 μl) is added, and the mixture is stirred overnight at 4°C. Water is added to hydrolyze unreacted succinyl chloride, and the solvents are then evaporated. The residue is extracted with chloroform, evaporated to dryness, and chromatographed on silica gel. The peak fractions (194 mg of 2β, 3β-isopropylidene ecdysone-22-hemisuccinate) are combined and crystallized, m.p. 160–170°C.

"2β, 3β-Isopropylidene-ecdysone 22-hemisuccinate (193 mg) in tetrahydrofuran (100 ml) and aqueous hydrochloric acid (10 ml, 1 M) is allowed to stand at room temperature for 6.5 h, and ammonium hydroxide solution (100 ml, 0.05 M) is then added. The tetrahydrofuran is then evaporated, and the aqueous residue is extracted twice with chloroform-ethanol (3 : 1, 100 ml). The combined extracts after washing with water are evaporated to dryness, and the residue (169 mg) is chromatographed on silica gel using chloroform-ethanol-acetic acid (89 : 10 : 1). The peak fractions are collected and recrystallized from ethyl acetate, m.p. 158–163°C. To characterize the acid, a portion can be methylated with diazomethane. Crystallization from ethyl acetate yields the methyl ester, m.p. 161–164°C" (Horn et al., 1976).

Succinylation of 20-hydroxyecdysone with succinyl chloride in tetrahydrofuran (Lauer et al., 1974) results in equal derivitization of all secondary hydroxyl groups. Unless these compounds are resolved prior to the protein coupling reaction, less predictable AB specificity may result.

Not only can ABs be produced that have the capacity to detect and quantify the metabolites of ecdysone (whether active or inactive), but they can also be directed against biosynthetic precursors of ecdysone—both those that have been identified and those whose existence is speculative. For instance, Δ^5, Δ^7-cholestadiene-3-ol (7-dehydrocholesterol), an intermediate in the synthesis of ecdysone from eholesterol (Warren et al., 1988) might be made immunogenic following derivitization and conjugation via the 3-hydroxyl group. ABs against a subsequent hypothetical oxidized product of 7-dehydrocholesterol (Rees et al., 1980; Milner et al., 1986), the Δ^5-epoxide, if sufficiently stable, could be produced in a similar manner to probe for this molecule. Alternatively, AB directed against the hydrated Δ^5-epoxide could be used to measure, following epoxide hydrolysis in situ, the derived 5,6-diol in a manner similar to that by which juvenile hormone concentrations are measured by RIA—i.e., assaying the diol of the hydrolyzed 10,11-epoxide (Strambi et al., 1984; see Chapter 7). Intermediates occuring after the establishment of the 5β(H)-6-keto-7-ene system such as the ketol, ketodiol, and triols (Hetru et al., 1978; 1982) can be derivatized through their ketone functions.

2. Synthesis of the Immunogen

The identity of the immunogenic protein to which the ecdysteroid hapten is linked does not usually affect the AB response or the specificity or sensitivity of the RIA utilizing these ABs. Thyroglobulin, having 10 times the molecular mass of bovine serum albumin (BSA) and having 400 free lysine amino groups (cf. 60 in BSA), can react with up to 10 times as much hapten linker molecules as BSA, but the AB response is similar (Hirn and Delaage, 1980). While large numbers of hapten groups linked to the surface of the protein may be counterproductive, too few (fewer than 3–10 per BSA molecule) may result in a different response—i.e., specific production of IgE ABs rather than IgG (Tung, 1983), or little or no response at all. UV spectroscopy of the conjugate or incorporation of radiotracers can be used to ensure adequate hapten incorporation.

Apparently, ecdysteroids are very immunogenic when linked to macromolecules. Small doses of immunogen (250 μg) have elicited adequate responses (Hirn and Delaage, 1980); those much above 1 mg per rabbit are probably unnecessary. Other proteins such as keyhole limpet hemocyanin and even purified major histocompatability antigens, thought to help control AB response, may also be used (see Chapters 1, 4.)

Two procedures are quite suitable for forming the link between the ε-amino groups on exposed lysine residues of the protein and the carboxylic acid function introduced into the hapten molecule. They can probably be used interchangeably in any particular situation.

A. Synthesis of 20-Hydroxyecdysone-Carboxymethyloxime (CMO) BSA Conjugate

"A mixed anhydride of 20-hydroxyecdysone-CMO is formed by mixing 20 mg 20-hydroxyecdysone-CMO and 50 μl tributylamine with 400 μl dioxane (Erlanger et al., 1967). Isobutylchloroformate (32 μmole) is added, and the mixture is allowed to incubate for 30 min. The resultant mixed anhydride is reacted with 32 mg BSA in 6.6 ml of a 1 : 1 solution of dioxane-water to which 300 μl 0.5 N NaOH had been added. The reaction proceeds for 5 h at 4°C with frequent mixing, after which the mixture is dialyzed overnight against running distilled water and then lyophilized. In alternative protocols, ethylchloroformate replaces isobutylchloroformate and dimethylformamide, or dimethylsul foxide may be used as solvent" (Change and O'Connor, 1979).

B. Ecdysone-22-Hemisuccinate-Thyroglobulin Conjugate

"A chilled solution of ecdysone-22-hemisuccinate (7.8 mg) in pyridine/water (1 : 1, 300 μl) is added to a vial containing 1-ethyl-3(3-dimethylaminopropyl)-carbodiimide hydrochloride (51.5 mg, Sigma Chemical Co.). The mixture is gently agitated in an ice bath until the carbodiimide dissolves (about 1 min). Thyroglobulin (9.9 mg, bovine type I, Sigma) dissolved in water (250 μl) is then added, and the mixture is rotated slowly for 3 min and allowed to stand overnight at room temperature. The reaction mixture is transferred to a dialysis tube (7 mm O.D), rinsed in with 30% pyridine (50 μl), and dialyzed against a fixed volume (100 ml) of stirred 30% aqueous pyridine which is changed at daily intervals. An examination of the material dialyzed out in the last dialysis should show it to be primarily unchanged ecdysone-22-hemisuccinate. For the purpose of raising ABs, it is not essential to have the Ag completely free of ecdysone-22-hemisuccinate, and the material after 2–3 days of dialysis is considered satisfactory for this purpose. The contents of the dialysis tube are transferred to a vial and lyophilized, the resulting amorphous powder being rather insoluble in water" (Horn et al., 1976).

In a two-step procedure that does not use a water-soluble carbodiimide reagent, an activated ester is first formed from the acid ecdysteroid derivative with N-hydroxysuccinimide and dicyclohexylcarbodiimide in an organic solvent, and the product is subsequently reacted with the carrier protein in aqueous alkaline solution (Lauer et al., 1974). Intramolecular cross-linking of the protein (or hapten) occurs occasionally with carbodiimide catalyzed reactions. The solubility characteristics of the hapten derivative influences the choice of conjugation procedure.

3. Immunization

Both soluble and insoluble ecdysteroid immunogens have elicited good responses (unpublished data). Between 0.25 and 4 mg protein-hapten conjugate is dispersed in 0.5 ml water or buffer, then homogenized with at least two volumes of Freund's adjuvant and injected into rabbits (see below). High-titer ABs are produced with either complete (containing killed mycobacteria) or incomplete adjuvant (Porcheron et al., 1976; Horn et al., 1976). The emulsion, with the appearance of stiff mayonnaise, must be made carefully, so that slow release of the immunogen for an extended period is assured. Aliquots are usually injected subcutaneously at 8–10 sites along the unshaven backs of at least three rabbits. Implantation of controlled drug delivery systems containing Ag or the use of insolubilized Ags made by the prior adsorption of Ag to aluminum hydroxide gel achieves the same prolonged release and adequate AB response. The tendency has been to use the least amount of immunogen necessary for an adequate response. As little as 20 μg of highly antigenic conjugate in Freund's complete adjuvant, if injected intradermally into 40–60 sites along the shaven back of a rabbit, has been successful (Vaitukaitis, 1981). Booster injections may be unnecessary to sustain the high titer of ABs. Like mycobacteria, *Bordetella pertussis* Ag is also used to elicit nonspecific proliferation of macrophages and lymphocytes—i.e., to prime the system. Other routes of rabbit immunization are normally not necessary, are usually quite painful, and can be dangerous to the health of precious high-titer rabbits. Reducing the requisite amount of Ag allows the use of rare, chemically or biochemically modified ecdysteroids as haptens.

From 8 to 10 weeks following initial subcutaneous injections, or whenever an adequate response has not been obtained after the primary immunization, a booster injection is administered, usually at half the dose of the primary injection. Repeated boosting at 1- to 2-month intervals thereafter may be helpful. Every 2–3 weeks after initial immunization or 8–12 days following a booster, blood is obtained by puncturing the central ear artery of a restrained rabbit with a 21-gauge needle. Rubbing the ear with xylene followed by alcohol makes the artery more prominent. The blood (30–60 ml) is allowed to clot at room temperature for 1 h and then overnight in the refrigerator. Serum is separated by centrifugation, partitioned into small aliquots, and either frozen at −70°C or lyophilized. Other animals, such as goats or sheep, can be used to produce large quantities of AB, although rabbits are normally quite adequate, and their IgG ABs bind well to protein A. Monoclonal antibodies can be produced following immunization of mice (see Chapter 3).

4. Antibody Characterization

Periodically throughout the immune response, the rabbit serum is analyzed for Ag-binding capability (AB dilution titer). This complex function of the concentrations of the heterogeneous population of anti-ecdysteroid ABs and their affinities for the radiolabel (tracer) is defined as the dilution factor (fold dilution or percent) of a specified volume of diluted ecdysteroid specific rabbit serum that will bind 30–50% of a convenient (10,000 dpm) amount of added radioligand (Fig. 6.4) or a comparable amount of fluorescent or enzyme-linked tracer (see below). The antisera are serially diluted with assay buffer, and a constant amount of radiolabeled Ag (either ^{3}H or ^{125}I) is added (see below). A wide range of buffers (borate, phosphate, citrate) have been used in the ecdysteroid RIA with equal success. The molarity can vary (0.01–0.1 M) as can the pH (from 6 to 8). Either normal rabbit serum as a carrier protein or BSA (to limit binding of AB and tracer to glass) is normally added to the buffer in addition to various preservatives—e.g., azide, merthiolate. Small,

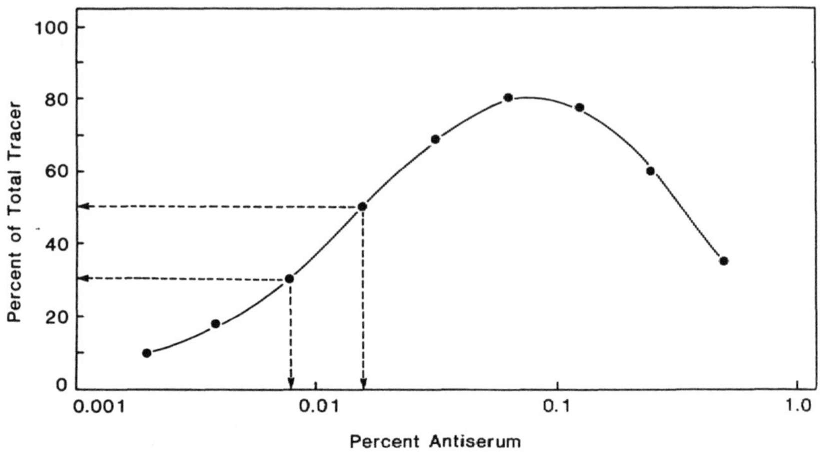

Figure 6.4. Antibody dilution titer (H-22 AB). Reduction of binding at high antiserum concentration is due to increased competition between tracer-labeled and nonlabeled AB for limited AB-binding sites on the protein A separation reagent.

disposable glass tubes (6 × 50 mm) are employed routinely with total incubation volumes of about 200 μl. Incubation can be carried out at room temperature (2 h) or at 4°C (overnight) to achieve a maximum equilibrium level of binding. Seperation of AB-bound label from free label (when necessary) can be accomplished by a variety of methods (see below).

AB titers as high as 1/200,000 have been achieved within a few weeks of boosting (Porcheron, 1979), and dilutions between 1/1,000 and 1/10,000 (0.1–0.01%) are common. The determinants of the AB titer, the affinity (association) constant (K_a), and the concentration of AB binding sites [n] can be determined from a Scatchard plot of the ratio of AB-bound ligand to free ligand vs. free ligand. The slope of the resulting straight line is equal to $[-K_a]$, and the intercept on the X axis is [n]. However, since the population of anti-ecdysteroid ABs can be very heterogeneous, curved lines usually result, and interpretation is more difficult. In general, the more curvature to the Scatchard plot, the less dynamic will be the binding response over any particular range of ecdysteroid concentrations; i.e., the standard curve will be less steep. The values of K_a for ecdysone range up to about 2×10^{10} L/mol (Chang and O'Connor, 1979; Hirn and Delaage, 1980).

An estimate of the maximum sensitivity of an RIA is $1/K_a$ (equal to K dissociation), or about 2 pg. The actual assay sensitivity for ecdysone or any other ecdysteroid (i.e., specificity) is determined from standard curves (Fig. 6.5). A working AB solution is prepared in buffer according to the AB dilution titer and should be stored frozen in aliquots that can be used within 1 or 2 months. Standard ecdysteroids dissolved in ethanol and quantified by UV absorption are subsequently diluted into assay buffer or unknown sample

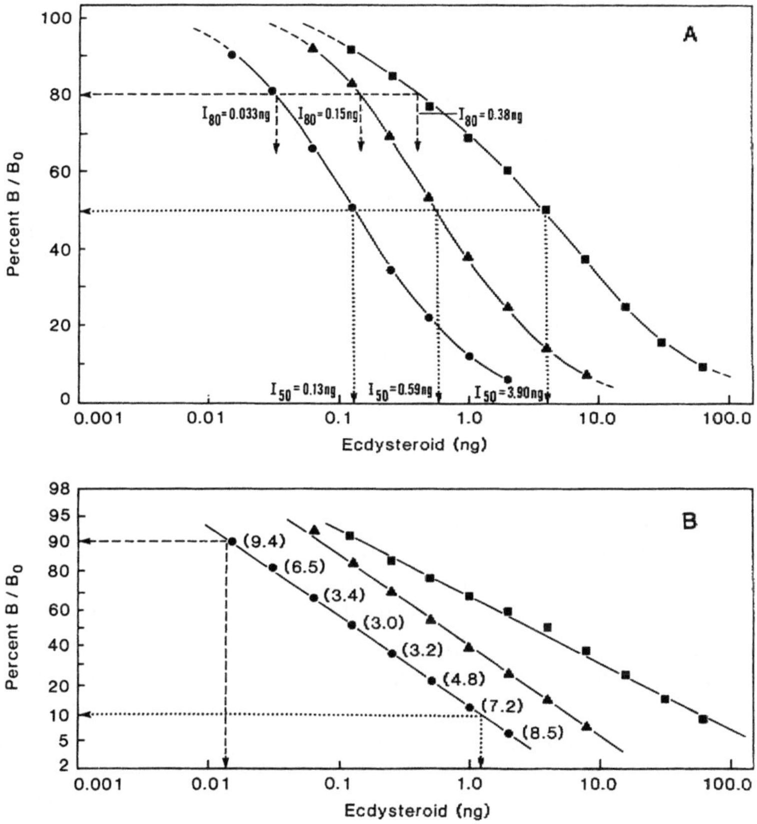

Figure 6.5. Standard ecdysteroid RIA competition curves. ●——●, ecdysone (E); ▲——▲, 20-hydroxyecdysone (20E); ■——■, ecdysteroid (UNK) unidentified. (A) Semilog plot. Relative cross-reaction (Xrxn) of an ecdysteroid defined as follows:

$$\text{Xrxn}(50) \text{ for } 20\text{E} = \frac{I_{50}(20\text{E})}{I_{50}(\text{E})} = 4.5$$

$$\text{Xrxn}(80) \text{ for } 20\text{E} = \frac{I_{80}(20\text{E})}{I_{80}(\text{E})} = 4.5$$

For the unknown ecdysteroid, Xrxn(50) = 30.0 whereas Xrxn(80) = 11.5, an indication that the competition curves are not parallel. Low concentrations of this ecdysteroid will be overestimated. (B) Linear logit-log plot. The useful range of the standard curve is between the I_{90} and I_{10}. Coefficient of variation of replicate standard B/B_o is below 10%.

matrix (i.e., hemolymph, Grace's media, or HPLC elution buffer) and stored frozen. Increasing amounts of ecdysone (10–2,000 pg, 10 μl) are then added to tubes together with the AB solution and the radiolabel, which has also been diluted into buffer and frozen in aliquots. The sequence of reagent addition can sometimes make a difference in the sensitivity and specificity exhibited by the assay, but the effect is variable with ABs elicited with immunogens made by conjugating low-molecular-weight haptens to proteins (Chard, 1982). After incubation and separation of bound and free label, specific AB-bound tracer is quantified.

Maximum sensitivity is defined as the smallest amount of added unlabeled ligand that can be distinguished from a sample containing no added unlabeled ligand. This is based on the overlap of 95% confidence limits as determined by statistical analysis of replicate standard assays. For almost any assay procedure, maximum sensitivity can be estimated as the amount of added ligand needed to reduce tracer binding to 90% of the maximum observed in the absence of added ligand (I_{90}). The maximum amount of ecdysone that can be reliably measured is the I_{10} (Fig. 6.5B). Samples containing ecdysteroids above this limit must first be diluted into this useful range of the standard curve (I_{90}–I_{10}). In fact, under the most stringent assay conditions, unknowns are assayed in triplicate at two separate dilutions within this range. Should it be necessary to decrease the sensitivity of the assay, as in the in vitro prothoracic gland assay for PTTH (Bollenbacher et al., 1983), the AB concentration can be increased and the specific activity of the tracer decreased (Warren et al., 1984). Increasing amounts of various other ecdysteroids are incubated with the ecdysteroid tracer and AB, and the resulting binding inhibition curves are compared to those of the unlabeled ligand (Fig. 6.5). The cross-reaction of a specific ecdysteroid is defined as the ratio of the respective amounts of the ecdysteroid or a standard, usually ecdysone, needed to inhibit maximum (B_0; zero standard) radiotracer binding by 50% (I_{50}). It is often advantageous to also compute this ratio using the I_{80} values of the respective two ecdysteroids:

$$\frac{I_{80}\ \text{(ecdysteroid)}}{I_{80}\ \text{(standard)}}$$

Significant variation from I_{50} calculated ratios is an indication that the two standard curves are not parallel, and this can lead to errors in quantifying cross-reacting ecdysteroids that have been resolved chromatographically but are present in low concentrations. Ideally, all individual ecdysteroids should be assayed at a concentration sufficient to reduce maximum tracer binding by 50% so that I_{50} cross-reaction data are valid. The relative specificities of various antisera preparations (different bleeds, rabbits, or immunogens) can be determined quickly from the analysis by RIA of a sequence of standard ecdysteroids of widely varying structure and polarity that have been adequately separated by HPLC. The cross-reaction of conjugated ecdysteroids can be determined by comparing their RIA response prior to, and following, total hydrolysis by suitable mixed esterase preparations—i.e., *Helix* extracts, potato acid phosphatase, and pig liver esterase (e.g., Warren and Gilbert, 1986).

It is seldom necessary, and sometimes not possible, to increase the sensitivity or specificity of the RIA by AB purification. Specific immunoabsorption can be used to separate ABs with different ligand affinities (Sage and O'Connor, 1976; Reum et al., 1982), but the highest-affinity ABs are the most difficult to elute and can be denatured in the process. Many chromatographic methods are available to purify heterogeneous polyclonal or monclonal ABs for studies requiring AB encapsulation or reaction to form solid-phase immunoreagents and immunolabels, such as for immunohistochemical and radioimmunometric applications (see Chapter 4).

5. The Ligand (Tracer)

The essential characteristic of a tracer is that it should behave in a manner similar to the unlabeled ligand in the assay system; i.e., it should have a similar affinity for the binder. In addition, it should be of the highest specific activity (usually related to increased assay sensitivity), chemically stable, easy to obtain or prepare, and measured conveniently by relatively economical automated equipment. Increasingly strict government safety regulations necessary to protect the workplace will certainly affect the type of tracer chosen.

Tracers can have either an internal label or an external label. In the case of the former, an atom in the ligand molecule is replaced by a radioactive isotope of that atom. Barring significant kinetic isotope effects, the tracer should behave like the unlabeled ligand. [^{3}H]Ecdysone of high specific activity (60 Ci/mmol) with tritium in the C-23 and C-24 positions can be obtained readily (New England Nuclear). It is relatively inexpensive (1¢ per/assay), stable in respect to radioactive decay and chemical degradation, and in any case, easily purified. With high-affinity ABs ($K_a = 1 \times 10^{10}$ L/mol), sensitivities of 10 pg ecdysone are routine and quite sufficient for most studies. The relative affinities of [^{3}H]ecdysone and ecdysone for binding to AB are indicated by the amount of unlabeled ecdysone (100 pg) needed to displace 50% of the bound [^{3}H]ecdysone (20 pg) from the binder (Warren et al., 1984). Other tritiated, high-specific-activity ecdysteroids have been prepared—e.g. ponasterone A and the ketol and ketodiols (Haag et al., 1986).

Because it cannot be substituted directly into the ecdysone molecule, radioiodine must be an external label. It can only be introduced into ecdysteroids that have been derivatized by covalent linkage to an iodine-reactive molecule—i.e., tyrosine, tyramine, or histidine residues. Such tracers are not identical to the unlabeled ligand and may behave differently, although in general this is not a serious problem. Radioiodine is a gamma emitter and can be utilized to label ecdysteroids to very high specific activity (2,000 Ci/mmol).

Inexpensive gamma counters can measure the activity of precipitated AB-bound tracer pellet without the need for added scintillators, removing the need for the tedious addition procedure required with beta emitters. The short, 60-day half-life of ^{125}I requires the synthesis of the radioligand every few months, but after the tyrosine or tyramine derivative of the ecdysteroid is made, iodination and purification are elementary and safe in laboratories equipped to handle the millicurie quantities of iodine involved. This derivative is usually made from the same hapten linker molecule (succinate or CMO) used to synthesize the immunogen.

A. Iodination

An ecdysone hemisuccinate (esterified at the C-2,22 or 26 position), inokosteronic acid, or 20-hydroxyecdysone CMO (1.5 μmol) dissolved in dioxane or dimethylformamide (100 μl) is treated with one equivalent each of tri-n-butylamine or triethylamine and isobutylchloroformate or ethylchloroformate for 15 min at 0°C (DeReggi et al., 1975; Porcheron et al., 1976). Reagent excess should not be necessary. The resulting mixed anhydride is then reacted at 0°C with tyramine (10 μmol) added in 20 μl of 0.01 N NaOH or tyrosine methyl ester (10 μmol) in dimethylformamide containing an equivalent amount of triethylamine. The reaction is immediate, and the product(s) are purified by column chromatography or TLC.

CMO derivatives of ecdysteroids are produced in both *syn* and *anti* forms, and this is true of their tyrosine or tyramine derivatives (Borst and O'Connor, 1974). Ideally, the predominant or sole oxime isomer that is used to produce the immunogen should reflect the stereochemistry of the similarly derived iodine label, although the reverse may prove useful in some instances—i.e., reducing specificity (see below). Although the method of Hunter and Greenwood (1962) has been used to iodinate *ortho* to the phenolic residues of the ecdysteroid derivatives, other methods are available (iodogen, lactoperoxidase, or iodine monochloride). Conversely, the phenolic derivative can be iodinated prior to reaction with the mixed anhydride activated hapten linker.

"A solution of chloramine-T (1 mg/ml) in dioxane is divided into aliquots of 10 μl, then dried at 40°C and stored at room temperature. For iodination, 20 μl of 10^{-4} M tyrosilated derivative in PBS (0.1 M, pH 7.0) and 1 mCi of ^{125}I iodide (IMS 30 from Amersham, or NEZ 033A from New England Nuclear) are added to the tube containing chloramine T. The reaction is left for 2 min and stopped by the addition of 20 μl of 0.6 mg/ml sodium metabisulfite in water. The mixture is loaded immediately on a Sephadex G-25 column (58 × 0.9 cm) and eluted with a 3 mM ascorbate buffer, pH 6.2, containing 10^{-2} M NaCl. The iodinated derivative is eluted as a sharp peak immediately after the free iodide. It must be diluted immediately to 200,000 cpm/ml in neutral buffer containing 2 mg/ml of serum albumin, divided into 2-ml aliquots, and stored at −20°C. Under these conditions, radiolysis is minimized, and the use of the iodinated derivative is only limited by the decay of ^{125}I. The column chromatographic purification of the iodinated derivative has many advantages over TLC (silica gel) purification—e.g., better protection against radiolysis and superior separation of iodinated derivatives from noniodinated molecules, so that a maximum specific radioactivity of about 2,000 Ci/mmole can be achieved" (Hirn and Delaage, 1980).

Since these iodinated tracers mimic the structure of the immunogen, they will certainly bind to the ABs derived from the immunogen. Unfortunately, they may bind too well, both to anti-ecdysteroid ABs and anti-linker ABs. On a molar basis, it requires 30 times as much 20-hydroxyecdysone to displace 50% of the [^{125}I]20-hydroxyecdysone CMO-tyramine tracer and 2,000 times as much to displace the [^{125}I]20-hydroxyecdysone-succinyl tyrosine methyl ester label from their respectively derived ABs (DeReggi et al., 1975; Porcheron et al., 1976; Delaage et al., 1982). The result can be a sensitivity much less than would be predicted by the specific activity of the label and the affinity constant of the AB. The effect can sometimes be reduced by employing in the tracer a linker distinct from that used in the immunogen—i.e., shorter or more bulky, yet attached in the same position to the ecdysteroid that is to be labeled. Such a tracer would presumably not bind to anti-linker ABs and might have an affinity constant for the anti-ecdysteroid ABs approaching that of the hapten. An increase in sensitivity would result, but specificity could be reduced, since other ecdysteroids might also more easily displace tracer from binder. External labels attached at ecdysteroid positions other than those used in the immunogen may also prove useful (Porcheron, personal communication). The use of iodinated tracers has resulted in a fivefold increase in sensitivity—i.e., down to 2 pg ecdysone.

B. Other External Labels

Other tracers can be used in ecdysteroid immunoassay. All are external labels and may elicit affinity problems similar to those observed with the iodine tracers. The same, or similar, ecdysteroid linker derivatives can be bonded to enzymes for use in an enzyme immunoassay (EIA). Following equilibration with AB and separation of bound from free, this tracer is quantified by incubating the pellet or supernatant with enzyme substrate, and the rate of production of UV or fluorescent product is measured (Porcheron, personal communication). In the case of enzyme-catalyzed chemiluminescence or bioluminescence reactions, light production is quantified (Reum, personal communication). Fluorescent molecules can also be attached to ecdysteroids, and their distribution in the AB-Ag reaction can be determined with great accuracy (fluorescence immunoassay, FIA; see Chapter 3). However, the sensitivity of these procedures is normally no greater than the RIA, nor are they more convenient, since the separation of bound and free tracer can still be required prior to tracer quantification.

Homogeneous fluorescent or enzyme immunoassays have been developed that do not require the physical separation of bound and free tracer (see Chapter 3). The specific inhibition (or enhancement) of the enzymatic activity or fluorescent quantum efficiency of the tracer when bound to AB has been used to determine the AB-Ag equilibria. Sensitivity of these homogeneous systems is usually lower than that of heterogeneous assays owing to nonspecific enzymatic or fluorescent activity in the unknown sample matrix. Exploitation of new techniques such as time-resolved fluorescence immunoassay and development of new equipment analogous to scintillation and gamma counters for multiple sample analysis, but containing enzyme rate recording spectrometers and spectrofluorimeters (see Chapter 3), will make these assays safe (no radioactivity) and routine but not necessarily inexpensive. A clinical need for an ecdysteroid immunoassay, such as the routine screening of human body fluids for immunoreactive ecdysteroids secreted by parasitizing helminths, may stimulate the development of this technology (Koolman et al., 1984; Koolman and Moeller, 1986).

6. Separation of Bound and Free

An immunoassay that incorporates a radiolabeled ligand is typically heterogeneous. To measure the distribution of ligand in the competitive binding reaction, bound tracer (either labeled ligand or labeled AB) must be separated from free label in solution. Many separation methods have been employed in published RIAs for ecdysteroids, and though seldom considered, they should result in the complete separation of the appropriate bound and free species while not disturbing the equilibrium of the reaction. Speed, simplicity, and cost will invariably determine the method of choice. Three basic mechanisms underlie all bound/free procedures: physical exclusion, fractional precipitation, and differential adsorption of either the free or AB-bound components.

A. Physical Exclusion

Equilibrium dialysis, devoid of artifacts, is theoretically the most elegant procedure in use (DeReggi et al., 1975). Tracer and standards or unknowns are added to one side of a semipermeable membrane and allowed to equilibrate with AB contained on the other side. Nothing is allowed to perturb the slowly attained equilibrium (18 h) in this one-phase system. In a variation of this procedure, following incubation (2 h) of tracer, standards (unknowns), and AB in the tops of Centricon-30 tubes (Amicon), a small sample of the free fraction is forced through a semipermeable membrane filter divider by a short, high-speed centrifugation step. It is inconvenient with present equipment to process large numbers of assays with these methods, and both the free and total fractions must be analyzed to calculate the amount bound. Gel filtration can also achieve separation of high molecular weight (AB-bound) from lower molecular weight (free tracer) species. Columns have been employed, but it is only necessary to add the gel (Sephadex or Biogel) to the assay tubes or incorporate it into the initial incubation media. As before, free ligand will distribute both inside and outside the gel matrix, whereas the bound complexes will be excluded (Garen et al., 1977).

B. Precipitation with Salts or Solvents

Precipitation with salts or organic solvents was among the earliest methods for the fractionation of biological molecules and is widely used for this purpose in RIA (Borst and O'Connor, 1974; Briers and DeLoof, 1981; Warren et al., 1984). Sufficient separation material is added to the incubation mixture to achieve a concentration at which the bound complexes are insoluble and precipitate, leaving the free fraction in solution. All the various agents (ethanol, isopropanol, dioxan, polyethylene glycol, sodium sulfite, ammonium sulfate, etc.) act in part by stripping the hydration shell from around the high molecular weight uncharged γ-globulins which become insoluble. The procedure does not affect the AB-Ag equilibrium. Indeed, with ammonium sulfate and polyethylene glycol (but not ethanol), the precipitation agent can be incorporated into the assay from the beginning, and the equilibrium is established between the two phases. Although inexpensive and of wide application, fractional precipitation can be labor-intensive. A washing step is normally required to reduce nonspecific binding due to entrapment below 10% of maximum binding. When the precipitating agent is to be added last, it must be mixed efficiently into each assay tube, one at a time, usually in the cold, to achieve reproducible results. Nevertheless, it is an invaluable procedure.

C. Adsorption

Nonspecific adsorption of free ligand by various particulates such as activated charcoal, sometimes coated with dextran, silicates (talc, Florisil), and hydroxyapatite, is rapid, simple, and inexpensive (Porcheron et al., 1976). Under the correct conditions, small crevices in the particles are thought to adsorb free ligand but exclude the larger bound complexes. A serious problem can develop with large assays when adsorbant progressively strips ligand from AB-binding sites, disturbing the steady state. These procedures require careful attention to time, temperature, ionic strength, and the protein concentration of the sample matrix.

The newest advances in RIA separation technology involve the specific adsorption of AB-bound ligand species from a mixture of bound and free ligand. In solid-phase AB systems, primary Ag-specific AB can be adsorbed onto, encapsulated within, or covalently attached to various polymers including plastic surfaces and glass, plastic, or gel beads (Möbus et al., 1987). The distribution of tracer between solid and liquid phases is determined following the physical separation of AB-bound label by decantation, centrifugation, or even magnetic attraction when the polymer contains ferric oxide.

D. Binding Proteins

In other methods, soluble or solid-phase AB-specific binding proteins are used to precipitate the primary AB-bound tracer complexes (Lazarovici et al., 1983). Goat or sheep anti-rabbit IgG or anti-rabbit-γ-globulin AB (second AB) can be used in soluble form where they bind to the F_c portion of the primary AB and AB-tracer molecules to form insoluble polymeric complexes. In the solid phase, bound second AB acts as an immunoglobulin sponge. Soluble second-AB methods can be lengthy procedures and require the incorporation of carrier γ-globulins into buffers and the use of additional second AB to enhance the size of the pellet formed after high-speed centrifugation. Normal rabbit serum (3–5%) must be added to buffers in fractional precipitation procedures for the same reason.

Protein A (Warren et al., 1984), a prominent component of the cell membrane of *Staphylococcus aureus* bacteria, is another binder of the F_c portion of IgG from many mammalian species, especially rabbit. For use as a binder, purified protein A can be bonded to various artificial solid supports or left in the membrane of heat-killed, formalin-treated bacteria. Like solid-phase second AB, it acts as a sponge for IgG, leaving non-bound tracer in solution. Binding is rapid at 4°C (15 min), and no extra carrier molecules are required in the assay medium, since the bacterial cell functions as an insoluble carrier. After low-speed centrifugation, the supernatant is easily aspirated or decanted from the compact cellular pellet, and a washing step is not needed to keep blanks below 10% of maximum tracer binding. If purchased from a commercial source, protein A reagent can be as expensive as second AB, but it is easy to prepare in large quantities in any laboratory equipped to conduct RIAs and can be stored frozen in working dilution for years without losing activity (Kessler, 1975).

Recent improvements in second-AB procedures have increased their efficiency and lowered their cost (Walgraeve et al., 1986). Preincubation of second AB with carrier γ-globulin (or primary AB) before addition to the assay mixture significantly reduces the equilibration time and requirement for second AB. The same effect is observed when second AB is first bound to the surface of *S. aureus* protein A–containing cells prior to addition to the assay. In addition, inclusion of polyethylene glycol in assays employing soluble second AB can greatly accelerate the formation of insoluble AB complexes and lessen assay time.

As in the case of some fractional precipitation methods, none of these specific AB adsorption procedures interfere with the AB-Ag reaction, and all can be incorporated into one-step assays. Standards or unknown samples are added to a preincubated mixture containing buffer, tracer, AB, and protein A (or second AB) which can be stored frozen in RIA tubes. After incubation overnight, the tubes are simply centrifuged and decanted, and the bound tracer is measured, most conveniently with gamma labels. The centrifuge is unnecessary with magnetic-coated surface or large bead-solid-phase immunobinding reagents, although extensive washing of the solid phase may be required to reduce blanks.

III. Sample Preparation and Assay

To avoid artifacts, it is important with any of the preceding assay procedures for the standards to be prepared in media as similar as possible to the composition of the sample; i.e., all assay tubes [blanks (no added AB), maximum binding (no added unlabeled ecdysteroid), standards, and unknowns] must contain the same components. Otherwise, both specific and nonspecific interference can occur. With protein A, immunoglobulins in the sample will cause problems (see Fig. 6.4), and with ammonium sulfate, lipids may interfere. Some interference with insect hemolymph titer assays by molecules other than ecdysteroids has been proposed. Endogenous ecdysteroid receptors and binding proteins present in the sample would disturb the AB-Ag reaction and may sequester bound hormone during alcohol denaturation and extraction of hemolymph and tissue samples (Feyereisen, 1977).

1. Titer of Immunoreactive Ecdysteroids

Media from the in vitro incubation of various tissues have been assayed directly, but centrifuged hemolymph should first be added to several volumes of cold methanol to denature the proteins present. After centrifugation, aliquots of the resulting supernatant are then evaporated in RIA tubes and analyzed. Depending on the developmental status of the animal, it may be necessary to wash the pellet once with 70% methanol to extract some polar metabolites more fully. If endogenous binding proteins are present, hemolymph extraction can be carried out at 60°C to facilitate disruption of this binding. Extreme pH conditions (below 5, above 9) should always be avoided with ecdysteroids owing to facile A-B ring isomerization and 14-hydroxyl elimination. Such changes would certainly affect AB binding. Other chaotropic agents may prove useful (urea, SDS, guanidinium hydrochloride, etc.). Tissue or whole-body preparations usually require more extensive extraction, and the incorporation of two-phase extractants—e.g., water-methanol-hexane or chloroform—will facilitate the removal of interfering nonpolar lipids and waxes.

2. Assay of Individual Ecdysteroids in Biological Material

The most important problem in ecdysteroid immunoassay, however, is the specific interference caused by more than one ecdysteroid in the sample. Even when 20-hydroxyecdysone is thought to be the major ecdysteroid and the only active hormone present in a sample, at least one chromatographic step prior to RIA is advised. Recent developments in TLC procedures have resulted in the convenient and rapid resolution of many ecdysteroids, both free and conjugated (Wilson, 1985; Wilson and Lafont, 1986). However, the tendency for the polar ecdysteroids to bind irreversibly to activated silica gel, both at the origin and after migration, makes consistent quantitative recovery difficult to achieve even with repeated extraction. Also, any silica gel carried over into the RIA will adsorb tracer and interfere. Recoveries of ecdysteroids from reverse-phase (and some normal phase) HPLC columns is consistently good, even when small amounts are injected (<50 ng total RIA-positive material). Most, if not all, known or suspected ecdysteroids contained in any particular extract can be resolved adequately with these techniques (Lafont et al., 1980; Dinan et al., 1981; Isaac et al., 1982). Usually, small aliquots (20 μl) of the HPLC eluant can be analyzed directly by various RIAs. Larger samples and those containing high concentrations of organic solvents must be reduced by evaporation prior to assay. Fractions containing conjugated ecdysteroids are evaporated and hydrolyzed by enzymes, and the free ecdysteroid(s) are then analyzed prior to and following further chromatography.

3. General Assay Procedure

[^{3}H]Ecdysone (62.7 pg, 18,000 dpm, 60 Ci/mmol, New England Nuclear) and the correct dilution of specific anti-ecdysteroid antisera in 100 mM borate buffer (0.2 ml, pH 8.5) containing 0.05% BSA (Sigma, fraction V) are added to aliquots (20 μl) of standard ecdysteroid solutions contained in glass tubes (6×75 mm) (Warren et al., 1984). Aliquots of HPLC elution fractions, or the evaporated residues following other chromatographic procedures or following extraction of biological material, replace the standard solution in other tubes. Replicate maximum binding (zero standard or B_0) tubes containing only tracer, AB, and buffer are included in each assay. All other AB-bound tracer counts (standards and unknowns) are normalized to these B_0 counts. Nonspecific binding (blank or infinite standard) replicates containing only buffer and tracer are also included, and all binding (maximum, standard, or unknown) is reduced by this value (Blk). Following overnight incubation at 4°C, a 5% (w/v) solution of heat-killed, formalin-treated *S. aureus* cells (20 μl) is added, and after 15 min at 4°C, the tubes are centrifuged (5,000g, 15 min). Supernatants are aspirated from the pellets containing AB-bound tracer, and pellets are then suspended in water (50 μl) and dissolved in scintillation cocktail (0.45 ml, Scintiverse I, Fisher) prior to counting (LKB Rack-beta).

4. Data Reduction and Quality Control

Large-capacity scintillation and gamma counters interfaced with computers have made RIA data processing almost effortless. Rather than raw data, finished tables and graphs complete with statistical analysis await the researcher the next morning. With small numbers of assays, manual data handling is almost as simple. Raw (cpm) nonspecific blank (Blk) binding is first subtracted from all other bound (B) counts. Blank corrected binding (B-Blk) is then expressed as a percent of blank corrected maximum binding (Bo-Blk)—i.e., B–Blk/Bo–Blk × 100 (% B/Bo). A linear standard curve usually results from plotting percent B/Bo vs. ecdysteroid concentration on logit-log paper, although many other plots can be used (Fig. 6.5). These calculations can be handled easily by pocket calculators. After a logit-log transformation and linear regression, unknown concentrations are obtained by interpolation. In equilibrium dialysis assays, bound/total can be related directly to ecdysteroid concentration, because there is no blank, and maximum binding approaches 100%.

Quality control in an RIA must be monitored constantly to assure consistent results. The coefficient of variation in intra- and interassay standard replicates should be under 10%. This will be reflected in consistent standard curve variables—i.e., the slope and standard I_{50} (or Y-axis intercept). Whenever new ecdysteroid standards (i.e., ecdysone) are prepared, it is important to ensure that these assay parameters remain constant. This is also necessary when new AB, label, or protein A working solutions are prepared or when other assay procedures are modified.

If clinical use of an ecdysteroid RIA becomes routine, it will become necessary to evaluate interassay performance more rigorously (Findlay et al., 1981; Warren et al., 1982). Separate low, medium, and high ecdysteroid standards prepared independently and dissolved in the sample matrix (i.e., human fluid) should be analyzed with every assay. In addition, replicate analysis of serum samples from patient pools that exhibit blank, low, and high titers is vital. The coefficient of variation of these interassay controls should also be within 10%.

IV. Conclusions

Although we have concentrated on the RIA of ecdysteroids, the principles discussed apply to all competitive protein binding assays. When care is taken, the qualitative and quantitative titer data generated with a combination of HPLC-RIA closely approximate those obtained via the more expensive and sophisticated analyses by GC/MS (compare Warren et al., 1986, and Thompson et al., 1987). For the insect biologist without routine access to a mass spectrometer, the RIA (or more specifically, a combination of RIAs) is the method of choice. However, it must be emphasized that a thorough knowledge of the principles of RIA and the characteristics of the AB are requisite for both the planning and execution of experiments involving the ecdysteroids of arthropods and other organisms.

Acknowledgments. We thank Susan Whitfield for graphics and Shelia King and Caroleigh Robinson for manuscript preparation. Research from our laboratory cited in this chapter was supported by MERIT grant DK-30118, grant DK-35347 from the National Institutes of Health, and grant DCB-8502194 from the National Science Foundation.

References

Ashburner, M. (1971). Induction of puffs in polytene chromosomes of *in vitro* cultured salivary glands of *Drosophila melanogaster* by ecdysone and ecdysone analogues. *Nature New Biol.* **230**:222–224.

Ashburner, M. (1973). Sequential gene activation by ecdysone in polytene chromosomes of *Drosophila melanogaster*. I. Dependence upon ecdysone concentration. *Dev. Biol.* **35**:47–61.

Bollenbacher, W.E., O'Brien, M., Katahira, E., and Gilbert, L.I. (1983). A kinetic analysis of the action of the prothoracicotropic hormone. *Mol. Cell. Endocrinol.* **32**:27–46.

Borst, D.W., and O'Connor, J.D (1972). Arthropod molting hormone: Radioimmune assay. *Science* **178**:418–419.

Borst, D.W., and O'Connor, J.D. (1974). The trace analysis of ecdysones by gas-liquid chromatography, radioimmunoassay and bioassay. *Steroids* **24**:637–656.

Briers, T., and DeLoof, A. (1981). Moulting hormone activity in the adult Colorado potato beetle, *Leptinotarsa decemlineata* Say in relation to reproduction and diapause. *Int. J. Invert. Reprod.* **3**:145–155.

Briers, T., Peferoen, M., and DeLoof, A. (1982). Ecdysteroids and adult diapause in the Colorado potato beetle, *Leptinotarsa decemlineata. Physiol. Entomol.* **7**:379–386.

Chang, E.S., and O'Connor, J.D. (1979). Arthropod molting hormones. In: *Methods of Hormone Radioimmunoassay, 2d Ed.* B.M. Jaffe and H.R. Behrman (Eds). Academic Press, New York, pp. 797–814.

Chard, T. (1982). An introduction to radioimmunoassay and related techniques. In: *Laboratory Techniques in Biochemistry and Molecular Biology*, Vol. 6 (II), T.S. Work and E. Work (Eds.). Elsevier, Amsterdam.

Cherbas, L., Yonge, C.D., Cherbas, P., and Williams, C.M. (1980). The morphological response of K_c–H cells to ecdysteroids: Hormonal specificity. *Wilhelm Roux's Arch.* **189**:1–15.

Delaage, M.A., Hirn, M.H., and DeReggi, M.L. (1982). Radioimmunoassay of ecdysteroids. In: *Methods in Enzymology* J.J. Langon and H. Van Vunakis (Eds.). Academic Press, New York, Vol. 84, pp. 350–358.

DeReggi, M.L., Hirn, M.H., and Delaage, M.A. (1975). Radioimmunoassay of ecdysone. An application to *Drosophila* larvae and pupae. *Biochem. Biophys. Res. Commun.* **66**:1307–1315.

Dinan, L.N., Donnahey, P.L., Rees, H.H., and Goodwin, T.W. (1981). High-performance liquid chromatography of ecdysteroids and their 3-epi, 3-dehydro and 26-hydroxy derivatives. *J. Chromatogr.* **205**: 139–145.

Erlanger, B.F., Bieser, S.M., Borek, F., Edel, F., and Leibermann, S. (1967). The preparation of steroid-protein conjugates to elicit anti-hormonal antibodies. In: *Methods in Immunology and Immunochemistry*. C.A. Williams and M.W. Chase (Eds.). Academic Press, New York, Vol. 1, pp. 149–150.

Feyereisen, R. (1977). A specific binding protein for the moulting hormone ecdysterone in locust haemolymph. *Experientia* **33**: 1111–1112.

Findlay, J.W.A., Warren, J.T., Hill, J.A., and Welch, R.M. (1981). Stereospecific radioimmunoassays for d-pseudophedrine in human plasma and their application to bioequivalency studies. *J. Pharm. Sci.* **70**: 624–631.

Fujimoto, Y., Miyasaka, S., Ikeda, T., et al. (1985). An unusual ecdysteroid (20S)-cholesta-7,14-diene-3β,5α,6α,20,25-pentaol (Bombycosterol) from the ovaries of the silkworm, *Bombyx mori*. *J. Chem. Soc. Chem. Commun.* 10–12.

Garen, A., Kauvar, L., and Lepesant, J.A. (1977) Roles of ecdysone in *Drosophila* development. *Proc. Natl. Acad. Sci. USA* **74**: 5099–5103.

Gilbert, L.I., Goodman, W., and Bollenbacher, W.E. (1977). Biochemistry of regulatory lipids and sterols in insects. In: *Biochemistry of Lipids II. T.W.* Goodwin (Ed.) University Park Press, Baltimore, Vol. 14, pp. 1–49.

Guédin-Vuong, D., Nakatani, Y., and Ourisson, G. (1985). Synthesis of C_{27}-substituted ecdysteroids. Potential tools for biochemical studies. *Tetrahedron Lett.* **26**: 2959–5962.

Haag, T., Meister, M.F., Hetru, C., et al. (1986). Synthesis of a labelled putative precursor of ecdysone-II (^{3}H)-3β-hydroxy-5β-cholest-7-ene-6-one: Critical re-evaluation of its role in *Locusta migratoria*. *Insect Biochem* **17**: 291–302.

Hetru, C., Lagueux, M., Bang, L., and Hoffmann, J.A. (1978). Adult ovaries of *Locusta migratoria* contain the sequence of biosynthetic intermediates for ecdysone. *Life Sci.* **22**: 2141–2154.

Hetru, C.C., Kappler, C., Hoffmann, J.A., Nearn, R., Bang, L., and Horn, D.H.S. (1982). The biosynthetic pathway of ecdysone: Studies with vitellogenic ovaries of *Locusta migratoria* (Orthoptera). *Mol. Cell. Endocrinol.* **26**: 51–80.

Hirn, M.H., and Delaage, M.A. (1980). Radioimmunological approaches to the quantification of ecdysteroids. In: *Progress in Ecdysone Research*. J.A. Hoffmann (Ed.). Elsevier/North-Holland, Amsterdam, pp. 69–82.

Horn, D.H.S. (1971). The ecdysones. In: *Naturally Occuring Insecticides*. M. Jacobson and D.G. Crosby (Eds.). Dekker, New York, pp. 333–455.

Horn, D.H.S., and Bergamasco, R. (1985). Chemistry of ecdysteroids. In: *Comprehensive Insect Physiology, Biochemistry and Pharmacology*. G.A. Kerkut and L.I. Gilbert (Eds.). Pergamon Press, Oxford, U.K., Vol. 7, pp. 185–248.

Horn, D.H.S., Wilkie, J.S., Sage, B.A., and O'Connor, J.D. (1976). A high affinity antiserum specific for the ecdysone nucleus. *J. Insect Physiol.* **22**: 901–905.

Hunter, W.M., and Greenwood, F.C. (1962). Preparation of iodine-131 labelled human growth hormone of high specific activity. *Nature* **194**: 495–496.

Isaac, R.E., Milner, N.P., and Rees, H.H. (1982). High-performance liquid chromatography of ecdysteroids and ecdysteroid-22-phosphates. *J. Chromatogr.* **246**: 317–322.

Kaplanis, J.N., Weirich, G.F., Svoboda, J.A., Thompson, M.J., and Robbins, W.E. (1980). Molting hormones of the tobacco hornworm and milkweed bug: Their chemistry and biochemistry. In: *Progress in Ecdysone Research.* J.A. Hoffmann (Ed.). Elsevier/North-Holland, Amsterdam, pp. 163–186.

Kellie, A.E. (1975). Radioimmunoassay of steroid glucuronides. In: *Radioimmunoassay of Steroid Hormones.* D. Gupta (Ed.). Verlag Chemie, Weinheim, F.R.G., pp. 165–183.

Kessler, S.W. (1975). Rapid isolation of antigens from cells with a staphylococcal protein A–antibody absorbent: Parameters of the interaction of antibody-antigen complexes with protein A. *J. Immunol.* **115:** 1617–1624.

Koolman, J. (1980). Ecdysteroids in the blowfly, *Calliphora vicina.* In: *Progress in Ecdysone Research.* J.A. Hoffmann (Ed.). Elsevier/North-Holland, Amsterdam, pp. 187–209.

Koolman, J., and Karlson, P. (1985). Regulation of ecdysteroid titer: Degradation. In: *Comprehensive Insect Physiology, Biochemistry and Pharmacology.* G.A. Kerkut and L.I. Gilbert (Eds.). Pergamon Press, Oxford, U.K., Vol. 7, pp. 343–362.

Koolman, J., and Moeller, H. (1986). Diagnosis of major helminth infections by RIA detection of ecdysteroids in urine and serum. *Insect Biochem.* **16:** 287–291.

Koolman, J., Walter, J., and Zahner, H. (1984). Ecdysteroids in helminths. In: *Biosynthesis, Metabolism and Mode of Action of Invertebrate Hormones.* J. Hoffmann and M. Porchet (Eds.). Berlin, Springer-Verlag, pp. 323–330.

Lafont, R. (1987). Metabolism and excretion of ecdysteroids. In VIII Ecdysone Workshop Abstracts, Marburg, F.R.G., p. 72.

Lafont, R., and Koolman, J. (1984). Ecdysone metabolism. In: *Biosynthesis and Mode of Action of Invertebrate Hormones.* J. Hoffmann and M. Porchet (Eds.). Springer-Verlag, Berlin, pp. 196–226.

Lafont, R., Somme-Martin, G., Mauchamp, B., Maume, B.F., and Delbeque, J.-P. (1980a). Analysis of ecdysteroids by high-performance liquid chromatography and coupled gas-liquid chromatography–mass spectrometry. In: *Progress in Ecdysone Research.* J.A. Hoffmann (Ed.). Elsevier/North-Holland, Amsterdam, pp. 45–68.

Lafont, R., Beydon, P., Somme-Martin, G., and Blais, C. (1980b). High-performance liquid chromatography of ecdysone metabolites applied to the cabbage butterfly, *Pieris brassicae. Steroids* **36:** 185–207.

Lafont, R., Beydon, P., Blais, C., et al. (1986). Ecdysteroid metabolism: A comparative study. *Insect Biochem.* **16:** 11–16.

Lauer, R.C., Solomon, P.H., Nakanishi, K., and Erlanger, B.F. (1974). Antibodies to the insect molting hormone β-ecdysone. *Experientia* **30:** 560–562.

Lazarovici, P., Shapira, D., and Shaaya, E. (1983). Determination of ecdysteroid titers by a double-antibody precipitation in the tropical warehouse moth, *Ephestia cautella* (Lepidoptera, Phycitidae). *Comp. Biochem. Physiol.* **74:** 525–529.

Milner M.J., and Sang J.H. (1974). Relative activities of α-ecdysone and β-ecdysone for the differentiation *in vitro* of *Drosophila melanogaster* imaginal discs. *Cell* **3:** 141–143.

Milner, N.P., Nali, M., Gibson, J.M., and Rees, H.H. (1986). Early stages of ecdysteroid biosynthesis. The role of 7-dehydrocholesterol. *Insect Biochem.* **16:** 17–23.

Möbus, P., Bidmon, H.-J., Reum, L., and Koolman, J. (1987) Solid-phase immunoassay for ecdysteroids. In VIII Ecdysone Workshop Abstracts, Marburg, F.R.G. p. 90.

Morgan, E.D., and Wilson, I.D. (1980). Progress in the analysis of ecdysteroids. In:

Progress in Ecdysone Research. J.A. Hoffmann (Ed.). Elsevier/North-Holland, Amsterdam, pp. 29–43.

Porcheron, P. (1979). Thèse de Doctorat d'Etat, Paris.

Porcheron, P., Foucrier, J., Gros, C., Pradelles, P., Cassier, P., and Dray, F. (1976). Radioimmunoassay of arthropod molting hormone: β-ecdysone antibodies production and [^{125}I]-iodinated tracer preparation. *FEBS Lett.* **61**:159–161.

Redfern, C.P.F. (1984). Evidence for the presence of makisterone A in *Drosophila* larvae and the secretion of 20-deoxymakisterone A by the ring gland. *Proc. Natl. Acad. Sci. USA* **81**:5643–5647.

Rees, H.H. (1985). Biosynthesis of ecdysone. In: *Comprehensive Insect Physiology, Biochemistry and Pharmacology*. G.A. Kerkut and L.I. Gilbert (Eds.). Pergamon Press, Oxford, U.K., Vol. 7, pp. 249–294.

Rees, H.H., Davies, T.G., Dinan, L.N., Lockley, J.S., and Goodwin, T.W. (1980). The biosynthesis of ecdysteroids. In: *Progress in Ecdysone Research*. J.A. Hoffmann (Ed.). Elsevier/North-Holland, Amsterdam, pp. 125–137.

Reum, L., and Koolman, J. (1979). Analysis of ecdysteroids by radioimmunoassay: Comparison of three different antisera. *Insect Biochem.* **9**:135–142.

Reum, L., Haustein, D., and Koolman, J. (1981). Immunoadsorption as a means for the purification of low molecular weight compounds: Isolation of ecdysteroids from insects. *Z. Naturf.* **36c**:709–797.

Reum, L., Kauser, G., Enderle, U., and Koolman, J. (1982). A steroid-binding protein from insect hemolymph isolated by photoaffinity labelling and immunoadsorption. *Z. Naturf.* **37**:967–976.

Richards, G. (1978). The relative biological activities of α- and β-ecdysone and their 3-dehydro derivatives in the chromosome puffing assay. *J. Insect Physiol.* **24**:329–335.

Richards, G. (1981). Insect hormones in development. *Biol. Rev.* **56**:501–549.

Robertson, J., and Pipa, R. (1973). Metamorphic shortening of interganglionic connections of *Galleria mellonella* (Lepidoptera) *in vitro*: Stimulation by ecdysone analogues. *J. Insect Physiol.* **19**:673–679.

Sage, B.A., and O'Connor, J.D. (1976). An affinity column for ecdysone binding proteins, *Anal. Biochem.* **73**:240–246.

Smith, S.L. (1985). Regulation of ecdysteroid titer: Synthesis. In: *Comprehensive Insect Physiology, Biochemistry and Pharmacology*. G.A. Kerkut and L.I. Gilbert (Eds.). Pergamon Press, Oxford, U.K., Vol. 7, pp. 295–342.

Soumoff, C., Horn, D.H.S., and O'Connor, J.D. (1981). Production of a new antiserum to arthropod molting hormone and comparison with two other antisera. *J. Steroid Biochem.* **14**:429–435.

Spindler, K.-D., Hamann, A., Spindler-Barth, M., Ihne, A., Beckers, C., and Emmerich, H. (1976). Derivatives of the insect moulting hormone for affinity chromatography and their biological activities. *Steroids* **27**:553–565.

Spindler, K.-D., Beckers, C., Groschel-Stewart, U., and Emmerich, H. (1978). A radioimmunoassay for arthropod moulting hormones introducing a novel method of immunogen coupling. *Hoppe-Seyler's Z. Physiol. Chem.* **359**:1269–1275.

Strambi, C., Strambi, A., DeReggi, M., and Delaage, M. (1984). Radioimmunoassays of juvenile hormones. State of the methods and recent data on validation. In: *Biosynthesis, Metabolism, and Mode of Action of Invertebrate Hormones*. J. Hoffmann and M. Porchet (Eds.). Springer-Verlag, Berlin, pp. 355–362.

Thompson, M.J., Svoboda, J.A., and Weirich, G.F. (1985a). Ecdysteroids in developing ovaries and eggs of the tobacco hornworm. *Steroids* **43**:333–342.

Thompson, M.J., Weirich, G., Rees, H., Svoboda, J., Feldlaufer, M., and Wilzer, K. (1985b). New ecdysteroid conjugate: Isolation and identification of 26-hydroxyecdysone-26-phosphate from eggs of the tobacco hornworm, *Manduca sexta* (L.). *Arch. Insect Biochem. Physiol.* **2**:227–236.

Thompson, M.J., Feldlaufer, M.F., Lozano, R., et al. (1987). Metabolism of 26-[^{14}C]-hydroxyecdysone-26-phosphate in the tobacco hornworm, *Manduca sexta* L., to a new ecdysteroid conjugate: 26-[^{14}C]-hydroxyecdysone-22-glucoside. *Arch. Insect Biochem. Physiol.* **4**:1–15.

Tung, A.S. (1983). Production, purification and characterization of antigen-specific murine monoclonal antibodies of IgE class. *Methods Enzymol.* **92**:47–65.

Vaitukaitis, J. (1981). Production of antisera with small doses of immunogen: Multiple intradermal injections. *Methods Enzymol.* **73**:46–51.

Walgraeve, H., Van Beek, E., Criel, G., Van Brussel, K., and DeLeenheer, A. (1986). Comparison of three separation systems for the radioimmunoassay of ecdysteroids in *Artemia* (Crustacea: Brachiopoda). *Insect Biochem.* **16**:41–44.

Warren, J.T., and Gilbert, L.I. (1986). Ecdysone metabolism and distribution during pupal-adult development of *Manduca sexta*. *Insect Biochem.* **16**:65–82.

Warren, J.T., Coker, G.G., Welch, R.M., Fowle, A.S.E., and Findlay, J.W.A. (1982). Plasma levels of a novel antidysrhythmic agent, meobentine sulfate, in humans as determined by radioimmunoassay. *J. Pharm. Sci.* **71**:665–670.

Warren, J.T., Smith, W., and Gilbert, L.I. (1984). Simplification of the ecdysteroid radioimmunoassay by the use of protein A from *Staphylococcus aureus*. *Experientia* **40**:393–394.

Warren, J.T., Steiner, B., Dorn, A., Pak, M., and Gilbert, L.I. (1986). Metabolism of ecdysteroids during the embryogenesis of *Manduca sexta*. *J. Liq. Chromatogr.* **9**:1759–1782.

Warren, J.T., Sakurai, S., Rountree, D.B., and Gilbert L.I. (1988). Synthesis and secretion of ecdysteroids by the prothoracic glands of *Manduca sexta*. *J. Insect Physiol.* (in press).

Wilson, I.D., (1985) Thin-layer chromatography of ecdysteroids. *J. Chromatogr.* **318**:373–377.

Wilson, I.D., and Lafont, R. (1986). Thin-layer chromatography and high-performance thin-layer chromatography of [^{3}H]-metabolites of 20-hydroxyecdysone. *Insect Biochem.* **16**:33–40.

Yund, M.A., and Osterbur, D.L. (1985). Ecdysteroid receptors and binding proteins. In: *Comprehensive Insect Physiology, Biochemistry and Pharmacology*. G.A. Kerkut and L.I. Gilbert (Eds.). Pergamon Press, Oxford, U.K., Vol. 7, pp. 473–490.

Chapter 7

Radioimmunoassay: Juvenile Hormones

Noelle A. Granger and Walter G. Goodman

I. Introduction

One of the major technological advances in analytical biochemistry during the past three decades has been the development of the radioimmunoassay (RIA). This relatively simple combination of binding and immunological techniques, in conjunction with the use of high-specific-activity radiotracers, has revolutionized the field of endocrine physiology. Because high specificity and sensitivity characterize the RIAs, this approach is particularly advantageous for the measurement of insect hormones, which normally occur in subnanogram concentrations.

The application of RIA technology to the measurement of the juvenile hormones (JHs) was a natural extension from the development of ecdysteroid RIAs (see Chapter 6). The family of JH molecules is collectively responsible for many critical physiological events during the postembryonic development of the insect, and there has long been a crucial need for a precise means by which to quantify JH.

Historically, analysis of these hormones has proven difficult, because in the insect the presence of "nanograms of labile lipid hormone concealed in milligrams of miscellaneous body lipids" (Judy, 1976) has made the application of conventional procedures for extraction, separation, and quantification virtually impossible. Moreover, JH was found to exist in Lepidoptera in several structurally distinct homologous forms (Meyer et al., 1968; Schooley et al., 1984), each with different biological activities as determined by bioassay, the most common method of JH quantification prior to the early 1970s. With increasing interest in the function of insect hormones, the inability of bioassays to distinguish quantitatively between the JH homologues pointed to the need for methods that would specifically quantify each of the hormones in a complex mixture.

II. History of the JH RIA

Two obstacles initially delayed the development of a JH RIA. First, no commercial source of the JH homologues, both radioinert and radiolabeled, was available. Second, since JH is not intrinsically immunogenic (see section III), it must be covalently attached to a molecule that can elicit an immunological response.

The first report of the generation of JH antisera (Lauer et al., 1974) was made soon after the tritiated JH I homologue became commercially available, and the synthetic routes developed in this study to produce the hormone: protein complex of JH III: human serum albumin (HSA) have subsequently been employed in the development of other RIAs. According to this procedure, JH III was hydrolyzed to the acid form and then attached through C-1 (carbon 1 of the chain; see Gilbert and Goodman, 1981, for nomenclature) to the protein carrier (see Tables 7.1, 7.2). Unfortunately, the serum generated to this complex displayed such low binding affinity that its use was precluded for the quantification of JH in biological samples (see Table 7.3). The low affinity of the antiserum may in part have been the result of immunization with high concentrations of antigen (Ag). Furthermore, the cross-reactivity of the antiserum with JH II obviated its use for the direct measurement of JH III in samples containing a mixture of homologues (see Table 7.5). The investigators concluded that increasing the time of immunization and the use of a radioligand with higher specific activity should increase affinity as well as sensitivity, and both of these predictions have since been borne out.

Table 7.1. Development of the Immunogen

Hapten	Protein carrier	Attachment site	mol Hapten/ mol carrier	Reference
JH III acid	Human serum albumin	C-1	20	Lauer et al., 1974
JH I, II, III acid	Human serum albumin	C-1	26	Baehr et al., 1976, 1979
JH I diol	Human serum albumin	C-1	17	Strambi et al., 1981
JH III acid	Thyroglobulin	C-1	104	Goodman, unpublished
JH III acid	Thyroglobulin	C-1	178	Granger and Goodman, unpublished

Table 7.2. Development of JH Antisera

Animal species (sex)	Amount of initial injection	Method of immunization	Immunization and bleeding schedule	Reference
Rabbit	0.4 mg	Toe pad injection	Inject days 0, 7, 14; bleed days 21 and subsequent for 1 month; boost day 49; bleed day 49 and subsequent for 1 month	Lauer et al., 1974
Rabbit (female)	0.6 mg	Intradermal injections on back	Inject day 0; boost at 5-week intervals; bleed 10 days after each booster	Baehr et al., 1976 Baehr et al., 1979
Rabbit	0.25 mg	Subcutaneous injections on back	Inject day 0; boost and bleed each week subsequent	Strambi et al., 1981
Rabbit (female)	0.3 mg	Subcutaneous injections on back	Inject day 0; boost days 30 and 44; bleed and boost every 24 days subsequent	Goodman, unpublished

Table 7.3. Sensitivity of the JH RIAs

Investigator	Antiserum	Lower limit of detection[a] (ng/assay)	50% Displacement[b] (ng/ml)	Radiolabeled ligand	Specific activity (Ci $mmol^{-1}$)
Lauer et al., 1974	JHIII	—	33.0	[^{3}H]JHI	14.1 Ci $mmol^{-1}$
	JHI	~0.030	1.0	[^{3}H]JHI	13.5 Ci $mmol^{-1}$
Baehr et al., 1976	JHI	~0.030	0.67	[^{125}I]JHI-histamine	~2000 Ci $mmol^{-1}$
Baehr et al., 1979	JHI	~0.020	0.75	[^{125}I]JHI-histamine	~2000 Ci $mmol^{-1}$
	JHII	~0.020	1.0	[^{125}I]JHII-histamine	~2000 Ci $mmol^{-1}$
	JHIII	~0.020	2.6	[^{125}I]JHIII-histamine	~2000 Ci $mmol^{-1}$
Granger et al., 1979, 1982	JHI	0.025	1.3	[^{3}H]JHI	11.3 Ci $mmol^{-1}$
Granger et al., 1982	JHIII	0.025	1.0	[^{3}H]JHIII	11.6 Ci $mmol^{-1}$
Strambi et al., 1981	JH diol	0.005	0.7	[^{125}I]Suc < JH-Gly-Tyr	~2000 Ci $mmol^{-1}$
Goodman et al., unpublished	JHIII	0.05	2.2	[^{3}H]JHII	58 Ci $mmol^{-1}$
Granger and Goodman, unpublished	JHII	0.025	2.0	[^{3}H]JHIII	11.0 Ci $mmol^{-1}$

[a] Least amount of JH that can be reproducibly detected by the assay.
[b] ng JH homologue causing 50% displacement of the radiolabeled ligand.

Table 7.4. Cross-Reactivities of JH I Antisera with JH Homologues and Structurally Related Compounds[a]

	Cross-reactivity (%)			
Hormone/compound	Baehr et al., 1976	Baehr et al., 1979	Granger et al., 1979	Strambi et al., 1981
JH I	100	100	100	16.7
JH II	37	7.3	12.6	—
JH III	0.4	0.8	1.7	—
JH 0	—	—	37	—
JH I acid	50	—	100	—
JH I diol	14	—	0.5	100
JH II diol	—	—	—	100
JH III diol	—	—	—	100
JH I isomers (17% *E,E,cis*)	—	—	37.2	—

[a] Modified from Granger and Goodman, 1983.

Baehr and his colleagues subsequently used the method of Lauer et al. (1974) for haptenic linkage of JH I, II, and III to HSA (Baehr et al., 1976, 1979). Using approximately half of the amount of immunogen (600 μg; see Tables 7.1, 7.2) for both the primary and booster injections and increasing the time between injections from 1 to 5 weeks, they obtained antisera exhibiting significantly higher binding affinities for their haptens. However, some cross-reactivity with other homologues was noted (see Tables 7.4, 7.5), thus initially precluding the accurate measurement of a single homologue (JH I) in a mixture. An [^{125}I]-radiolabeled JH derivative was synthesized for use in this assay and was found to enhance its sensitivity slightly. With either the tritiated or radioiodinated JH, binding affinities of the antisera were high enough to permit their use in the quantification of JH in biological samples (see Table 7.3) (Caruelle et al., 1979; Deleurance et al., 1979, 1980; Lensky et al., 1979; Baehr et al., 1982; 1986; Porcheron et al., 1982; Poras et al., 1983; Ismail et al., 1984). Although these particular antisera were initially thought to be sufficiently specific to permit the accurate identification and quantification of JH in hemolymph and whole-body extracts, evidence from other sources has not always confirmed the results (see Section VIII.1).

An alternative derivatization method was utilized by Strambi et al. (1981) for the generation of antisera to JH diol. This approach was undertaken because of two salient characteristics of the JH diol: its increased solubility in aqueous solutions, and its ease of separation from lipids, which are the main source of nonspecific interference in JH RIAs. With this approach, the epoxide ring of JH I was opened to form the diol, which was then succinylated and conjugated to HSA. The antibodies (ABs) raised recognized the diols of each JH homologue equivalently (see Table 7.4), and the antiserum displayed

Table 7.5. Cross-Reactivities of JH III Antisera with JH Homologues and Structurally Related Compounds[a]

Hormone/compound	Cross-reactivity (%)				
	Lauer et al., 1974	Baehr et al., 1979b	Granger et al., 1982	Goodman et al., unpublished	Granger and Goodman, unpublished
JH III	100	100	100	100	100
JH II	50	0.9	0.9	71.0	4.0
JH I	—	0.9	0.6	51.8	2.0
JH 0	—	—	0.6	34.1	—
JH III acid	—	—	77.1	7.3	—
JH III diol	—	—	12	0	—
Methyl farnesoate	—	—	7.9	0	—
Methyl epoxy stearate	—	—	0	—	—
Methyl linoleate	—	—	0	0	—

[a] Modified from Granger and Goodman (1983).

a sufficiently high affinity to yield a particularly useful assay when an iodinated radioligand was also used (see Table 7.3). The sensitivity of the assay permitted its use in the measurement of JH in hemolymph and whole-body extracts (Strambi, 1981; Strambi et al., 1981, 1984; Vancassel et al., 1984; deKort et al., 1985) and JH synthesis in vitro (Strambi, 1981; Strambi et al., 1984). A drawback to this assay, however, is that JH must be derivatized to the diol prior to the assay, since the hormone itself is not recognized by the antiserum.

An RIA has recently been developed by Goodman (unpublished observations), using rigorously defined derivatization and conjugation procedures (see Tables 7.1, 7.2). As in previous assays (Lauer et al., 1974; Baehr et al., 1976), JH III was derivatized through the C-1 position, but thyroglobulin rather than HSA was used as the carrier protein. The relative binding affinity achieved was comparable to that of antisera previously raised to JH derivatized through C-1, when tritiated JH was used as the radioligand in binding studies (see Table 7.3). Interestingly, this antiserum cross-reacted equally well with I, II, and III when enantiomerically enriched (>97% 10*R*, 11*S*) (Prestwich and Wawrzenczyk, 1985) JH II was used as the tracer. More importantly, the antiserum can be used to measure JH in extracts of hemolymph; preliminary titer results have been found to agree with results obtained by gas chromatography–mass spectrometry (GC-MS) (Baker et al., 1987). The use of this same Ag has also resulted in the generation of antisera specific for JH III (Granger, unpublished observations).

The characteristics of the recently developed assays indicate that many of the past problems with JH RIAs have been overcome and that these assays will take their place with other RIAs as indispensable analytical tools.

III. Immunogen Synthesis

In general, compounds with a molecular weight of less than 1,000 make poor Ags. The JHs are thus too small to be intrinsically immunogenic and must be covalently attached to protein molecules to stimulate a specific antigenic response. Preparing the immunogen, which involves derivatizing the hormone and linking it to the carrier molecule, can pose a number of problems for the investigator interested in generating JH antisera for RIAs.

1. Choice of Carrier Protein

Although virtually any protein can be used as a carrier, the protein of choice should be immunogenic and should remain soluble during the conjugation reaction (Erlanger, 1980). Insoluble conjugates can be used; however, characterization of the precipitate is more difficult to perform. The most frequently used carrier proteins are globulin fractions, such as serum albumins. Also used are the hemocyanins from molluscs, ovalbumins and thyroglobulin. There is disagreement as to whether different proteins make a significant difference in the generation of specific antisera. Some investigators have suggested that the choice of protein has little to do with the resulting antisera (Niswender et al., 1976; Erlanger, 1980). However, Raz et al. (1975) demonstrated that antisera raised to a homologous series of prostaglandins with thyroglobulin as the carrier displayed greater specificity than antisera generated with serum albumin as the carrier. Jaffe and Behrman (1979) also noted significant differences in AB titer when different carriers are used. Nevertheless, the differences noted in these two reports may simply reflect the biological diversity of the immunized population, rather than greater immunogenicity of a specific protein. It is clear that more systematic data are needed before the choice of a protein carrier for the JH immunogen can be optimized.

2. Derivatization of the JH Molecule and Covalent Attachment to the Carrier Protein

The molecular structure of the Ag and its structurally similar congeners often dictate the site of attachment as well as the type of conjugation reaction. For example, it has been known for some time that AB specificity is directed primarily at the portion of the hapten molecule most distant from the site of attachment to the protein carrier. Thus, functional groups that make the congeners structurally different from one another must not be involved in the conjugation if the antiserum is to display any specificity. Although JH does not have a diversity of possible attachment sites, there are two functional groups at opposite ends of the molecule that have been exploited for derivation and conjugation: the methyl ester and the epoxide function.

The first step in the derivatization process requires a highly purified Ag, since the hapten will only be as good as the starting material. Although highly purified JH is commercially available as the naturally occurring geometrical isomer (10,11-*cis*), it is not yet available as the naturally occurring enantiomer (10*R*, 11*S*, JH I and II; 10*R* JH III). Thus the availability of highly purified, optically pure JH for the generation of antisera remains a problem; and investigators wishing to use the natural enantiomer will need to obtain it from biological sources or from chemical synthesis until significant quantities are produced commercially. Despite the high degree of purity of the commercially produced geometrical isomer, it should be purified further to eliminate trace quantities of other geometrical isomers and any contaminants. There are several published methods for analytical or semipreparative purification of JH by high-performance liquid chromatography (HPLC), using normal or reverse-phase separation (Schooley, 1977; Goodman et al., 1978a; Mauchamp et al., 1981). Quantification of the hormone is then made spectrophotometrically:

$$\varepsilon^{21}_{\text{methanol}} = 14{,}800$$

(Goodman et al., 1978a).

Covalent attachment of the hapten at C-1 was first reported by Lauer et al. (1974), and this approach has been used subsequently by other investigators with some modifications (Baehr et al., 1976; Goodman, unpublished observations). The first step in derivatization of the hormone for attachment to the carrier protein is the production of milligram amounts of JH acid by base hydrolysis. This is normally performed using NaOH under mild alkaline conditions. Lithium hydroxide has also been used as the base, with apparently more quantitative conversion of JH to the acid (W.-S. Eng and G.D. Prestwich, personal communication). The reaction is pH-, temperature-, and time-dependent, requiring careful monitoring to ensure a high yield. Another procedure for generating JH acid is enzymatic hydrolysis using hemolyph esterases (Metzler et al., 1972). Although this procedure is selective and efficient, it is not practical for hapten synthesis, since hemolymph contains a high level of lipids that would need to be removed for recovery of the acid.

A simple procedure for obtaining JH acids in relatively high yield has been reported by Goodman and Adams (1984). With this procedure, 10–30 mg of the purified hormone is converted to JH acid at a slightly elevated temperature (40°C) and under mildly alkaline conditions (pH ~ 12, methanol: 1 M NaOH, 1:1, v/v). After titration of the reaction mixture to pH 5, the acid is extracted with chloroform: toluene (9:1, v/v) and then purified on an analytical or semipreparative LiChrosorb Diol column with isocratic conditions (chloroform preserved with 1% ethanol). The purified JH acid is quantified

$$\varepsilon^{21}_{\text{acetonitrile}} = 11{,}840$$

and can be stored at −20°C neat for up to 2 weeks without significant breakdown. Long-term storage of JH acid is not recommended, since it results in the appearance of unidentified peaks when the acid is chromatographed as previously described. This procedure yields 70–75% JH acid from the starting material, and the acid has been demonstrated to be free of other contaminants.

After the removal of the carbomethoxy function, the N-hydroxysuccinimide ester of the hormone is formed using dicyclohexylcarbodiimide (DCC) and N-hydroxysuccinimide (NHS) (Lauer et al., 1974; Baehr et al., 1976). Use of DCC may result in low yields of JH-NHS, and a water-soluble form of carbodiimide, 1-ethyl-3-(3-dimethylaminopropyl)-carbodiimide (EDC), has recently been substituted (Goodman, unpublished observations). This compound permits a two-step synthesis of the immunogen rather than the three-step procedure employed by investigators using DDC, and the thin-layer chromatographic (TLC) separation of the reactive ester is eliminated. Thus, further loss of the product and contamination from the TLC plates are avoided.

The earlier syntheses of immunogen (Lauer et al., 1974; Baehr et al., 1976) were performed at pH 9 to 9.5 to ensure that attachment sites, primarily the epsilon amino groups of lysine residues, were ionized. In a more recent procedure (Goodman, unpublished observations), the reaction is first carried out at pH 6 to favor the coupling of JH acid to NHS. After a 6 h incubation of JH acid with EDC and NHS, a catalyst, dimethylaminopyridine, is added with the carrier protein to the incubation mixture, and the pH is brought to 9. The addition of the catalyst, an organic base, results in a doubling in the yield of hapten linked to the protein (Goodman, unpublished observations). After incubation, the low-molecular-weight by-products are removed by exhaustive dialysis, and the degree of incorporation is determined by the radioassay of the internal standard (see subsequent discussion).

Strambi et al. (1981) utilized the epoxy function at C-10 to C-11 as an alternative site for covalent attachment. The epoxide function is opened under mildly acidic conditions (dioxane: 0.25 M H_2SO_4, 15 h, 40°C). The diol is then purified by TLC (chloroform: methanol: water, 80:20:2, v/v/v) and converted to succinylated JH by incubating the hormone with succinic anhydride and triethylamine (40 h, 38°C). The product is purified from the reactants by ion exchange and gel filtration chromatography. The purified succinylated JH is then dissolved in water with HSA and EDC at pH 5.5 and incubated for 4 h. After incubation, the conjugate is repurified by gel filtation, concentrated, dialyzed, and radioassayed to determine the degree of incorporation. One advantage of this synthesis is that the portion of the JH molecule that displays the enantiomeric configuration may not be recognized as an epitope owing to its proximity to the attachment site (Strambi et al., 1984). Thus, attachment through C-10 reduces the need for optically pure JH but may not eliminate it completely, since the chirality of C-10 and C-11 is still retained.

The isoprenoid side chains present other possible sites of attachment, which could theoretically generate some rather interesting antisera. Unfortunately, no linkages have yet been made at those sites, owing to the difficult chemical syntheses involved. Recently a 12-hydroxylated trisubstituted (*Z*)-olefin has been used as a precursor in the development of enantiomerically enriched radiolabeled JH I of high specific activity (Prestwich and Wawrzenczyk, 1985). This route should permit the introduction of a hydroxyl moiety into a chiral intermediate such as C-7 or C-12. Covalent attachment of intermediates of this type, especially through the C-7 position, should result in antisera with which all of the homologues are equally immunoreactive, thus enabling a determination of total JH titers.

The desired degree of antigenic attachment to the carrier protein is an important aspect to be considered during the planning stages for the synthesis of the immunogen. Too few haptens per protein molecule result in unusable antiserum (Niswender and Midgley, 1970), while too many appear to be equally disadvantageous (Carraway, 1979). Most investigators agree that a ratio of one hapten per three attachment sites will yield a high-affinity antiserum (Niswender et al., 1976). Since the serum albumins have 60 lysine residues per molecule available for attachment, and thyroglobulin over 400 (Chard, 1978, 1982), the protein-hapten ratio of the current JH RIAs falls close to the ideal epitope density (see Table 7.1).

There are several methods for determining hapten density. Lauer et al. (1974) used 2,4,6-trinitrobenzenesulfonic acid (TNBS) to determine the degree of attachment; others have determined density by isotope dilution. For isotope dilution analysis, a small amount of radiolabeled JH is mixed with the radioinert hormone, modified accordingly, and covalently attached to the carrier protein. After the reaction is complete, the immunogen is dialyzed exhaustively, and a small sample is radioassayed to determine the number of molecules of hormone covalently attached. The data can also be used to determine the efficiency of the derivatization and attachment.

IV. Antiserum production

The production of antiserum involves several important considerations: the animal species to be immunized, the immunogen dose, the adjuvant, the route of immunization, the schedule of immunization, and schedule of antisera collection (Chard, 1978, 1982).

The choice of animal species for immunization is primarily a function of financial considerations, animal accommodations, and the volume of antiserum required for long-term studies. No matter what species is selected, investigators in the U.S. should be aware of recent changes in federal guidelines dealing with vertebrates and should be familiar with the booklet entitled "Guide for the Care and Use of Laboratory Animals," provided by the NIH. In view of the rather limited number of investigators requiring JH antisera, the use of rabbits for antiserum production is most practical. Since most investigators use randomly bred rabbits that display considerable immunological diversity, as many animals as is practical should be immunized in order to obtain at least one antiserum with good specificity and affinity. It is recommended that 3–6 rabbits be immunized initially and monitored until useful antiserum producers are identified. In our studies (unpublished), six female New Zealand white rabbits approximately 2.5 kg or 4–6 months old were used. Because rabbits producing useful antisera are thus a critical resource, they should be free of the bacterium *Pasturella* and maintained in a *Pasturella*-free facility.

The amount of immunogen to be administered to an animal for optimal antiserum production is not determined on the basis of a simple dose-dependent relationship. Although the literature indicates that the initial dose for rabbit immunization can vary from approximately 20 μg to 1 mg, most investigators use amounts less than 300 μg per animal. The rationale for using lower doses of immunogen derives from the fact that low doses of immunogen stimulate B cells that have high affinity for the hapten; higher amounts of immunogen apparently stimulate lower-affinity B cells, and may result in immunological tolerance (Eisen, 1974). Table 7.2 illustrates this point: animals immunized with more than a milligram of immunogen produced an antiserum with rather low affinity and/or low titers, whereas animals immunized with lower amounts produced antisera with much higher affinities and/or higher titers.

Virtually all immunization protocols call for complete Freund's adjuvant as the vehicle of choice for administration of the immunogen. Complete Freund's adjuvant is a mixture of detergent, mineral oil, and dead mycobacteria; introduction of the immunogen to the immune system as an emulsion in this adjuvant results in steady, long-term exposure of the immunogen rather than one large challenge. Since a water-in-oil emulsion is more effective for slow release than is oil-in-water, the proportion of adjuvant to antigen should ideally be 2 to 1. Emulsification of the immunogen in the adjuvant is also critical; when properly done, the adjuvant-immunogen mixture should remain as a single droplet when gently placed on an aqueous surface. There are several methods for emulsification that avoid significant loss of the immunogen. Immunogen dissolved in saline can be added dropwise to complete Freund's adjuvant with simultaneous mixing in a Vertis homogenizer (Niswender et al., 1976). Emulsification can also be accomplished with the use of two syringes. With this method, the immunogen is drawn into one syringe and the adjuvant into another. Air is expelled, and the two syringes are connected by a double-headed Luer-lok needle. The syringes are then pushed back and forth, forcing the two components to emulsify. In most immunization schedules, complete Freund's adjuvant is used for the initial immunization, and incomplete Freund's adjuvant (without the mycobacteria component) is used for boosters. The dead mycobacteria provide a general nonspecific stimulus to the immune system, which is not necessary in subsequent immunizations (Chard, 1978, 1982).

The primary immunization of the animal can be performed by any of a number of different methods (Garvey et al., 1977); however, the simplest—intradermal and subcutaneous injections—remain the method of choice for most investigators (see Chapter 4). Intradermal injection requires that the immunogen be placed between the layers of skin, a technique that may prove somewhat difficult for both novice immunologist and rabbit. As described by Vaitukaitis (1971), the back of the rabbit is shaved, and the surface is scrubbed with 70% alcohol. The skin is then stretched between the thumb and index finger of one hand, while the syringe barrel is held by the other at an angle of about 30° to the skin surface, with the beveled edge of the needle facing up. The tip of the needle should be inserted just under the outer layer of skin, and the syringe lowered to a position almost parallel to the skin surface. The needle tip is then inserted about a millimeter beyond the bevel, and a small amount (25 μl) of immunogen is slowly injected. Thirty to fifty such injection sites are made across the back, with 0.5 ml of crude *Bordetella pertussis* vaccine injected at a separate site.

The must commonly used method of injection is subcutaneous. For this method, the back is shaved and scrubbed with 70% alcohol. A tent of skin is raised with the thumb and index finger, and the needle is inserted beneath the skin, such that the needle lies parallel to the underlying muscle. The immunogen is then injected, and the needle is quickly withdrawn. Mutiple injections are made across the back. The area of the injections can be expected to abcess and ulcerate but will eventually heal.

Booster, or secondary, immunizations are performed as described above, except that the immunogen is emulsified in incomplete Freund's adjuvant. A booster can also be given by intravenous injection. For this, the immunogen is dissolved in sterile saline rather than adjuvant, since Freund's adjuvant given intravenously is lethal. With the rabbit restrained, the skin over the external marginal ear vein is shaved, and xylene is rubbed over the shaved area to dilate the vein. After being wiped with alcohol, the ear is held in one hand with the thumb clamping the ear against the palm. The syringe is held in the other hand, bevel up and pointing in the direction of blood flow. After the needle is pushed through the skin and into the vein, the syringe is gradually lowered to a parallel position. The needle is held in place by a thumb over the point of entry, and the contents are injected. The thumb should remain in place after the needle is withdrawn, until the blood has clotted.

The blood must subsequently be drawn from an immunized animal, and as with injection of the immunogen, care, patience, and practice are essential. There are several procedures for bleeding a rabbit, but the most commonly used is venous puncture. For this procedure the rabbit is restrained, and the skin over the inner marginal ear vein is shaved and then vigorously rubbed with a pad soaked in xylene. The vein is cut transversely about 2 cm from the distal end, and drops of blood are collected in a centrifuge tube. If the blood flow slows or stops, massaging the ear may reinitiate the flow. This procedure can be shortened by inserting a vacuum bleeding apparatus (Belco Glass Co.) over the ear and applying a slight vacuum; the blood from the cut vein collects in the attached centrifuge tube.

Collected blood should be allowed to clot at room temperature for 1–2 h. The clot is then carefully separated from the wall of the tube with a metal spatula, and the tube is cooled to 4°C for 24 h to permit clot contraction. The serum is then decanted into centrifuge tubes and spun at 1,000*g* for 30 min. After centrifugation, the serum is carefully decanted from the cellular pellet and stored at −20°C in aliquots of a useful volume.

A survey of the literature indicates that there are numerous schedules for boosting and bleeding immunized animals. Differences in Ags and in immunological responses between individuals preclude the detailing of a specific timetable (Table 7.2). A general protocol that has proved effective and convenient (unpublished observations) is to inject the immunogen and wait a month before injecting the first booster. A small amount of serum is drawn 2 weeks after the booster and examined for activity. Thereafter, the titer of the serum is monitored monthly. When the titer falls, another booster is given. The reaction of each animal is variable, and the production of useful antiserum may take some time. Goodman et al. (unpublished observations) found that one animal immunized with JH-thyroglobulin (104 JH/thyroglobulin) immunogen and boosted on a monthly schedule did not produce usable antiserum (1:2,000 dilution) until day 135 after immunization. The same carrier with 178 JH haptens per molecule elicited a much quicker response (Granger and Goodman, unpublished observations). Previous studies using a JH-HSA immunogen reported usable antisera within 60 days of the primary injection (Lauer et al., 1974; Baehr et al., 1976; Strambi et al., 1981).

Because the specificity and affinity of an antiserum are critical to the development of a successful RIA, these parameters should be monitored for each animal at each bleeding. Moreover, the titer of a specific antiserum first rises and then falls dramatically after each boost; thus to optimize the serum titer, it is important that blood be drawn at the time of the rise. Specificity of the antiserum has also been reported to change during an immunization program (Abraham and Grover, 1971; Eisen, 1974). Antiserum drawn later in the immunization schedule usually displays a higher affinity for the ligand than that drawn earlier; however, the increase in affinity is often offset by an increase in cross-reactivity.

V. Tracer Ligand

Determining the distribution of bound and unbound ligand is of critical importance to any binding assay. Since RIAs are performed at concentrations that are orders of magnitude below the direct detection limits of a hormone, the hormone must be labeled for detection. Any type of label can be used, including radioisotopes, enzymes, phage, and fluorescent and luminescent agents, as long as the labeled hormone is homogeneous and easily detected and behaves identically to the unlabeled hormone in the assay system. For example, Baehr et al. (1987) have developed an enzyme immunoassay (EIA) for JH that uses as the tracer ligand acetylcholinesterase covalently attached to a modified JH molecule. At the present time, however, only the radiolabeled tracers are both highly labeled and readily available; thus discussion will focus on the use of this type of tracer in a JH immunoassay.

There are two types of radiolabeled tracer: Those in which the radiolabeled atom is an integral part of the molecule are termed internally labeled tracers, and those that have the radiolabel covalently attached to a modified portion of the molecule are termed externally labeled tracers. Both internally and externally radiolabeled JHs have been used in RIAs. The first radiolabeled JH to be used in an RIA was a commercially available tritiated JH I, which had a tritium atom incorporated into the side chain of C-7. More recently, the replacement in JH I, II, and III of a hydrogen at the C-10 position with a tritium atom has offered new opportunities for the development of highly specific assays. Unfortunately, none of these labeled JHs are of particularly high specific activity, nor are they enantiomerically pure. Moreover, the labeling pattern of the available JHs makes them susceptible to proton exchange under mildly acidic conditions. Thus, they are unfavorable internal markers for monitoring reactions performed under acidic conditions, such as diol formation. Recently, Prestwich and Wawrzenczyk (1985) synthesized high-specific-activity JHs (58 Ci mmol^{-1}) which are enriched (95%) for their naturally occurring 10*R*, 11*S* enantiomers. Since the positions of the radioisotopic labels in these JHs are at C-12, proton exchange is presumably less likely to occur. The development of these enantiomerically enriched, high-specific-activity JHs should pave the way for the widespread use of JH RIAs.

JH has also been externally labeled using the radioisotope ^{125}I (Baehr et al., 1976; Strambi et al., 1981). In contrast to the tritiated form of JH, the iodinated form permits use of a considerably smaller mass of radioactive Ag and results in a more sensitive assay system. Baehr et al. (1976) iodinated JH by conjugating it to histamine labeled with ^{125}I; the resulting JH derivative had a specific activity close to the theoretical limit of $\sim$2,000 Ci mmol^{-1}. The method employed utilized chloramine T, which is a potent oxidizing agent capable of converting iodide to a more reactive form, and followed the procedures outlined for synthesis of the immunogen. With this relatively simple protocol, JH is first converted to the acid, then derivatized to the reactive JH-NHS ester (see section III). The JH-NHS is then covalently attached to histamine, and the histamine is iodinated using chloramine T. Iodination is achieved by mixing radiolabeled sodium iodide, chloramine T, and JH-NHS in a reaction vessel for 1 min and then stopping the reaction by the addition of sodium metabisulfite. The resulting tracer is purified by TLC. With a very similar method, Strambi et al. (1981) iodinated tyrosine rather than histamine as the source of the radiolabel for JH. A comparison of the assay using the iodinated tracer with that using the tritiated form indicates that sensitivity is increased with the [^{125}I]-JH, although not as much as one might anticipate. In one case, the I_{50} (the amount of unlabeled ligand resulting in 50% displacement of the label) was reduced from 300 to 200 pg (Baehr et al., 1976); in the other (Strambi et al., 1981), the lowest detectable quantity was reduced to 5 pg/tube. Since the lowest detectable quantity with assays employing the tritiated hormone is 20 pg (see Table 7.3), the use of iodinated tracers may prove useful; and with the recent synthesis of a JH I analogue internally labeled with ^{125}I (Eng and Prestwich, 1986), the availability of internally iodinated JHs for use as tracers appears imminent.

The chemistry of the JH molecule makes the labeled JHs of high specific activity particularly susceptible to radiodecay. Damage to the tracer can be incurred from the spontaneous disintegration of the radioisotope. Furthermore, radiation emanating from the decay can damage nearby hormone molecules. It is therefore recommended that the highly labeled hormones be stored diluted and at low temperature in distilled and degassed toluene or toluene-heptane (1 : 1, v/v) to minimize the damage (Goodman, unpublished; Prestwich, personal communication). Strambi et al. (1981) found that dilution of their externally labeled, radioiodinated JH tracer into buffer containing hemoglobin (2 mg/ml) preserved full immunoreactivity up to 6 months. The radiopurity of stored, labeled hormone should be checked monthly, and the hormone should be repurified by HPLC if impurities amounting to more than a few percent of the total have accumulated.

VI. Separation of Bound and Unbound Ligands

The separation of bound and unbound hormone in an RIA can be accomplished by many different procedures, but not all of them are applicable to JH RIAs. Various techniques that have been used to separate bound and unbound JH include secondary antibody precipitation (Lauer et al., 1974), dextran-coated charcoal adsorption (Goodman et al., 1976; Kramer et al., 1976), polyethylene glycol precipitation (Baehr et al., 1976, Koeppe et al., 1981), hydroxylapatite adsorption (Goodman et al., 1978), equilibrium dialysis (Goodman and Gilbert, 1978; Strambi et al., 1981), and ammonium sulfate precipitation (Granger et al., 1979).

Ammonium sulfate (Granger et al., 1979, 1981; Goodman, unpublished observations) and polyethylene glycol (Baehr et al., 1976) are both currently used to separate bound and unbound fractions in JH RIAs, with the same physicochemical principles inducing AB precipitation in each case. With this type of assay termination, the AB-hormone complex is precipitated while the unbound hormone remains in solution. The precipitated AB is then centrifuged to form a compact pellet, the supernatant is removed, and the pellet is resuspended in scintillation fluid for radioassay (see Fig. 7.1 for details). Both ammonium sulfate and polyethylene glycol disrupt the water layer around the AB molecule, so that the hydrophobic domains on the surface of one molecule interact with like domains on the surfaces of other molecules, leading to the formation of large aggregates, which then precipitate. The unbound JH remains in solution, and several wash-centrifugation cycles remove any remaining unbound JH from the precipitate. The advantages of this type of separation are low cost, simplicity, and speed. Moreover, the procedure is carried out in a single disposable tube, thus eliminating potential contamination and the need to transfer samples.

Another method of separating bound from free hormone that merits some discussion is equilibrium dialysis (Strambi et al., 1981). In this procedure, the radiolabeled tracer is placed in one chamber of a commercially produced equilibrium dialysis unit, and diluted antiserum is placed in the other. The two

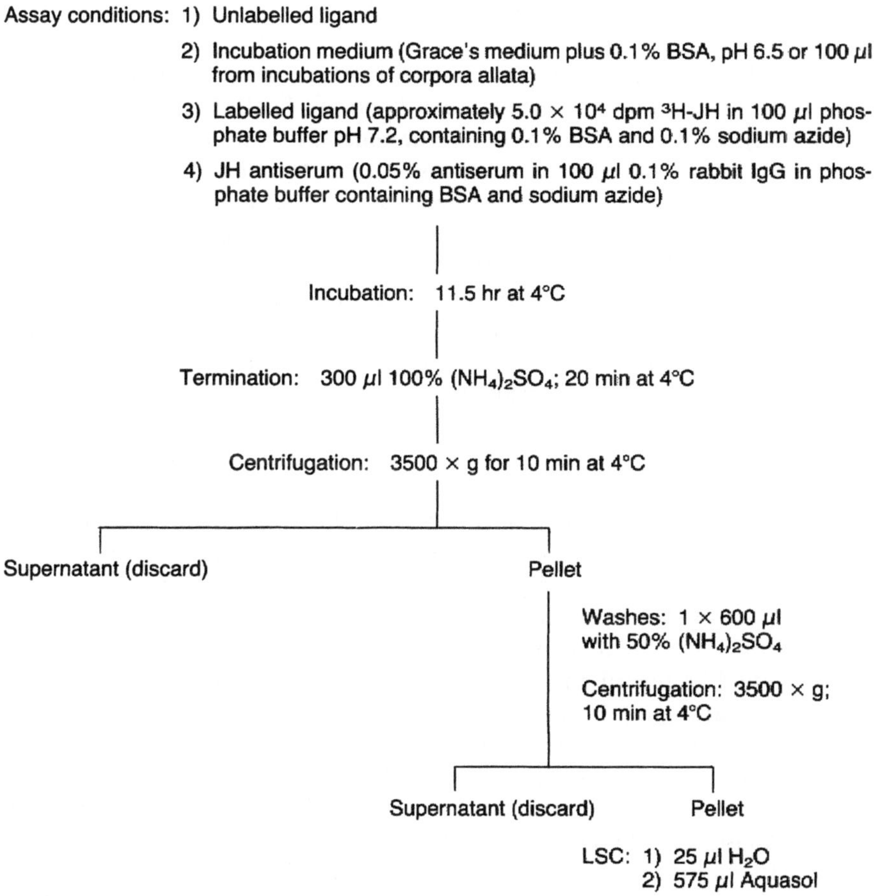

Figure 7.1. Protocol for a JH radioimmunoassay (modified from Granger et al., 1981).

chambers are then clamped together with a semipermeable membrane separating them, to allow free diffusion of the hormone between the two chambers. The AB is too large to cross the membrane and thus is limited to one side of the unit. Equilibrium conditions are eventually established (22 h), with the concentration of unbound JH diol equal in both chambers. The unit is then opened, and a sample from each chamber is radioassayed. To determine the amount of hormone bound, counts from the chamber containing no antiserum are subtracted from those in the chamber containing antiserum. Although the method clearly establishes binding equilibrium, the process is tedious, more costly than ammonium sulfate or polyethylene glycol precipitation, and lengthy. Moreover, contamination is a possible problem, since the chambers are reused.

VII. Preparation of Samples for RIA

The ease of sample preparation for an RIA remains one of the most attractive aspects of this technology, since frequently no purification is necessary prior to the assay. Unfortunately, this is not the general situation with the JH RIA. The nature of the hormone and the high concentrations of interfering substances in biological samples have made at least some sample preparation imperative in most cases. The guidelines offered by Chard (1978, 1982) provide a generally applicable basis for the development of a sample purification procedure: (1) the procedure should be rapid and simple, since there are often many samples to be assayed at one time; (2) the procedure should concentrate the ligand into a volume considerably smaller than the original volume in order to increase sensitivity; (3) the process should not concentrate or introduce nonspecific interfering substances; (4) recovery of the starting material should be greater than 50% and, most importantly, should be reproducible; (5) the procedure should not cause degradation of the hormone; and (6) the adsorbents used in the purification should be readily available and not vary in performance from batch to batch. There are two additional caveats specifically pertaining to the preparation of JH samples. First, the purification protocol should be sufficiently rigorous to strip the hormone from hemolymph and cellular binding proteins, and second, the method of purification should be reverified each time JH is titered in tissue from a different developmental stage. This last is very important, since the biochemical composition of the insect changes dramatically during development.

Stock hormone solutions used in the RIA for the competition curves are generated by spectroscopy (Goodman et al., 1978). Storage of JH has been described elsewhere (Schooley, 1977), but some mention of the stability of stock JH solutions must be made. Many laboratories routinely use JH dissolved in methanol or ethanol for stock concentrations. Stored at −20°C, very little breakdown occurs over a 3-week period as determined by normal-phase HPLC; however, long-term storage in methanol is not recommended, since contaminants begin to appear (Goodman, unpublished). However, milligram quantities of JH can be stored at −20°C for long periods of time, neat or in HPLC-grade hexane, with little for no breakdown (Goodman, unpublished).

In bioassay and physicochemical assays, the addition of radiolabeled JH prior to sample purification offers an excellent means of monitoring hormone recovery. However, the presence of radiolabel other than the tracer in the RIA can prevent a precise determination of the amount of hormone bound. This problem can be circumvented in several different ways. If sufficient biological material is available, it can be divided and radiolabel added to one portion to monitor the efficiency of the purification procedure. If iodinated JH derivatives are to be used as the tracer, the tritiated form of the hormone can be employed to determine the recovery, provided that no tritium-proton exchange occurs during sample preparation.

Sample preparation presents a real dilemma for the JH investigator. On the one hand, a single-step extraction of JH for RIA analysis has generally proven insufficient for the removal of sources of nonspecific inhibition present in insect tissue. On the other hand, more complex procedures of purification often introduce nonbiological factors that may inhibit binding nonspecifically or reduce hormone recovery to unacceptable levels. These factors fall into a broad category termed nonimmunological inhibition, and an understanding of these factors is necessary if accurate, sensitive, and specific assays are to be routinely performed.

In aqueous solutions, JH is highly surface active (Law, 1980) and has a marked propensity for binding to untreated glass and plasticware. All glassware coming in contact with aqueous solutions of JH must be coated with polyethylene glycol to reduce nonspecific binding (Goodman et al., 1976; Giese et al., 1977). At no time should the unbound hormone come in contact with plasticware. Except for Hamilton microsyringes, storage volumetric flasks, and reactivials, all glassware should be disposable. When glassware (except syringes) is to be reused for JH studies, it must be rinsed in methanol, then scrubbed thoroughly with Alconox (Alconox Inc., New York) and rinsed extensively. The glassware should then be soaked overnight in 1 M HCl, rinsed in 1 M NaOH, rinsed again several times in double-distilled water, and then heat-treated for at least 4 h at 500°C. Syringes should be rinsed exhaustively in several organic solvents to remove contaminants and should be segregated from those in general use to avoid cross-contamination. Syringes that are used to transfer microgram quantities of the hormone should never be used for the transfer of smaller amounts. The apparatus for evaporating solvent carriers, usually a battery of long cannulas attached to a nitrogen source, should be wiped down before each use to remove any hormone that has been inadvertently splashed onto the needle (deKort et al., 1985).

Solvents and chromatographic material represent another potential source of nonimmunological inhibition. HPLC grade solvents should be routinely used for extraction and standard-curve development. Chromatographic substrates such as TLC plates or Sep-Paks (Waters Assoc., Milford, MA) should be washed thoroughly prior to use. Indeed, the substrates themselves should be monitored to ensure that they are not contributing nonspecific inhibitors to the sample. For a concise discussion of nonspecific factors influencing the RIA, the reader is referred to Abraham and Grover (1971). More extensive reviews can be found in Chard (1978, 1982).

Several procedure for extraction of JHs from hemolymph and whole bodies have been interfaced with a JH RIA (Baehr et al., 1979; Strambi et al., 1981, 1984). One that has proven particularly effective (Goodman, unpublished observations) is a general procedure for the extraction of JH from both hemolymph and whole-body extracts developed by Bergot et al. (1981a). According to this protocol, approximately equal portions of hemolymph are added incrementally to acetonitrile at room temperature, with mixing after each addition. The material is briefly centrifuged to sediment the precipitate.

The supernatant is rapidly transferred to a separatory funnel containing 40 ml pentane and 150 ml 4% aqueous NaCl. The tube contents are then extracted three times with 5 ml of acetonitrile. The combined supernatants remaining in the separatory funnel are shaken, the aqueous layer removed, and the organic phase washed with 50 ml of brine. The pentane phase is transferred to a conical bottom vial, and the solvent is evaporated on a rotary evaporator. The oily pentane-soluble material is redissolved in pentane and transferred to a reactivial. The solvent is then evaporated under N_2, and the extract is redissolved in a small amount of pentane before being loaded onto a normal-phase TLC plate (silica gel G; E.M. Merck) that has been prewashed once in ethanol and once in developing solvent. The plate is developed in hexane : chloroform : ethyl acetate (7 : 7 : 1, v/v/v). The JH-containing fractions are scraped from the plate into Pasteur pipettes packed at the bottom with approximately 1 cm of delipidated cotton. JHs are then eluted with 4 ml ethyl acetate and dried under N_2. The extract is redissolved in hexane, transferred to the appropriate assay tubes, and processed for RIA. With this procedure, the final recovery of JH added to hemolymph from wandering-stage *Manduca* larvae (day 5, last instar) is 60%. Since JH is not normally present in the hemolymph at this stage (Baker et al., 1987), the hemolymph extract thus serves as a good biological blank to monitor nonimmunological inhibition.

Whole insects (5–15 g) are anesthetized and placed in a 100-ml blending flask containing 40 ml acetonitrile and an amount of Celite equal to one-third the sample weight of the insects. The contents are homogenized in a chilled high-speed blender for 3–5 min. The homogenate is then filtered through a fritted glass funnel, and the retained Celite cake is rehomogenized and filtered. The combined filtrates are evaporated under vacuum to 20 ml and then decanted into a separatory funnel for the pentane-brine partition as described in the previous paragraph. The final steps in processing the extract are the same as those for hemolymph.

VIII. Establishment of Reliability

Once the antiserum has been developed and the methods of extraction have been defined, the investigator must then consider three important parameters: accuracy, sensitivity, and specificity.

1. Accuracy

Of central importance to any assay not based on physicochemical principles is its accuracy. In the case of the JH RIA, the assay-determined value must reflect the JH physically present in the sample rather than a combination of JH and nonspecific factors. Thus it was of considerable concern (see Tobe and Feyereisen, 1983) when comparisons of RIA-derived JH hemolymph titers with titers determined by GC-MS (Bergot et al., 1981) revealed that the RIA measurements were consistently 10- to 100-fold greater. Moreover, cross-reactivity problems apparently led to difficulties in the identification of the correct homologue. For example, changes in the titers of JH I, II, and III were measured by RIA in the hemolymph of *Locusta migratoria*, *Leptinotarsa decemlineata*, and *Schistocerca gregaria* (Deleurance et al., 1979, 1980; Papillon et al., 1980; Caruelle et al., 1981), whereas only JH III was detected by physicochemical procedures (Bergot et al., 1981; Rembold, 1981; deKort et al., 1982).

In the past, simultaneous comparisons of RIA and physicochemical data to determine assay accuracy have been done with difficulty, since such studies require the close collaboration of the RIA operator with one of the few laboratories routinely performing GC-MS analyses of JH titers. In one such study (deKort et al., 1985), RIA-derived JH titers reported for the hemolymph of *Leptinotarsa* adults agreed remarkably well with titers previously determined by GC-MS (deKort et al., 1982). In an effort to standardize JH quantification between laboratories, the various methods used by five different laboratories to quantify JH were directly compared (J. Koeppe, unpublished observations). Each laboratory titered samples that came from a single source and that included JH standards, freeze-dried aliquots of hemolymph, and freeze-dried medium from incubations of corpora allata. As subsequently reported at the Fourth International Conference on the Juvenile Hormones (Koeppe, unpublished), the various assays, which were not identified but which included both RIA and physicochemical methods, showed as much agreement among themselves as a previous comparison between RIA and GC-MS (Granger and Goodman, 1983). The values for all the determinations ranged from 0.05 to 2.7 times the actual concentrations. In only three instances, each determined by a different method and laboratory, was the given amount ($\pm 10\%$) actually the value arrived at.

2. Sensitivity

A second parameter of central importance to the RIA is sensitivity, defined as the lowest concentration of unlabeled ligand yielding displacement that is statistically significant from the value obtained without unlabeled ligand (Chard, 1978, 1982). Thus, sensitivity is, in part, a measure of both the affinity of the antiserum and the variation in samples of the same concentration. Minimal sensitivity has often been interpreted as the first concentration of unlabeled ligand with a displacement value different from that of no ligand, but this is only correct if the two points are significantly different.

Optimization of sensitivity is generally achieved by using $8-15 \times 10^3$ cpm radiolabeled ligand per assay and determining the minimum concentration of AB needed to give 30–50% binding of the radioligand (Parker, 1976). The result gives the minimum concentration of radiolabeled Ag and AB needed to provide a level of AB-bound radioactivity from which displacement can be reproducibly measured. The apparent sensitivity of a system can be manipulated to some extent by the use of nonequilibrium assay conditions, small incubation volumes, or long counting time, but the use of a radiolabeled ligand of high specific activity is the most critical factor determining the sensitivity of an assay. The RIAs employing commercially available radiolabeled hormones (JH I, II, III) have two problems that impair both their sensitivity and accuracy: the relatively low specific activity of these tracers (11–13 Ci $mmol^{-1}$), and their racemic compositions. Thus, the extent to which the natural enantiomer in a sample of tissue extract or incubation medium displaces the racemic labeled JH may not accurately reflect the concentration of JH in the sample.

A comparison of the sensitivities of existing JH RIAs is presented in Table 7.3. Interestingly, with the exception of the Lauer et al. (1974) study, these data are not vastly different for the different assays, despite the fact that the specific activities of the radiolabeled ligands used varied considerably. Thus, the sensitivities of those assays using the commercially available tritiated JHs (Baehr et al., 1976; Granger et al., 1979, 1982; Goodman, unpublished observations; Granger and Goodman, unpublished observations) are not significantly less than those using the [^{125}I]JH at specific activities reported at or close to the theoretical limit of ~2,000 Ci $mmol^{-1}$ (Baehr et al., 1976, 1979; Strambi et al., 1981, 1984). For the studies by Baehr et al. (1976, 1979), JH was iodinated through the attachment of an iodinated histamine, and it is possible that significantly better sensitivity was not achieved because the hormone was racemic and highly derivatized. It is also possible that the exceptionally high specific activity of this radioisotope increased the radiochemical decay of the nuclide and caused radiolytically induced chemical degradation. The RIA of Strambi et al. (1981, 1984) also used an iodinated radioligand, succinyl-JH-glycyltyrosine, which is similar to the derivatized JH used for conjugation. In this case, the sensitivity of the assay was somewhat improved from that achieved by the other assays.

Recently, enantiomerically enriched (>95% 10*R*,11*S*) JHs and their antipodes, both labeled (58 Ci $mmol^{-1}$) and unlabeled, have been synthesized by Prestwich and his colleagues (Prestwich and Wawrzenczyk, 1985; Prestwich et al., 1987). Granger and Janzen (unpublished) have compared the sensitivities of JH I RIAs using either (10*R*,11*S*)[^{3}H]JH I (58 Ci $mmol^{-1}$) with (10*R*,11*S*) JH I as the competing unlabeled ligand or racemic [^{3}H]JHI (New England Nuclear, 13.5 Ci $mmol^{-1}$) with unlabeled racemic JH I (Fig. 7.2).

With the natural hormones, the assay had a lower limit of detection of 5 pg and and I_{50} of 0.18 ng, compared to 25 pg and 0.50 ng, respectively, for the assay with racemic JHs. Thus, the approximate four-fold increase in specific activity enhanced the sensitivity of the assay about two- to three-fold. Furthermore, 10 times less antiserum (0.005%) was required for the assay using the natural hormones. The use of a higher specific activity radiolabeled JH can

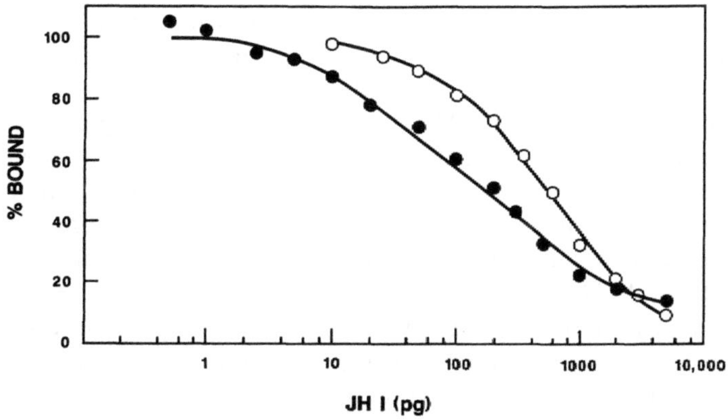

Figure 7.2. JH I RIA standard curves generated with (10*R*,11*S*)-[^{3}H]JHI (58 Ci mmol^{-1}) and (10*R*,11*S*) JHI as the unlabeled ligand (●) or with racemic [^{3}H]JH I (New England Nuclear; 13.5 Ci mmol^{-1}) and racemic JH I as the unlabeled ligand (o). Antiserum used was DB 195 (Granger et al., 1979).

thus significantly increase the sensitivity of the RIA, although it appears that there are limits to the degree of enhancement.

3. Specificity

The problem of assay specificity is another important parameter that must be thoroughly understood by the RIA operator if a quantitative assay is to be performed. Specificity is determined by the degree to which the antiserum binds substances other than the hapten. All nonspecific binding falls into one of two categories: specific nonspecificity and nonimmunological inhibition (Chard, 1978). Specific nonspecificity in a JH RIA can be evaluated by running standard curves with radiolabeled Ag and physicochemically similar Ags that might be suspected of cross-reactivity. Nonimmunological inhibition refers to interference by substances that do not directly interact with the AB but nevertheless affect binding. For example, denaturation of the AB by substances in a biological extract would result in less binding of the labeled ligand, but for a reason completely unrelated to displacement. For a detailed examination of specificity and its interpretation, the reader is referred to Chard (1978, 1982).

The specific nonspecificity of the JH RIA is affected to some extent by the fact that ABs to hydrophobic haptens such as the JHs tend to be more immunologically cross-reactive with structurally related haptens than are those to more polar compounds (Parker, 1976). Furthermore, some lepidopterans have circulating titers of two or more of the five structural homologues of JH thus far identified in this order (Schooley et al., 1984). Therefore, unless the operator isolates each of the homologues prior to the assay, the usefulness of a particular JH antiserum for measuring absolute hormone concentrations may be determined largely by its specificity for a single homologue in a mixture.

The specificities of existing JH I antisera have been compared previously (Granger and Goodman, 1983) (Table 7.4). This comparison revealed that the JH antisera raised against an immunogen derivatized at C-1 exhibited considerable specificity for the ethyl group at C-7, for the configuration around the epoxide, and for certain geometrical isomers (Baehr et al., 1976, 1979; Granger et al., 1979). Lack of recognition of the methyl ester also characterized these antisera, with different lots displaying up to 100% cross-reactivity with JH I acid (Baehr et al., 1976; Granger et al., 1979). This particular lack of specificity is proving to be an extremely useful feature of the assay, since there are stages in the development of both *Manduca* and *Galleria* when the corpora allata produce JH acid (Bhaskaran et al., 1986; Granger and Janzen, unpublished observations; Sehnal and Rembold, unpublished observations). The other major JH metabolite, JH I diol, cross-reacted with the antisera to a far lesser extent. The antiserum of Strambi (1981, 1984), raised against a succinylated and HSA-coupled diol of JH I, recognized the diols of JH I, II, and III equivalently. The only other substances shown to have significant levels of cross-reactivity with this antiserum were JH I, iso-JH II diol, and JH III methoxyhydrin; the diol of JH 0 was 60-fold less immunoreactive than the other diols. These binding studies demonstrate that JH can only be measured by this assay after derivatization to the diol form; however, the sensitivity of the assay and its equal recognition of the different diols are extremely useful features for the measurement of total JH or any one of the homologues, provided it is purified prior to the assay.

Antibodies are known to discriminate between optical and geometric isomers, and the question of these types of discrimination by antisera generated to racemic JH has been considered previously (Granger and Goodman, 1983). It is known, for example, that antisera generated to racemic JH contain subpopulations of ABs that can recognize the configuration around the epoxide. Optical specificity has recently been examined in more detail for two different antisera, and different results were obtained. One antiserum (Goodman, unpublished observations) was found to recognize the unnatural (10*S*,11*R*) enantiomer of JH I better than the naturally occurring (10*R*,11*S*) enantiomer, indicating enantiomeric specificity. By contrast, antiserum from a different source (DB 195, Granger et al., 1979; Granger and Janzen, unpublished observations) was found to bind both the 10*R*,11*S* and 10*S*,11*R* enantiomers and the racemic mixture equally (Fig. 7.3). This apparent lack of enantiomeric specificity is further supported by the results of a comparison of the amounts of JH I or JH I acid synthesized in vitro by last-instar *Manduca sexta* corpora allata, measured by RIAs using either racemic labeled (13.5 Ci $mmol^{-1}$) and unlabeled JH I or (10*R*,11*S*) labeled (58 Ci $mmol^{-1}$) and unlabeled JH I (Table 7.6). No statistical differences were observed between

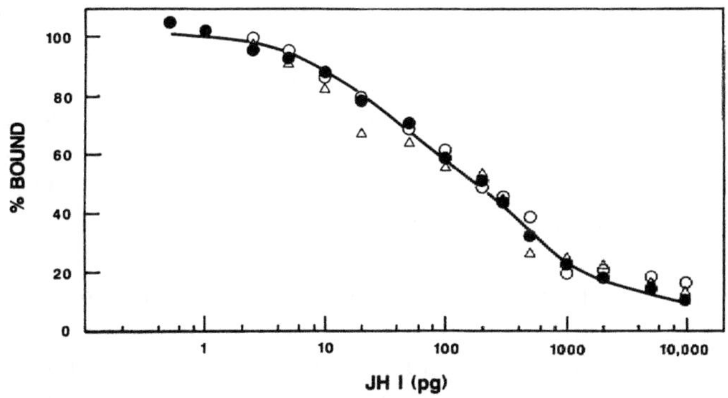

Figure 7.3. Competitive binding analysis of the specificity of a JH I antiserum (DB 195) for JH I enantiomers. The labeled ligand was (10*R*,11*S*)-[^{3}H]JH I (58 Ci $mmol^{-1}$). The unlabeled ligands were (10*S*,11*R*) JH I (Δ), (10*R*,11*S*) JH I (•), and racemic JH I (o).

Table 7.6. Amounts of JH I/JH I Acid Synthesized In Vitro by *Manduca sexta* Corpora Allata from Different Stages of the Last Instar as Measured by RIAs For JH I Using Either Racemic or (10*R*,11*S*) [^{3}H]JH I

Stage of development[a]	[^{3}H]JH I[b]	JH I equivalents[c]
V/0	Racemic	0.88 ± 0.14
	10*R*,11*S*	0.73 ± 0.16
V/4	Racemic	0.44 ± 0.13
	10*R*,11*S*	0.40 ± 0.05
V/6	Racemic	1.49 ± 0.09
	10*R*,11*S*	1.55 ± 0.17
V/7	Racemic	1.75 ± 0.13
	10*R*,11*S*	1.81 ± 0.17

[a] V/0 = day 0, fifth instar; V/4 = day 4, fifth instar; V/6 = day 6, fifth instar; V/7 = day 7, fifth instar.
[b] Either racemic [^{3}H]JH I (13.5 Ci $mmol^{-1}$) and racemic JH I as the unlabeled ligand or (10*R*, 11*S*) [^{3}H]JH I (58 Ci $mmol^{-1}$) and (10*R*, 11*S*) JH I as the unlabeled ligand were used in the standard curves.
[c] ng 2 pair CA^{-1} 6 h^{-1} ± SEM.

the amounts of JH I/JH I acid measured by these two assays in incubations of glands from several different stages during the last instar. Nevertheless, an antiserum to racemic JH can theoretically recognize the optical isomers differentially (see above), and, depending on assay conditions, its use could lead to inaccurate quantification of the natural hormone. One means of circumventing the need for antiserum to the natural enantiomer would be to use the radiolabeled and radioinert forms of the 10*R*,11*S* enantiomer. In this assay, AB subpopulations recognizing the 10*S*,11*R* isomer would be unable to bind the radiolabeled natural enantiomer.

Varying degrees of specificity for the JH homologues, degradative products, and analogues have been found for JH III antisera covalently attached at C-1 (Table 7.4). Two of these studies demonstrate that the choice of the radiolabeled ligand for an RIA can significantly affect the cross-reactivities observed. When [^{3}H]JH I was used as the labeled ligand with the JH III–directed antiserum of Lauer et al. (1974), the antiserum bound JH I and JH II equivalently and displayed 50% cross-reactivity with JH III. When [^{3}H]JH III was used, significant cross-reactivity was observed only with JH II (50%) and a succinylated methyl ester of JH III (83%). The results of a recent specificity study (Goodman, unpublished observations) indicate that the radiolabeled ligand can be chosen to exploit substantial cross-reactivity between homologues in order to determine total JH. When high-specific-activity JH II (58 Ci mmol^{-1}) was used as the radiolabeled ligand with a JH III–directed antiserum, JH I, II, and III were bound equivalently. Lauer et al. (1974) also observed that when the same antiserum was used with a different radiolabeled ligand, its degree of specificity for different homologues was significantly altered.

Nonspecific immunological inhibition is a more difficult parameter to assess. In the case of the JH RIA, it is the probable source of most of the problems with quantification, since this type of nonspecificity generally results in an overestimation of the hormone (see section VIII.1). There are several reasons for nonspecific nonimmunological inhibition, all of which can be traced to the extracted non-JH material in a biological sample: destruction or sequestration of the antibody, tracer, or unlabeled ligand; and interference with the AB-Ag interaction (see section VII) (Chard, 1978, 1982). The most commonly used approach for detecting nonspecificity of this type is a test for parallelism between dilutions of the biological extract and the standard.

IX. Analysis of RIA Data

Until the availability of the personal computer (PC) for routine data reduction in the laboratory, the vast majority of RIA results were calculated by manual interpolation. With this convention, which is still generally the first used by the novice, the RIA standard curve is expressed as a semilogarithmic plot of percentage tracer bound against concentration of unlabeled ligand (see Fig. 7.4A). The dependent variable can also be expressed as the ratio of bound to unbound ligand (see Fig. 7.4B), unbound to bound, or as one of several other variations (Chard, 1978, 1982), the choice of which is determined mainly on the basis of personal taste or experience. The quantity represented by each of the unknown values is then estimated from the standard curve by simple interpolation from percent tracer bound (bound/unbound) in the presence of the unknown (ordinate) to concentration of ligand (abscissa).

Another method of plotting the typically nonlinear RIA standard curve

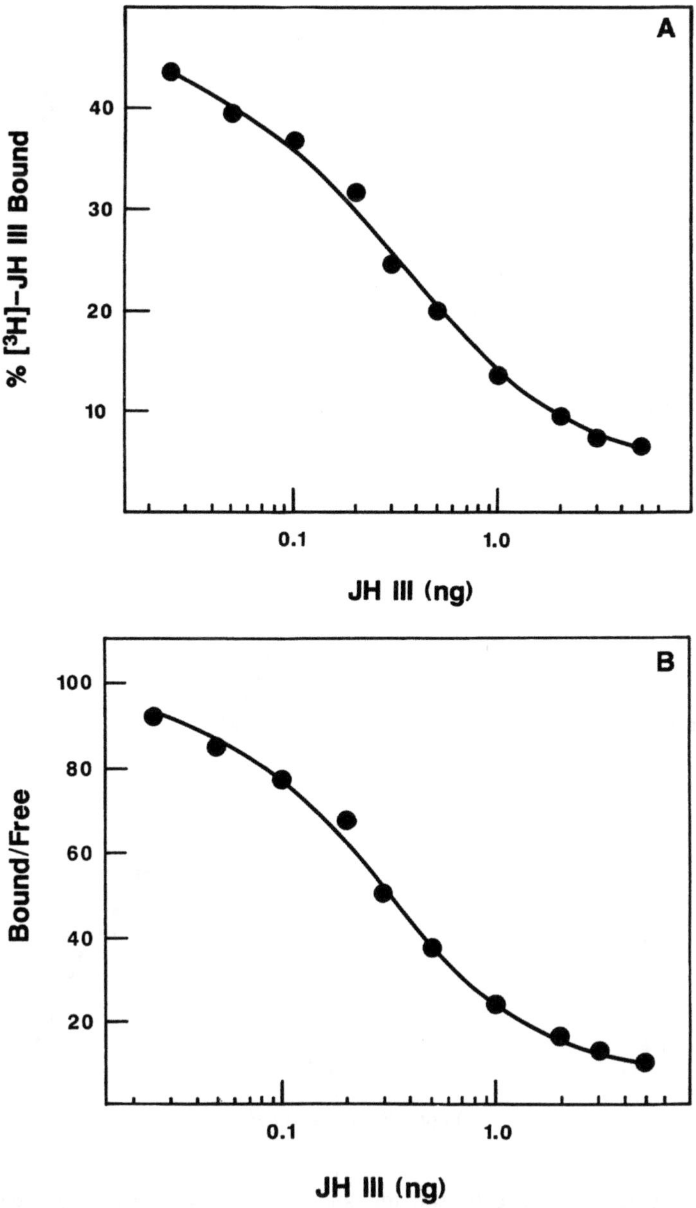

Figure 7.4. JH III RIA standard curve expressed as a semilogarithmic plot of percentage tracer bound against concentration of unlabeled JH III (A); the ratio of bound to free ligand against the concentration of unlabeled JH III (B); and a log-logit plot (C) (Rodbard and Lewald, 1970). The log-logit plot is a straight line, computer-derived from a logit transformation of the standard curve data: logit b = $\log_e$ where b is the proportion of tracer bound expressed as a percentage of the zero standard (+ = datum point; • = datum point falling on the derived plot).

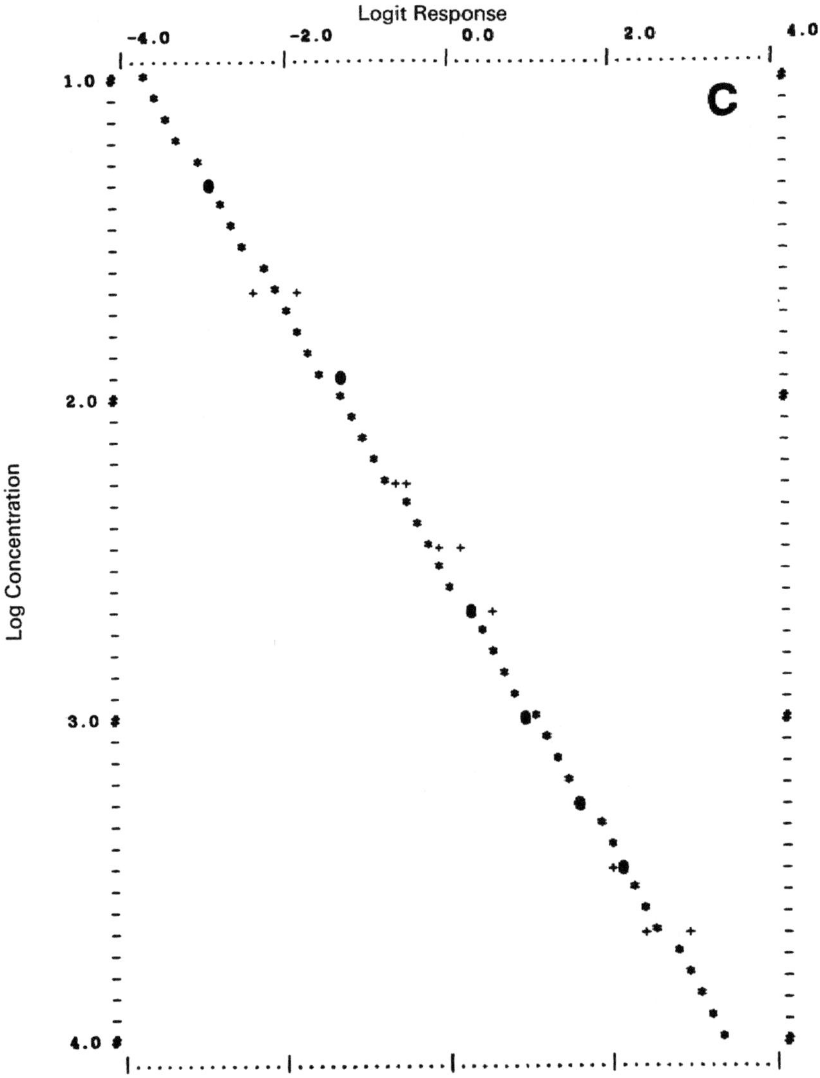

Figure 7.4 (*continued*)

involves fitting the curve to the measured standard points by means of a weighted least-squares method (Naus et al., 1977). However, the most accurate interpolation of unknown values is achieved when the dose response curve is converted from sigmoidal (or hyperbolic) to a straight line. There are several transformational formulae that can be used to achieve this conversion, the most widely used being the logit (logit-log, log-logit) transform introduced by Rodbard et al. (1969; see also Rodbard and Lewald, 1970) (Fig. 7.4C). The linearization of the standard curve may then be smoothed by a spline function

(Rodbard, 1971; Marschner et al., 1975). Because of the lengthy computations involved, these more complex data reductions were not routinely used until the advent of the PC. The increased precision of RIA data analysis achieved with these techniques, particularly for values at the extremes of the standard curve, clearly demonstrates that RIA measurements can be significantly affected by the method used for data reduction.

Regardless of which approach is taken for data analysis, the accuracy of the end results will be directly affected by both intra- and interassay variability. Intraassay variability can be attributable to errors in the technical operation of the assay—i.e., in the manipulations involved at the different steps (pipetting, mixing). This type of error can be assessed by replicate measurements of each sample. Points on the standard curve also need to be run in replicate, because the precision of an assay varies with the concentration of the ligand being measured. Since the greatest precision is usually obtained in the central part of the curve, it is recommended to run points in replicate that include the extremes of the standard curve.

By contrast, interassay variability is derived from several sources: operator error, different operators performing the same assay, and differences in assay components between assays. Interassay variability can be assessed by the measurement with each assay of "internal control" samples—carefully stored samples of a biological standard at high, intermediate, and low concentrations. The coefficients of variation for the values from a series of assays can then be used to demonstrate differences between assays. Another means of assessing this type of variability is to compare standard-curve statistics: the zero standard, assay blank, intercepts on the standard curve, and slope of the standard curve. For example, the RIA data analysis program used by Granger and colleagues since 1981 routinely includes a quality control summary for the previous 8–10 assays of such data as the concentrations of standard displacing 20%, 50%, and 80% of the labeled ligand; the Y intercept; and the slope of the curve (log-logit plot). The means and standard deviations for the intercepts are automatically computed, allowing a direct comparison of the new assay with those previously run. An assay can then be rejected if the intercept values of the new assay are significantly different—i.e., 2 standard deviations from the mean.

In summary, it is necessary to monitor continually the precision of a JH RIA by some means of quality control. Without an assessment of inter- and intraassay variability, even the best analytical program for data reduction can not improve assay precision.

X. Future Prospects of JH RIAs

Although the problems faced by investigators seeking to measure JH in biological samples have yet to be fully resolved, recent developments offer considerable encouragement. JH RIAs now lend themselves well to the analysis of JH biosynthesis in vitro. At least one JH RIA has proven to be reliable, if somewhat laborious, for the measurement of JH in tissue samples, and new assays are being developed. However, if the assays are to find widespread applicability, improvements in their sensitivity and specificity are necessary. What are the prospects?

Given the nature of the hapten and the similar sensitivities of existing antisera obtained with different derivatization sites (methyl ester vs. epoxide function), varied immunization schedules, and individualized assay protocols, it is unlikely that the current levels of sensitivity will be substantially improved without considerable effort. There are several directions that might be profitably explored to achieve better sensitivity. These include the synthesis of high-specific-activity radiolabeled JHs and the development of very high affinity monoclonal ABs. The development of commercially available high-specific-activity hormones such as those generated by Prestwich and Wawrzenczyk (1985) will certainly improve sensitivity; however, the stability of internally labeled JH at higher specific activities is still open to question. As noted earlier, use of radiolabeled iodinated tracers with specific activities exceeding 2,000 Ci $mmol^{-1}$ only modestly enhanced the sensitivity of the assay. Thus, to increase sensitivity significantly, higher-affinity ABs are needed. Hybridoma technology thus might resolve this problem through the development of very high affinity monoclonal ABs.

Improving antiserum specificity by reducing or eliminating nonspecific nonspecificity and specific nonspecificity is also clearly critical for the future JH RIAs. The ideal JH antiserum is one that requires no purification of the JH prior to assay and that measures a single homologue in a mixture of homologues. The importance of this latter characteristic has become evident with the knowledge that the biological activities of the homologues are quite different (Granger and Bollenbacher, 1982) and that their syntheses may be differentially regulated (Granger et al., 1981, 1984, 1986). Antisera whose characteristics approach the ideal may in fact already exist, needing only to be identified by a thorough investigation of the effects of various radiolabeled homologues and analogues on its specificity, which the choice of radiolabeled ligand can profoundly affect (see section VIII.3). Significant improvement in antiserum specificity may also be achieved with the development of monoclonal ABs (mABs) to the JHs.

The production of mABs to the JHs thus deserves careful consideration, since the products could potentially affect improvements in both sensitivity and specificity. Certainly the availability of a virtually immortal cell line that produces an unvarying and highly specific population of ABs would be a tremendous advantage in light of the existing situation with polyclonal ABs: the frequent need to generate immunogen, the lengthy immunization schedules, and the laborious binding studies to obtain antisera that may vary in specificity and affinity from laboratory to laboratory. The question yet unanswered is whether the specificity of a given mAB and the sensitivity of the resulting RIA would be any greater than that achieved with a given subpopulation of ABs in a polyclonal serum.

The first comparisons have recently been made between the assay characteristics of monoclonal and polyclonal ABs for the same hormone. The results of these comparisons have revealed that the affinities of murine mABs for the steroid's testosterone (Fantl and Wang, 1983) and progesterone (Brochu et al., 1983) and for human chorionic gonadotropin (Wiley, 1985) showed no better overall affinity for their haptens than corresponding rabbit polyclonal antisera. The mABs to testosterone and progesterone similarly demonstrated no general improvement in specificity when compared with the corresponding rabbit antisera. Furthermore, extreme variability in specificity from one mAB to another was found, apparently because of their recognition of small, discrete epitopes. As with polyclonal antisera, the choice of an appropriately labeled ligand proved critical to the achievement of assay sensitivity and accuracy with these particular hybridoma lines. On the basis of these results, it appears that replacement of polyclonal antisera with monoclonal antisera would not necessarily be expected to improve the performance of an RIA for steroid hormones. For the same reasons, the use of mABs to the JHs may not effect significant improvements in the specificity and sensitivity of a JH RIA. Thus, the remaining clear advantage of the use of hybridoma technology for the generation of JH antisera appears to be the availability of a continuous source of ABs with unchanging specificity.

There are other immunological approaches to the quantification of JH that could utilize either polyclonal or monoclonal ABs. For example, enzyme immunoassays (EIA) utilizing JH antisera may be a useful future approach to the measurement of JH, since they offer the possibility of increased sensitivity. Moreover, since exceedingly low levels of enzyme activity can be measured by such simple techniques as colorimetry, the EIA is more universally practicable and less costly. A fluorescence immunoassay (FIA) that utilizes a fluorescently labeled JH could also become a principal means of nonisotopic measurement of JH (see Chard, 1978, 1982; see Chapter 3). Since the lower limit of detection of a fluorescent tracer can extend into the picomolar range with the proper equipment for detection, this type of assay is directly applicable to the measurement of circulating titers of JH in the insect.

Regardless of the type of immunoassay for JH, the same rigorous attention to detail and thorough understanding of proper controls will be necessary if the subnanogram amounts of JH normally found in tissue extracts are to be measured accurately. If the appropriate precautions are taken in the design of the assay, the use of immunological approaches to the measurement of JH (RIA, EIA, FIA) should become indispensable for studies of the biochemistry and physiology of this hormone.

Acknowledgments. The authors thank Dr. G.D. Prestwich for his generous and continued provision of JH I and II enantiomers, both radiolabeled and radioinert, for their studies of antiserum specificity and sensitivity. This work was supported by NIH grants GM 34257 (W.G.G.) and NS 14816 (N.A.G.) and by the College of Agricultural and Life Sciences, University of Wisconsin, Madison. The literature search was completed in October 1986.

References

Abraham, G.E., and Grover, P.K. (1971). Covalent linkage of hormonal haptens to protein carriers for use in radioimmunoassays. In: *Principles of Competitive Protein-Binding Assays*. W.D. Odell and W.H. Daughaday (Eds.). J.B. Lippincott, Philadelphia, pp. 134–157.

Baehr, J.C., Pradelles, P., and Dray F. (1976). A simple and sensitive radioimmunoassay of insect juvenile hormone using an iodinated tracer. *FEBS Lett.* **69**:123–128.

Baehr, J.C., Pradelles, P., and Dray, F. (1979). A radioimmunological assay for naturally occurring insect juvenile hormone using iodinated tracers: Its use in the analysis of biological samples. *Ann. Biol. Anim. Biochem. Biophys.* **19**:1827–1836.

Baehr, J.C., Caruelle, J.P., Porcheron, P., and Cassier P. (1981). Quantification of juvenile hormones using a radioimmunological assay in the analysis of biological samples. In: *Juvenile Hormone Biochemistry: Action, Agonism and Antagonism*. G.E. Pratt and G.T. Brooks (Eds.). Elsevier/North Holland, Amsterdam, pp. 46–57.

Baehr, J.C., Cassier, P., Caussanel, C., and Porcheron, P. (1982). Activity of corpora allata, endocrine balance and reproduction in female *Labidura riparia* (Dermaptera). *Cell Tissue Res.* **225**:267–282.

Baehr, J.C., Caruelle, J.P., and Poras, M. (1986). The activity of denervated corpora allata in a diapausing strain of *Locusta migratoria*: In vivo and in vitro studies. *Int. J. Invert. Rep. Dev.* **10**:143–150.

Baehr, J.C., Casas, J., Messegeur, A., Pradelles, P., and Grassi, J. (1987). Enzyme immunoassays of JH3 and JH3–diol using acetylcholinesterase as label: An alternative to radioimmunoassay. *Insect Biochem.* (in press).

Bhaskaran, G., Sparagana, S.P., Barrera, P., and Dahm, K.H. (1986). Change in corpus allatum of function during metamorphosis of the tobacco hornworm, *Manduca sexta*: Regulation at the terminal step in juvenile hormone biosynthesis. *Arch. Insect Biochem. Physiol.* **3**:321–338.

Baker, F.C., Tsai, L.W., Reuter, C.C., and Schooley, D.A. (1987). In vivo fluctuations of JH and JH acid, and ecdysteroid titer and JH esterase activity, during development of fifth stadium *Manduca sexta. Insect. Biochem.* **17**:989–996.

Bergot, B.J., Baker, F.C., Cerf, D., Jamieson, G., and Schooley, D.A. (1981a). Qualitative and quantitative aspects of juvenile hormone titers in developing embryos of several species: Discovery of a new JH-like substance extracted from eggs of *Manduca sexta*. In: *Juvenile Hormone Biochemistry: Action, Agonism and Antagonism*. G.E. Pratt and G.T. Brooks (Eds.). Elsevier/North Holland, Amsterdam, pp. 33–45.

Bergot, B.J., Schooley, D.A., and deKort, C.A.D. (1981b) Identification of JH III as the principal juvenile hormone in *Locusta migratoria. Experientia* **37**:909–910.

Bitsch, C., Baehr, J.C., and Bitsch, J. (1985). Juvenile hormones in *Thermobia domestica* females: Identification and quantification during biological cycles and after precocene application. *Experientia* **41**:409–410.

Brochu, M., Veilleux, R., Lorrain, A., and Belanger, A. (1984). Monoclonal antibodies for use with ^{125}iodine-labeled radioligands in progesterone radioimmunoassay. *J. Steroid Biochem.* **21**:405–411.

Carraway, R. (1979). Neurotensin and related substances. In: *Methods of Hormone Radioimmunoassays*. B. Jaffe and H. Behrman (Eds.). Academic Press, New York, pp. 139–169.

Caruelle, J.P., Baehr, J.C., and Cassier P. (1979). Synthèse d'hormones juvéniles in vitro par les corpora allata de larves du 5e stade de *Locusta migratoria migratorioides* (R. et F.). *C. R. Acad. Sci. Paris* **288**:1107–1110.

Chard, T. (1978). An introduction to radioimmunoassay and related techniques. In: *Laboratory Techniques in Biochemistry and Molecular Biology*. T.W. Work and E. Work (Eds.). Elsevier North-Holland, Amsterdam, pp. 291–534.

Chard, T. (1982). An introduction to radioimmunoassay and related techniques. In: *Laboratory Techniques in Biochemistry and Molecular Biology*. T.W. Work and E. Work (Eds.). Elsevier North-Holland, Amsterdam. Vol. 6 Pt. 2, pp. 1–284.

Dahm, K.H., Bhaskaran, G., Peter, M.G., Shirk, P.D., Sesham, K.R., and Röller, H. (1976). On the identity of the juvenile hormone in insects. In: *The Juvenile Hormones*. L.I. Gilbert (Ed.). Plenum Press, New York, pp. 19–47.

deKort, C.A.D., Bergot, B.J., and Schooley, D.A. (1982). The nature and titre of juvenile hormone in the Colorado potato beetle, *Leptinotarsa decemlineata*. *J. Insect Physiol.* **28**:471–474.

deKort, C.A.D., Koopmanschap, A.B., Strambi, C., and Strambi A. (1985). The application and evaluation of a radioimmunoassay for measuring juvenile hormone titres in Colorado beetle hemolymph. *Insect Biochem.* **15**:771–775.

Deleurance, S., Baehr, J.C., Porcheron, P., and Cassier, P. (1979). Dosages radioimmunologiques des ecdystéroides et des hormones juvéniles au cours des stades larvaires III et IV, nymphal et imaginal de *Leptinotarsa decemlineata* Say (Coleoptera, Chysomelides). *C. R. Acad. Sci. Paris* **289**:1089–1092.

Deleurance, S., Baehr, J.C., Porcheron, P., and Cassier, P. (1980). Diapause et évolution des taux des ecdystéroides et des hormones juvéniles chez les nymphes et les imagos des *Leptinotarsa decemlineata* Say. *C. R. Acad. Sci. Paris* **290**:367–370.

Eisen, H.M. (1974). *Immunology: An Introduction to Molecular and Cellular Principles of the Immune Response*. Harper and Row, New York.

Eng, W.-S., and Prestwick, G.D. (1986). Synthesis of radiolabelled juvenile hormone analogs and chiral homologs. *Bull. Soc. Chim. Belg.* **95**:895–914.

Erlanger, B.F. (1980). The preparation of antigenic-hapten carrier conjugates: A survey. *Methods Enzymol.* **70**:85–104.

Fantl, V.E., and Wang, D.Y. (1980). Characteristics of monoclonal antibodies raised against testosterone. *J. Steroid Biochem.* **19**:1605–1610.

Garvey, J.S., Cremer, N.E., and Sussdorf, D.H. (1977). *Methods in Immunology*, 3d Ed. W.A. Benjamin, Reading, MA.

Giese, C., Spindler, K.D., and Emmerich, H. (1977). The solubility of insect juvenile hormone in aqueous solutions and its adsorption by glassware and plastics. *Z. Naturforsch.* **32c**:158–160.

Gilbert, L.I., and Goodman, W. (1981). Chemistry, metabolism and transport of hormone controlling insect metamorphosis. In: *Metamorphosis*. L.I. Gilbert and E. Frieden (Eds.). Plenum Press, New York, pp. 139–175.

Goodman, W.G., and Adams, B. (1984). Semipreparative synthesis and purification of juvenile hormone acids by high-performance liquid chromatography. *J. Chromatogr.* **294:**447–451.

Goodman, W., and Gilbert, L.I. (1978). Juvenile hormone binding protein titers during the fourth larval instar of *Manduca sexta*. *Gen. Comp. Endocrinol.* **35:**27–34.

Goodman, W.G., Bollenbacher, W.E., Zvenko, H.L., and Gilbert, L.I. (1976). A competitive binding protein assay for juvenile hormone. In: *The Juvenile Hormones*. L.I. Gilbert (Ed.). Plenum Press, New York, pp. 75–95.

Goodman, W., Schooley, D.A., and Gilbert, L.I. (1978a). Specificity of the juvenile hormone binding protein: The geometrical isomers of juvenile hormone I. *Proc. Natl. Acad. Sci. USA* **75:**185–189.

Goodman, W., O'Hern, P., Zaugg, R., and Gilbert, L.I. (1978b). Purification and characterization of the hemolymph juvenile hormone binding protein from *Manduca sexta*. *Mol. Cell Endocrinol.* **11:**225–242.

Granger, N.A., and Bollenbacher, W.E. (1981). Hormonal control of insect metamorphosis. In: *Metamorphosis*. L.I. Gilbert and E. Friedan (Eds.). Plenum Press, New York, pp. 105–136.

Granger, N.A., and Goodman, W.G. (1983). Juvenile hormone radioimmunoassays. Theory and practice. *Insect Biochem.* **13:**333–340.

Granger, N.A., and Janzen, W.P. (1987). Inhibition of *Manduca sexta* corpora allata in vitro by a cerebral allatostatic neuropeptide. *Mol. Cell. Endocrinol* **49:**237–248.

Granger, N.A., Bollenbacher, W.E., Vince R., Gilbert, L.I., Baehr, J.C., and Dray, F. (1979). In vitro biosynthesis of juvenile hormone by the larval corpora allata of *Manduca sexta*: Quantification by radioimmunoassay. *Mol. Cell. Endocrinol.* **16:**1–17.

Granger, N.A., Bollenbacher, W.E., and Gilbert, L.I. (1981). An in vitro approach for investigating the regulation of the corpora allata during larval-pupal metamorphosis. In: *Current Topics in Insect Endocrinology and Nutrition*. G. Bhaskaran, S. Friedman, and J.G. Rodriguez (Eds.). Plenum Press, New York, pp. 83–105.

Granger, N.A., Niemiec, S.M., Gilbert, L.I., and Bollenbacher, W.E. (1982). Juvenile hormone III biosynthesis by the larval corpora allata of *Manduca sexta*. *J. Insect Physiol.* **28:**385–391.

Granger, N.A., Mitchell, L.J., Janzen, W.P., and Bollenbacher, W.E. (1984). Activation of *Manduca sexta* corpora allata in vitro by a cerebral neuropeptide. *Mol. Cell. Endocrinol* **37:**349–358.

Ismail, S., Baehr, J.C., Porcheron, P., and Fuzeau-Braesch, S. (1984). The role of juvenile hormones in the larval diapause of the cricket *Gryllus campestris* (Orthoptera). *Comp. Biochem. Physiol.* **77A:**383–387.

Jaffe, B., and Behrman, H. (1979). Prostaglandins and prostaglandin metabolites. In: *Methods of Hormone Radioimmunoassays*. M. Jaffe and H. Behrman (Eds.). Academic Press, New York, pp. 19–42.

Judy, K.J. (1976). Summary of session on the biosynthesis and metabolism of juvenile hormone. In: *The Juvenile Hormones.* L.I. Gilbert (Ed.). Plenum Press, New York, pp. 96–100.

Judy, K.J., Schooley, D.A., Durham, L.L., Hall, M.S., Bergot, B.J., and Siddall, J.B. (1973). Isolation, structure and absolute configuration of a new natural insect juvenile hormone from *Manduca sexta. Proc. Natl. Acad. Sci. USA* **70**:1509–1513.

Koeppe, J.K., Kovalick, G.E., and LaPointe, M. (1981). Juvenile hormone interactions with ovarian tissue in *Leucophaea maderae.* In: *Juvenile Hormone Biochemistry: Action, Agonism and Antogonism.* G.E. Pratt and G.T. Brooks (Eds.). Elsevier/North-Holland, Amsterdam. pp. 215–231.

Kramer, K., Dunn, P.E., Peterson, R.C., and Law, J.H. (1976). Interaction of juvenile hormone with binding proteins in insect hemolymph. In: *The Juvenile Hormones.* L.I. Gilbert (Ed.). Plenum Press, New York, pp. 327–341.

Lauer, R.C., Solomon, P.H., Nakanishi, K., and Erlanger, B.F. (1974). Antibodies to insect C_{16}-juvenile hormone. *Experientia* **30**:558–559.

Law, J.H. (1980). Lipid-protein interactions in insects. In: *Insect Biology in the Future.* M. Locke and D.S. Smoth (Eds.). Academic Press, New York, pp. 295–309.

Lensky, Y., Baehr, J.C., and Porcheron, P. (1978). Dosages radio-immunologiques des ecdysones et des hormones juveniles au cours du developpement postembryonnaire chez les ouvrières et les reines d'Abeille. *C. R. Acad. Sci. Paris* **287**:821–824.

Marschner, I., Erhardt, F., Henner, J., and Scriba, P.C. (1975). A modular analyzer system for double antibody radioimmunoassays. *Z. Klin. Chem. Klin. Biochem.* **13**:481–488.

Mauchamp, B., Lafort, R., and Krien, P. (1981). Analysis of juvenile hormones by high-performance liquid chromatography coupled with mass spectrometry. In: *Juvenile Hormone Biochemistry: Action, Agonism and Antagonism.* G.E. Pratt and G.T. Brooks (Eds.). Elsevier/North-Holland, Amsterdam, pp. 21–31.

Metzler, M., Meyer, B., Dahm, K.H., Röller, H., and Siddall, J.B. (1972). Biosynthesis of juvenile hormone from 10-epoxy-7-ethyl-3,11-dimethyl-2,6-tridecadienoic acid in the adult *Cecropia. Z. Naturforsch. Pt. B* **29**:321–322.

Meyer, A.S., Schneiderman, H.A., Hanzmann, E., and Ko, J.H. (1968). The two juvenile hormones from the *Cecropia* silk moth. *Proc. Natl. Acad. Sci. USA* **60**:853–860.

Naus, A.J., Kuppens, P.S., and Borst, A. (1977). Calculation of radioimmunoassay curves. *Clin. Chem.* **23**:1624–1627.

Niswender, G.D., and Midgeley, A.R. (1970). Hapten-radioimmunoassay for steroid hormones. In: *Immunologic Methods in Steroid Determination.* F.G. Person and B.V. Caldwell (Eds.). Appleton-Century-Crofts, New York, pp. 149–173.

Niswender, G., Akbar, A.M., and Nett, T.M. (1976). Radioimmunoassay procedures for quantification of steroid hormones. In: *Hormones in Human Blood.* H.N. Antonides (Ed.). Harvard University Press, Cambridge, MA, pp. 751–755.

Papillon, M., Porcheron, P., and Baehr, J.C. (1980). Effects of the rearing temperature upon growth and hormonal balance in *Schistocerca gregaria* during the last two larval instars. *Experientia* **36**:419–422.

Parker, C.W. (1976). *Radioimmunoassay of Biologically Active Compounds.* Prentice-Hall, Englewood Cliffs, NJ.

Poras, M., Baehr, J.C., and Cassier, P. (1983). Control of corpus allatum activity during the imaginal diapause in females of *Locusta migratoria* L. *Int. J. Invert. Reprod. Dev.* **6:** 111–122.

Porcheron, P., Papillon, M., and Baehr, J.C. (1982). Hormonal levels and protein variations during sexual maturation of *Schistocera gregaria*: Effect of rearing temperature. *Experientia* **38:** 970–972.

Prestwich, G.D., and Wawrzenczyk, C. (1985). High specific activity, enantiomerically enriched juvenile hormones: Synthesis and binding assay. *Proc. Natl. Acad. Sci. USA* **82:** 5290–5294.

Prestwich, G.D., Eng, W.-S., Boehm, M.F., and Wawrenczyk, C. (1987). High specific activity tritium- and iodine-labelled JH homologs and analogs for receptor binding studies. *Insect Biochem.* **17:** 1033–1037.

Raz. A., Schwartzmann, M., Kenig-Wakshal, R., and Perl, E. (1975). The specificity of antisera to conjugates of prostaglandins E with bovine serum albumin and thyroglobulin. *Eur. J. Biochem.* **53:** 145–149.

Rembold, H. (1981). Modulation of JH III titer during the gonadotropic cycle of *Locusta migratoria*, measured by gas chromatography–selected ion monitoring mass spectrometry. In: *Juvenile Hormone Biochemistry: Action, Agonism and Antagonism.* G.E. Pratt and G.T. Brooks (Eds.). Elsevier/North-Holland, Amsterdaum, pp. 11–20.

Rodbard, D. (1971) Statistical aspects of radioimmunoassays. In: *Principles of Competitive Protein Binding Assays.* W.D. Odell and W.H. Daughaday (Eds.). J.B. Lippincott, Philadelphia, pp. 204–253.

Rodbard, D., and Lewald, J.E. (1970). Computer analysis of radioligand assay and radioimmunoassay data. *Acta Endocrinol. Suppl.* **147:** 79–103.

Rodbard, D., Bridson, W., and Rayford, P.L. (1969). Rapid calculation of radioimmunoassay results. *J. Lab. Clin. Med.* **74:** 770–781.

Schooley, D.A. (1977). Analysis of the naturally occurring juvenile hormones: Their isolation, identification and titer determination at physiological levels. In: *Analytical Biochemistry of Insects.* R.B. Turner (Ed.). Elsevier, New York, pp. 241–287.

Schooley, D.A., Baher, F.C., Tsai, L.W., Miller, C.A., and Jamieson, G.C. (1984). Juvenile hormones 0, I, and II exist only in Lepidoptera. In: *Biosynthesis, Metabolism, and Mode of Action of Invertebrate Hormones.* J. Hoffman and M. Porchet (Eds.). Springer-Verlag, Berlin, pp. 373–378.

Strambi, C. (1981). Some data obtained by radioimmunoassay of juvenile hormones. In: *Juvenile Hormone Biochemistry: Action, Agonism and Antagonism.* G.E. Pratt and G.T. Brooks (Eds.). Elsevier/North-Holland, Amsterdam, pp. 47–57.

Strambi, C., Strambi, A., DeReggi, M.L., Hirn, M.H., and DeLaage, M.A. (1981). Radioimmunoassay of insect juvenile hormones and of their diol derivatives. *Eur. J. Biochem.* **118:** 401–406.

Strambi, C., Strambi, A., DeReggi, M., and DeLaage, M. (1984). Radioimmunoassays of juvenile hormones. State of the methods and recent data on validation. In: *Biosynthesis, Metabolism, and Mode of Action of Invertebrate Hormones.* J. Hoffman and M. Porchet (Eds.). Springer-Verlag, Berlin, pp. 355–362.

Tobe, S.S., Feyereisen, R. (1983). Juvenile hormone biosynthesis: Regulation and assay. In: *Endocrinology of Insects.* G.H. Downer and H. Laufer (Eds.). Alan R. Liss, New York, pp. 161–178.

Vaitukaitis, J., Robbins, J.B., Nieschlag, E., and Ross, J.T. (1971). A method for producing specific antisera with small doses of immunogen. *J. Clin. Endocrinol. Metab.* **33**:988–991.

Vancassel, M., Foraste, M., Strambi, A. and Strambi, C. (1984). Normal and experimentally induced changes in hormonal haemolymph titers during parental behavior of the earwig *Labidura riparia*. Gen. Comp. Endocrinol. **56**:444–456.

Willey, K.P. (1985). Simple and complex antibody reactions in radioimmunoassay and the prediction of assay characteristics. *J. Immunol. Methods* **84**:343–358.

Chapter 8

Radioimmunoassays and ELISAs: Peptides

Rainer Keller

I. Introduction

It is hardly an exaggeration to state that the introduction of the radioimmunoassay (RIA) more than 25 years ago has been the single most important methodological advance in endocrinology in the past 50 years. The ability to measure notoriously small amounts of hormone in blood samples has decisively shaped our current concepts regarding endocrine mechanisms, especially in vertebrates. The impact of this technique on invertebrate endocrinology is of a more recent date. The introduction of the ecdysone RIA 16 years ago (Borst and O'Connor, 1972; see Chapter 6) has had enormous consequences for the development of arthropod endocrinology.

In general, however, not many immunoassays for invertebrate hormones are available at present. In particular, assays for invertebrate peptide hormones have come in use only recently. Progress in this field has for a long time been lacking, because invertebrate peptides could rarely be isolated in sufficient amounts and/or sufficiently pure form to produce specific antibodies. This situation has already changed and is bound to change even more in the near future by methodological advances in peptide separation techniques and analysis. By high-pressure liquid chromatography (HPLC), the purification to homogeneity of minute amounts of peptides from typically small amounts of starting material is no longer the serious problem it used to be. HPLC-purified peptides in subnanomole amounts can now be sequenced by modern manual microsequencing techniques, the automated gas phase sequencer, or FAB mass spectroscopy. Once the structure is known, the peptides, especially smaller ones, can easily be made by automated or half-automated peptide synthesis. Antisera can then be raised to the synthetic peptides, and immunoassays can be developed. At present, the limiting factor appears to be the availability of suitable bioassays to track down biologically interesting peptides and to assist in the purification procedure. It is not known how many different peptide hormones exist in arthropods and other invertebrates. However, considering the predominance of neuroendocrine control in the lower phyla, a considerable number can be expected.

Sensitive immunoassays will in many cases be crucial to establishing that an identified neuropeptide indeed has a hormonal role. One example may illustrate this. The prothoracicotropic neurohormone (PTTH) of insects was discovered 65 years ago (Kopec, 1922), and it is only very recently that the primary structure of a brain factor displaying PTTH activity in the bioassay has been completely elucidated (Nagasawa et al., 1986). Final proof that this neuropeptide functions as a PTTH under physiological conditions wil require the demonstration that it appears in the hemolymph during critical stages of development.

In the following article, four immunoassays for arthropod neuropeptides will be described. The descriptions are, perhaps with the exception of the ELISA for crustacean black pigment dispersing hormone, detailed enough to be used as recipes. Moreover, it is hoped that these first examples will stimulate further attempts to develop immunoassays for other arthropod peptides.

II. Crustacean Hyperglycemic Hormone (CHH)

Since the discovery of a hyperglycemic principle in the eyestalks of decapod crustaceans more than four decades ago (Abramowitz et al., 1944), a considerable amount of work has been carried out on different aspects of the physiology and biochemistry of this neurohormone (for a recent review see Keller et al., 1985). Its primary structure has yet to be elucidated, and knowledge concerning the physiological significance of CHH remains limited. Its role in maintaining hemolymph glucose homeostasis is clearly established, but it may have other functions that are not, or are only indirectly, related to carbohydrate metabolism. Recent evidence suggests, for example, that it may act as a secretagogue on hepatopancreas tissue, causing release of digestive enzymes (Keller and Sedlmeier, 1987). Thus, CHH may be a hormone with a broader spectrum of actions, but further work is clearly required to elucidate this suggestion.

CHH has been purified to homogeneity from several crustacean species, and amino acid compositions have been determined (Keller et al., 1985). The best-studied CHH, that of *Carcinus maenas*, is a peptide of 57 amino acid residues with two disulfide bridges, a blocked N terminus and an isoelectric point of 4.6. Because of the presence of four Tyr residues, iodination can readily be carried out. CHHs from different crustacean species differ in their amino acid composition but share certain important features like molecular mass, presence of two disulfide bridges, and the blocked N terminus. These species- or systematic group-related differences are most markedly displayed in cross experiments on interspecific biological activity; e.g., it has been shown that the hormone from astacurans has no or very little hyperglycemic activity in brachyurans, and vice versa (Keller et al., 1985). Hormones from different species also show marked differences in their binding to a particular antiserum. These differences in cross-reactivity have to be taken into account in the development of RIAs for CHHs.

1. Preparation of Antigen

A good CHH preparation is best obtained by use of isolated sinus glands (SG) as starting material. SGs can be dissected from exposed eyestalk ganglia without much difficulty and can be collected in distilled water or 2 N acetic acid and stored at −20°C until enough have been accumulated. The compact SG of brachyurans is particularly easy to collect, but the more diffuse organ of the Astacura also does not pose serious problems.

The collection of SG, although time-consuming, yields a starting material that contains the hormone in a highly enriched state, as more than 99% of the CHH present in the whole eyestalk are stored in this very small neurohemal organ. Thus pure hormone can be obtained by use of a simple separation procedure in good yield. The use of whole eyestalks as starting material necessitates a lengthy purification procedure that inevitably causes great losses. SGs are thoroughly extracted with 2 N acetic acid (Webster and Keller, 1986). After Speedvac (Savant Instruments) drying, the residue is dissolved in a small volume of 2 N acetic acid which is applied directly to an HPLC column.

Pure hormone is obtained by use of a simple two-step HPLC procedure (Keller and Kegel, 1984; Webster and Keller, 1986). In brief, it involves a first chromatographic step with manual collection of the very prominent CHH peak and rechromatography with a shallower gradient of the same solvents on the same or a different column. With *Carcinus* material, we have consistently found two peaks of biological activity, one main peak and a smaller prepeak with a slightly lower retention time. These two peaks are indistinguishable in amino acid composition and biological activity, and the nature of the difference is unknown at present. The hormone can be easily identified by bioassay (Kleinholz and Keller, 1973).

Since one SG (of *Carcinus*) contains approximately 150–180 pmole (1 μg) of hormone, a considerable amount of pure CHH can, in theory, be isolated from limited amounts of starting material. In practice, however, losses during purification are inevitable because of the strong binding of CHH to reverse-phase supports. In our experiments, approximately 3.5–4.5 nmole CHH was obtained from 100 *Carcinus* SG and approximately 1 nmole from 100 *Orconectes* SG.

2. Immunization

CHH is large enough to be immunogenic without conjugation to a carrier protein (see Chapters 1, 7). For immunization of two white New Zealand rabbits that had been tested for the presence of crossreacting antibodies (ABs) beforehand and were found to be negative, 54 nmole (360 μg) *Carcinus* CHH was dissolved in 2.0 ml of 50 mM phosphate-buffered saline (0.9% NaCl, PBS), pH 7.4, and thoroughly mixed with 2.0 ml of Freund's complete adjuvant.

Each animal received two subcutaneous injections of 500 μl each and two intradermal injections of the same size in the dorsal skin, in the region of the scapula and the femoral joint, respectively. Thirty days later, each animal received 55 μg CHH in Freund's incomplete adjuvant as a booster injection. The final bleeding was carried out after another 30 days. The serum was divided into smaller portions for storage at −80°C.

The two rabbits developed comparable titers of ABs, and the antisera were suitable for RIA at dilutions of 1:30,000 to 1:60,000 (approximately 50% binding of tracer; Fig. 8.1). Employing the same immunization schedule, an

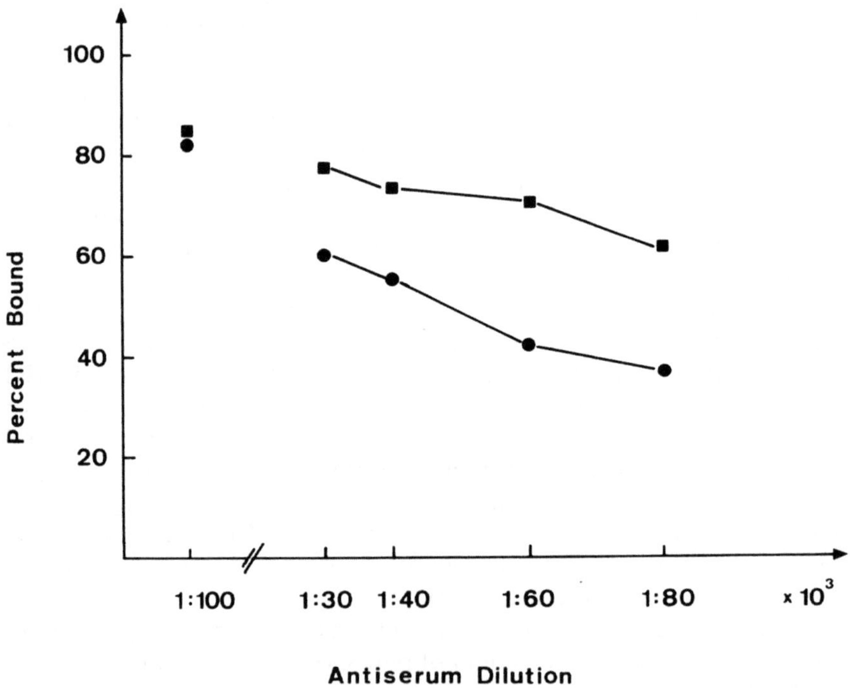

Figure 8.1. Binding of [^{125}I]CHH (*Carcinus*) by dilutions of antiserum from two immunized rabbits. The serum represented by the lower curve was selected for routine use.

antiserum against *Orconectes limosus* CHH was produced with only 60 μg CHH per animal as a first injection and 35 μg as the booster injection.

3. Preparation of Labeled Antigen

The iodination of *Carcinus* CHH is carried out according to Frechet et al. (1971) as follows. To the tracer, 1 mCi $Na^{125}I$ in 5 μl 0.1 N NaOH (New England Nuclear), 50 μl 0.3 M phosphate buffer, pH 7.4, and 1 μg CHH (150 pmole) in 10 μl distilled water are added. After addition of 0.8 μg chloramin T in 20 μl 0.3 M phosphate buffer, pH 7.4, and rapid mixing, the sample is incubated for 2–3 min at room temperature. Then 1 μg sodium metabisulfite in 10 μl and 140 μg potassium iodide in 140 μl, both dissolved in the above phosphate buffer, are added.

The iodinated CHH is separated from free ^{125}I by passage through a small Sephadex G-25 column, equilibrated with PBS containing 1.0% bovine serum albumin (BSA), which is also used as eluant. The labeled antigen (Ag) is divided into smaller portions and stored at −20°C. It is usable for 4–6 weeks. After this period, it is generally no longer possible to obtain enough good tracer by rechromatography on Sephadex G-25.

4. Radioimmunoassay

Antiserum, tracer, standard, and samples to be measured are taken up and diluted in assay buffer: 50 mM phosphate pH 7.4; 0.9% NaCl; 1.0% BSA; 0.03% EDTA. The tubes with a final assay volume of 400 μl are made up as follows:

1. For determination of nonspecific binding (NSB): 200 μl assay buffer and 100 μl of 1:100 diluted human serum are combined.
2. For determination of zero binding (B_0): 200 μl assay buffer and 100 μl antiserum dilution (1:7,500–1:15,000) are combined.
3. For the standard curve and test samples: 100 μl assay buffer, 100 μl CHH standards or unknown samples, and 100 μl antiserum dilution (1:7,500–1:15,000) are mixed.

All tests are set up in duplicate or triplicate. After preincubation for 4 h at 4°C, 10,000 cpm [^{125}I]CHH in 100 μl assay buffer is added and mixed, and incubation is continued for 16 h at 4°C.

The separation of bound and free hormone is achieved with polyethyleneglycol 6,000 (PEG). To every tube, 50 μl of undiluted human serum is added, followed by 400 μl of 30% PEG in 50 mM phosphate buffer, pH 7.5. After thorough mixing on a vortex mixer, the tubes are incubated for 30 min at 4°C. The precipitate is then sedimented in a refrigerated centrifuge at approximately 4,800 *g* for 10 min. The supernatants are aspirated off and discarded, and the pellets are resuspended in 400 μl 15% PEG. After centrifugation and removal of the supernatants, the tubes with the pellets are counted.

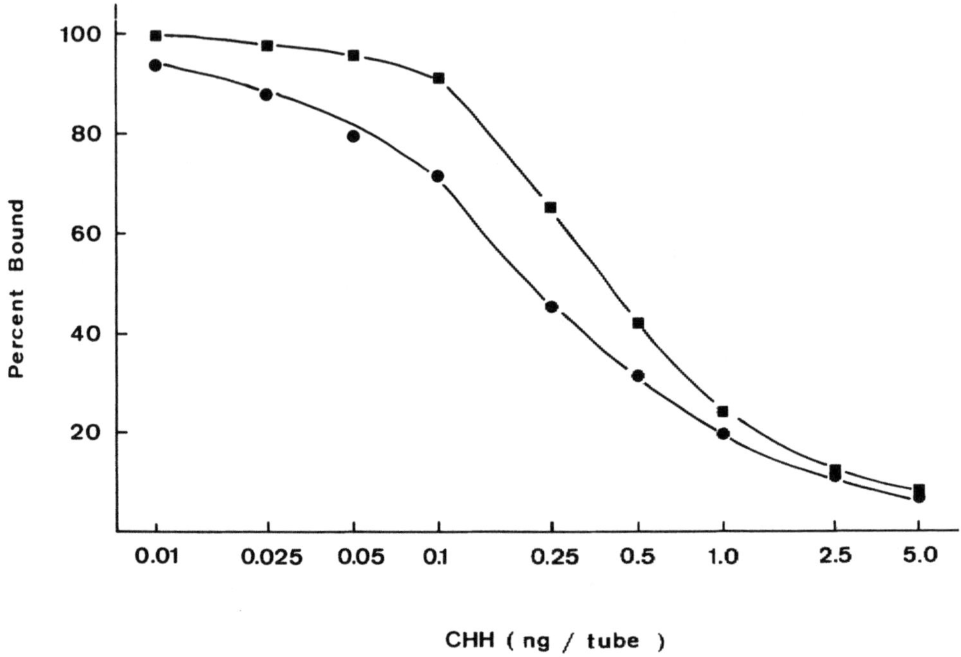

Figure 8.2. RIA for *Carcinus* CHH. The two standard curves were obtained with the two sera shown in Figure 8.1 (the symbols are the same). The RIA represented by the lower curve was routinely used, since it permits the determination of smaller amounts. The RIA represented by the upper curve is more accurate in a range of higher concentrations.

Figure 8.2 shows typical standard curves for *Carcinus* CHH, as obtained with the sera of the two rabbits. The lower limit of detection is approximately 6 fmole with one of the sera. CHH from *Orconectes limosus* has been tested by RIA (data not shown). It clearly cross-reacts, but with low potency. For 50% displacement of [^{125}I]*Carcinus* CHH, approximately 1,000 times more *Orconectes* CHH is required. Thus, the RIA based on the antiserum against *Carcinus* CHH is rather insensitive for *Orconectes* CHH.

5. Application of the RIA

Thus far, the RIA has demonstrated its usefulness in quantitative determinations of CHH in different parts of the central nervous system of *Carcinus* and of hormone levels in the hemolymph. Measurements of extractable hormone in nerve tissues showed a highly localized distribution; of the total hormone present in eyestalks, cerebral and thoracic ganglia, commissures, and pericardial organs combined, approximately 99% is localized in the eyestalks.

In the eyestalks, the bulk of the hormone (150–180 pmole) is stored in the neurohemal organ, the SG, whereas the groups of CHH-producing perikarya in the medulla terminals X-organ contain only approximately 1.1 pmole. The other optic ganglia in the eyestalk contain very small amounts (Keller et al., 1985). Among the other tissues, only the pericardial organs possess significant amounts of CHH (approximately 0.6 pmole). The physiological significance of its presence in, and possible release from, these neurohemal organs is unknown at present.

Measurements of resting levels of hemolymph CHH in *Carcinus* have yielded values of between 30 and 180 fmole/ml. In animals stressed by handling, low osmolarity, elevated temperature, or hypoxia, considerably higher values, occasionally exceeding 1.4 pmole/ml were obtained (unpublished results). This supports the long-standing notion that CHH may have a physiological role in adaptation to stress. In crabs that were eyestalk-ablated 24 h prior to hemolymph sampling, no CHH was detectable by RIA. This indicates that there is continuous release of CHH from the SG, presumably to maintain and regulate resting glucose levels, as hypoglycemia following eyestalk removal has often been reported (see Kleinholz and Keller, 1979).

Taking into account the lower limit of detection in the RIA (approximately 6 fmole), hemolymph samples of 100–200 μl are required for a single determination (in *Carcinus*).

III. Proctolin

The pentapeptide proctolin, Arg-Tyr-Leu-Pro-Thr, was originally discovered as a myotropic hindgut factor in *Periplaneta*, and large-scale purification from this species has resulted in its isolation and structural elucidation (Starratt and Brown, 1975). Subsequent research has suggested the presence of proctolin or a proctolinlike immunoreactive substance not only in insects in general, but in other invertebrates, especially arthropods, as well.

In crustaceans, proctolin has now been firmly established as an autochthonous neuropeptide (Schwarz et al., 1984; Stangier et al., 1986), and this appears to be true also for chelicerates, especially *Limulus* (Benson et al., 1981). It may also occur in vertebrates (Holets et al., 1984). However, final proof of the existence of proctolin, by isolation and complete structural identification, has thus far, except for insects, been provided only for crustaceans (Schwarz et al., 1984; Stangier et al., 1986).

From numerous studies on arthropod neuronal and muscle preparations, a general role of proctolin as a locally acting neurotransmitter and/or modulator has been deduced, but some results obtained in insects and *Limulus* and, in particular, crustaceans, point to a neurohormonal role of the peptide. Its presence in neurohemal structures (i.e., the crustacean pericardial organs) has been particularly suggestive.

Final proof will depend on the application of an RIA to demonstrate the presence of circulating proctolin. By immunocytochemistry, proctolinlike reactivity has recently been demonstrated in another neurohemal structure, the neurosecretory vena cava system of *Octopus* (Martin and Voigt, 1987), and it will ultimately also require the application of an RIA to establish whether it is released into the hemolymph. (First, of course, it must be confirmed that this immunoreactive substance is indeed proctolin, which would make this the first finding in mollusks).

In addition to its obvious value in studies on a possible neurohormonal role of proctolin, an RIA will undoubtedly be a useful tool in attempts to ascertain that proctolinlike immunoreactivity is indeed authentic proctolin, especially in animals (e.g., *Limulus*) in which the evidence is still incomplete. Also, it will assist in screening a larger number of animal species to see how widely distributed proctolin is, and will enable quantitative measurements in different parts of the nervous system as a supplement to immunocytochemical mapping of positive neurons.

The procedures for the development and application of an RIA to be described are those of Schwarz et al. (1984). Proctolin RIAs have also been described by Bishop et al. (1981) and Kingan and Titmus (1983).

1. Preparation of Antigen

The pentapeptide proctolin is too small to induce an immunoresponse and must be coupled to a larger carrier protein. Two common methods have been employed to produce conjugates that proved to be immunogenic—conjugation by carbodiimide to BSA or thyroglobulin, and conjugation by glutaraldehyde to BSA.

The carbodiimide cross-linked proctolin BSA conjugate was prepared by mixing 250 μl of an aqueous solution of 1-ethyl-3-(3-dimethylaminopropyl) carbodiimide (80 mg/ml) with 0.5 ml of water containing 2.0 mg of proctolin (Peninsula Laboratories), 4.0 mg BSA, and approximately 20,000 cpm [^{125}I]proctolin. This solution was incubated overnight and then dialyzed against 1 L of 0.9% NaCl for 24 h at 4°C. The Ag was finally diluted to 20 ml with distilled water (Ag 1).

The same method was used in modified form (Skowsky and Fischer, 1972) to prepare a proctolin-thyroglobulin conjugate. Two hundred fifty microliters of an aqueous solution of l-ethyl-3-(3-dimethyl-aminopropyl) carbodiimide (10 mg/ml) was mixed with 750 μl of water containing 5 mg of proctolin, 42 mg of thyroglobulin, and 20,000 cpm of [^{125}I]proctolin. After incubation for 18 h at room temperature, 1 ml of 1 M hydroxylamine was added to the solution, and the incubation was continued for another 5 h. The Ag was dialyzed against 2 L of distilled water for 48 h at 4°C (Ag 2).

Glutaraldehyde cross-linked proctolin BSA conjugate was prepared by mixing 10 μl of 50% glutaraldehyde into 200 μl of 0.1 M sodium phosphate (pH 7.4) containing 2.0 mg proctolin, 4.0 mg BSA, and 20,000 cpm [^{125}I] proctolin. Overnight incubation at room temperature was followed by dialysis against 1 L of 0.9% NaCl. Finally, the Ag was diluted to 2.0 ml with distilled water. (Ag 3).

As judged from the incorporation of tracer, the coupling yield was 50% in the glutaraldehyde-BSA conjugation and 60–65% in the carbodiimide-BSA and carbodiimide-thyroglobulin conjugations, respectively.

2. Immunization

Ag 1, diluted to 20 ml with distilled water, was emulsified with an equal volume of Freund's complete adjuvant. Each animal received an initial injection of 1.0 ml of the emulsion into the muscles of the hindlegs and 0.5 ml injections once a month thereafter. The AB titer was checked in blood samples taken 1 week after each injection from the lateral ear vein or central ear artery.

Of eight New Zealand white rabbits thus injected, one produced antibodies to proctolin (antiserum D). Its binding characteristics are shown in Figure 8.3. The same injection and bleeding schedule was used for Ag 3. Three rabbits were injected, and one of them produced ABs (antiserum A; see Fig. 8.3 for

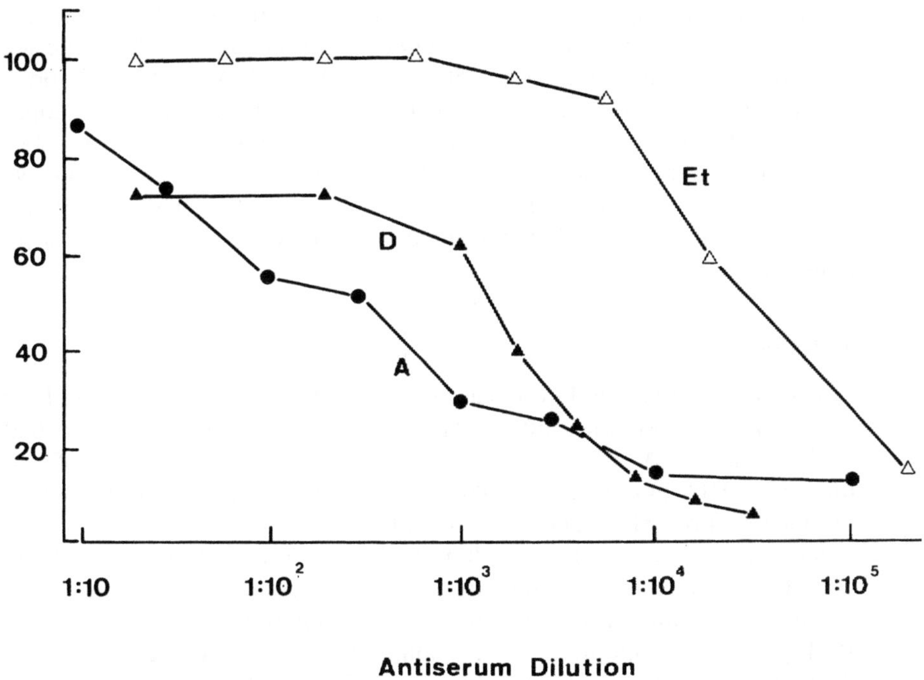

Figure 8.3. Binding of [^{125}I]proctolin by dilutions of three different antisera. Serum D was used for the RIA (Schwarz et al., 1984).

binding characteristics). Ag 2 was administered according to a different schedule.

For one rabbit, Ag at a concentration of 1.5 mg of conjugated proctolin per milliliter was used, and a second animal received material that was diluted 10-fold. The Ag solutions were emulsified with equal volumes of Freund's complete adjuvant, and five injections of 0.6 ml of the mixture were given at 10-day intervals. Thereafter, booster injections of the same size were given every 15–20 days.

Both rabbits developed ABs to proctolin at a rather high titer. The binding characteristics of one of the two sera, antiserum Et, are shown in Figure 8.3. It is not known whether this particular successful immunization was due to the injection schedule or to the different Ag.

3. Preparation of Labeled Antigen

For iodination of proctolin, 0.5 μg of the peptide in 10 μl 0.1 M sodium borate (pH 10) was combined with 0.5 mCi of $Na^{125}I$ in 5.0 μl 0.1 N NaOH, and the reaction was started by addition of μl of dilute hypochlorite solution (20 μl of Clorox (Procter and Gamble) in 13.3 ml of sodium borate buffer, pH 10).

The reaction was allowed to proceed for 15 sec at room temperature and was then stopped by addition of 20 μl of a sodium metabisulfite solution (10 mg/ml borate buffer). After 2 min, 20 μl of NaI (100 mg/ml 60% acetic acid) was added. Again, 2 min later, 50 μl of 0.25% BSA in 0.05 M sodium phosphate buffer, pH 7.6, was added. The reaction mixture was then applied to a small Sephadex 6–10 column, first washed with RIA buffer (see below), and then equilibrated with 10% acetic acid. Elution was carried out with 10% acetic acid, and 0.5 ml fractions were collected. Those containing [^{125}I]proctolin were combined, diluted with an equal volume of 60% propanol, and stored at 4°C.

4. Radioimmunoassay

For the RIA, antiserum D was selected. A 1 : 500 dilution was prepared in 50 mM sodium phosphate buffer (pH 7.6) containing 0.1% BSA and 0.2% sodium azide (RIA buffer). To 100 μl of the diluted antiserum, 50 μl of tracer (diluted in RIA buffer to give 3,000–10,000 cpm/50 μl) and 50 μl of either a sample or a proctolin standard were added.

Subsequent incubation for 18 h at 4°C was terminated by the addition of 100 μl of 2% dextran-coated charcoal (in 10% fetal calf serum and 6.6 mM sodium phosphate buffer, pH 7.6) and centrifugation at 2,000g for 12 min. Both pellets and supernatants were counted. In this RIA, the lower limit of detection was 10 fmole of proctolin, and 70 fmole reduced AB binding of the [^{125}I]proctolin by 50%.

The following substances were tested in the assay and were not found to cross-react: α-MSH, Leu- and Met-enkephalin, somatostatin, eledoisin, eledoisin-related peptide, neurotensin, oxytocin, FMRF-amide, physalaemin, substance P, 5HT, and octopamine.

5. Applications of the RIA

The RIA has been used successfully to measure the content of proctolin in different parts of the nervous system of insects and crustaceans (Bishop et al., 1981; Kingan and Titmus, 1983, Schwarz et al., 1984). In the lobster, these studies have been corroborated by a detailed immunocytochemical mapping of proctolinlike reactive neurons (Siwicki and Bishop, 1986). Results of the two studies are in good agreement and show that, as in insects, proctolin is present in nearly every portion of the nervous system.

By far the highest concentrations (59–500 pmole/mg of protein; Schwarz et al., 1984) were found in the lobster pericardial organs. There is no structure in insects that contains such high concentrations. This, together with the demonstration of an abundance of immunoreactive fibers and terminals in this neurohemal tissue, is the strongest evidence for a neurohormonal role of proctolin. In addition, release of proctolin from pericardial organs in vitro under depolarizing conditions (high K^+) and in the presence of Ca^{2+} was shown by the RIA.

In view of these results it is surprising, however, that the crucial test—i.e., demonstration of circulating proctolin in the hemolymph by the RIA—has thus far yielded only negative results (Kingan and Titmus, 1983; Schwarz et al., 1984). This is obviously not due to an insensitivity of the RIA. Schwarz et al. (1984) have extracted a larger volume of lobster hemolymph and have not detected any proctolin reactivity in an assay that would have detected levels as low as 4×10^{-12} M. As an effective concentration in the hemolymph, at least about 10^{-10} M may be postulated, as this is the lowest demonstrated threshold for an effect on a crustacean tissue, the dactyl opener muscle of the walking leg (Schwarz et al., 1984).

Thus, the RIA data do not yet support a neurohormonal role for proctolin in the sense that it can act on peripheral tissues via the general circulation. Effective concentrations may be built up in close proximity to the release sites only—i.e., in the pericardial cavity. There, proctolin may act on the heart before it is diluted to ineffective and undetectable levels in the general circulation. Alternatively, proctolin may only be released in appreciable amounts under certain, heretofore unknown physiological conditions.

IV. Adipokinetic Hormone (AKH)

The adipokinetic hormone, especially AKH I, has become known as an important metabolic hormone that is responsible for the mobilization of diglycerides from the fat body of migratory locusts as fuel for sustained flight (see Beenakkers, 1983, for review; see also Chapters 3 and 4). It is synthesized, stored, and secreted by the glandular lobe cells of the corpus cardiacum. It is a decapeptide, whose primary structure, pGlu-Leu-Asn-Phe-Thr-Pro-Asn-Trp-Gly-Thr-NH_2, was elucidated by Stone et al. (1976).

In addition, AKH II, an octapeptide, is present in the corpus cardiacum (Siegert et al., 1985) and is most likely synthesized by the same cells as AKH I. AKH II displays sequence homology to AKH I in the N-terminal portion and is also a lipid-mobilizing hormone, although its physiological significance is less clear than that of AKH I. It is present in smaller amounts than AKH I.

Considering the significance of AKH I, the development of an RIA has been desirable for a long time. Hemolymph levels of the hormone could thus far be investigated only by bioassay, which was time-consuming and inevitably inaccurate. An RIA, if sensitive enough, can be expected to be of great help in detailed studies on the dynamics of AKH secretion in relation to the onset and maintenance of flight and on other details of endocrine control mechanisms in which AKH I is involved. A specific RIA may also prove useful in screening larger numbers of species in order to study the distribution of AKH I in insects and other arthropods.

The development of an RIA has been hampered by the fact that AKH I is too small to elicit an immunogenic response and by its unsuitability for conjugation to carrier proteins due to the blocked N- and C-terminal ends and the lack of other suitable functional groups. Also, radiolabeling is not possible without modification. Thus far, antisera have been produced by use of a conjugate of a [Tyr^1]AKH I analogue to thyroglobulin (Schooneveld et al., 1983) and of the N-terminal tetrapeptide of native AKH I via a diaminohexane spacer to thyroglobulin (Schooneveld and Veenstra, 1985), but these antisera have been used for immunocytochemical studies only (see Chapter 4).

Since AKH I has been shown to be just one member of a larger family of related peptides, which consists at present of eight different hormones and which is likely to grow in the future, specificity of an RIA is a matter that should be carefully considered. The following RIA is based on an antiserum that appears to be specific for the full length of the decapeptide, but its cross-reactivity with other members of the family has not yet been investigated.

The aforementioned problems in Ag preparation and preparation of a radiolabeled ligand have been circumvented by the procedure of Moshitzky et al. (1987), described below.

1. Preparation of Antigen

To introduce a functional group for conjugation, the AKH-I analogue [Glu^1]AKH-I was synthesized by conventional Merrifield solid-phase synthesis. Final purification of the analogue was achieved by partition chromatography on a 15 × 450 mm column of Sepharose CL-6B with the solvent system 1-butanol:acetic acid:water (9 : 1 : 10 v/v/v). The homogeneity of the peptide was checked by TLC on silica gel in 1-butanol: acetic acid: water (4:1:1 v/v/v) with chlorine-tolidine as detecting reagent.

2. Immunization

The analogue (5 mg) was conjugated to BSA by the glutaraldehyde procedure (Kishida et al., 1975). A mixture of 1 mg hapten and 4 mg *Mycobacterium tuberculosis* (Difco) in Freund's complete adjuvant and PBS (2:1, v/v) was initially given by multiple-site subcutaneous injections. Three 1-month-old male New Zealand white rabbits were thus injected. Booster injections of 300 μg hapten/animal without *M. tuberculosis* were given 3 weeks later, and animals were bled 10 weeks after the initial injections.

3. Preparation of Labeled Antigen

The derivative 4-hydroxyphenylpropionyl-[Glu^1]AKH-I was prepared by reacting 5 mg of [Glu^1]AKH-I and 2 mg Bolton-Hunter reagent (N-succinimidyl-3-(4-hydroxyphenyl) propionate; Pierce) in 1 ml of 0.1 M borate buffer for 2 h at 24°C. The reaction product was isolated by reverse-phase HPLC on a Vydac C 18 (300 Å) column (4.6 × 250 mm) using isocratic elution with 20% 2-propanol in 0.1% trifluoroacetic acid.

Iodination of the derivative was carried out as follows: $Na^{125}I$, 1 mCi in 5 μl of 0.1 M NaOH (New England Nuclear), 5 μg of the derivative in 5 μl PBS, and 25 μl 0.5 M phosphate buffer, pH 7.5, were mixed. The reaction was initiated by addition of 6 μl of a 0.35% chloramin T solution in phosphate buffer and mixing. After 30 sec, 20 μl of a 10% solution of β-mercaptoethanol was added, and the reaction mixture was chromatographed on a Sephadex LH 20 column (9 × 200 mm) in 1 M pyridine acetate, pH 5.5, containing 3% ethanol and 1% β-mercaptoethanol.

Fractions containing the second peak of radioactivity were collected and stored at 4°C. Prior to the RIA, aliquots of the labeled analogue were purified by reverse-phase HPLC on an Alltech C-6,000 column (4.6 × 250 mm, particle size 10 μm, 300 Å, Alltech) using isocratic elution with 30% 2-propanol in 0.1% trifluoracetic acid.

4. Radioimmunoassay

AB titer and optimal working dilution of the antiserum were determined as follows: [^{125}I]AKH-I analogue (10,000 cpm) in 50 μl of assay buffer (0.05 M phosphate buffer, pH 7.4; 0.25% BSA, 0.5% β-mercaptoethanol, 0.05% sodium azide, and 0.001% polylysine (MW = 4,000); Sigma), 50 μl of assay buffer, and 50 μl of diluted antiserum in assay buffer were combined, mixed, and incubated overnight at 4°C. Blanks and dilutions of normal rabbit serum were included as controls.

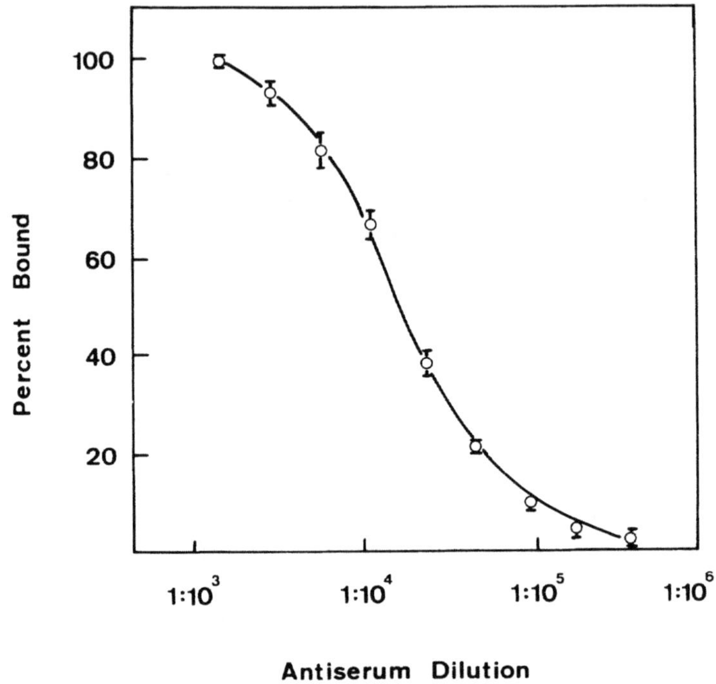

Figure 8.4. Binding of [^{125}I]4-hydroxyphenyl propionyl-(Glu^1)-AKH-I([^{125}I]-AKH I analogue) by dilutions of antiserum (Moshitzky et al., 1987).

For separation of bound and free ligand, 0.7 ml of a suspension of dextran-coated charcoal was added to each tube. After 5–10 min at 4°C, the tubes were centrifuged at 13,000*g* for 5 min at 4°C. Supernatant aliquots of 0.5 ml were taken for counting. The dextran-coated charcoal was prepared by adding 1% dextran T-70 (Pharmacia), 0.5% BSA, and 2% Norit A decolorizing carbon (Pfanstiehl Labs, Waukegan, IL) to 0.05 M phosphate buffer (pH 7.4). Assays with different dilutions of antiserum showed that approximately half-maximal binding (40%) was obtained by a final antiserum dilution of 1: 36,000 (Fig. 8.4).

The RIA was carried out by mixing AKH I standards or test samples in 50 μl of assay buffer with 50 μl of AKH I serum, diluted with assay buffer. The tubes were then incubated for 12–24 h at 4°C. The tracer (10,000 cpm) was then added in 50 μl of assay buffer, and the incubation was continued for 12–24 h at 4°C. The separation of bound and free ligand with dextran-coated charcoal was carried out as previously described. A typical standard curve is shown in Figure 8.5. The amount of unlabeled AKH I required for half-maximal inhibition of tracer binding is 44 pg per tube. The lower limit of detection is approximately 3 pg (3.3 fmole).

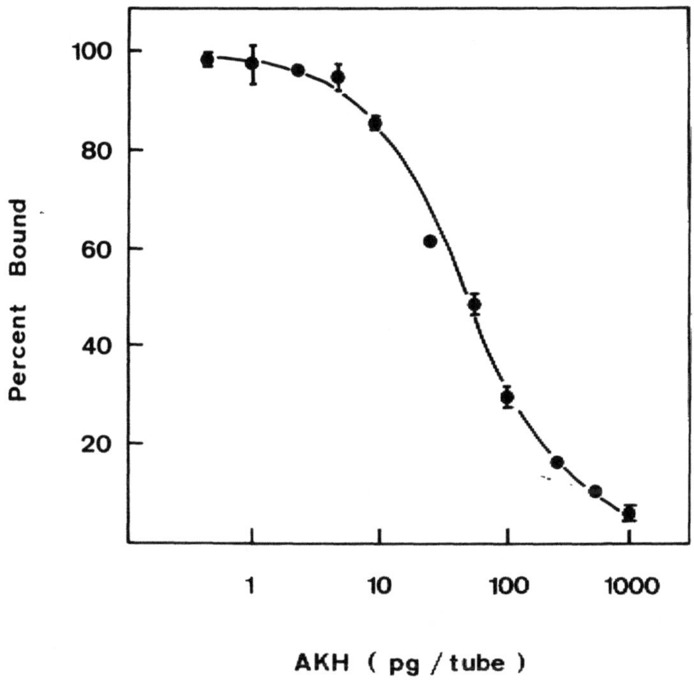

Figure 8.5. Standard curve of the AKH I RIA, showing the inhibition of binding of [^{125}I]AKH I analogue by increasing amounts of authentic AKH I (Moshitzky et al., 1987).

Some common neuropeptides were tested in the RIA and were found not to cross-react significantly: proctolin, substance P, ACTH, α-MSH, β-endorphin. Also, a synthetic analogue of the structurally related crustacean red pigment–concentrating hormone (RPCH) [Tyr^4]RPCH was tested, and no cross-reactivity was found.

5. Application of the RIA

The efficacy of the AKH I RIA has been demonstrated by the determination of the AKH I content of locust corpora cardiaca. After HPLC separation of methanol extracts of corpora cardiaca, an amount of 0.45 ± 0.18 μg per corpus cardiacum (cc) was found (Moshitzky et al., 1987). This agrees well with previous calculations which were based on bioassay procedures and which yielded a value of 0.2–0.5 μg/cc (Stone et al., 1976). In addition, it was shown that the locust brain contains a radioimmunoreactive substance that co-chromatographs with authentic AKH-I. Its amount is less than 10% of that found in the corpus cardiacum.

From bioassay data it may be estimated that AKH I levels in the circulating hemolymph in vivo range between 10^{-10} and 10^{-8} M (Stone et al., 1976). This would correspond to approximately 1–100 fmoles 10 μl of hemolymph. Since the lower limit of detection in the RIA is approximately 3.3 fmoles, it is realistic to assume that the RIA is sensitive enough to measure circulating hormone in hemolymph samples (e.g., a few microliters) that can be drawn from locusts without much difficulty. The RIA should therefore prove extremely useful for studies on the dynamics of AKH I secretion in vivo, especially in relation to development, environmental and metabolic stress, etc.

V. Crustacean Black Pigment–dispersing Hormone (BPDH)

The black pigment–dispersing neurohormone (BPDH) of the crustacean eyestalk is one of the neuropeptides involved in the regulation of epidermal chromatophores and other pigmentary effectors. For example, BPDH is identical to the distal retinal pigment hormone (DRPH) that controls the light adaptation of the crustacean compound eye. The physiology and biochemistry of this hormone have been studied rather extensively, and a detailed review is available (Rao, 1985). The hormone is also of potential interest to insect endocrinologists, since there is evidence for the occurrence of related substances in insects (Rao, 1985). The primary structure of two different, but homologous BPDHs, both octadecapeptides, has been elucidated (Fernlund, 1976; Rao et al., 1985; Kleinholz et al., 1986).

The release of BPDH appears to be regulated in a complex way by several mechanisms, including circadian and circatidal rhythms, illumination, and aminergic input. To unravel these mechanisms in detail, a method for detecting circulating hormone in small hemolymph samples would be of great help. Quackenbush and Fingerman (1985) have described such an enzyme-linked immunosorbent immunoassay (ELISA). For the development of this assay, an Ag of proven homogeneity was not available either for immunization or for use as standard substance. It can therefore, at the present stage, not be considered to be a completely valid assay in the sense that it yields values that are correct in absolute terms. It is, however, felt that the procedure is valid and that the assay could be upgraded by using synthetic BPDH. Moreover, it may serve as an example of how an immunoassay can be developed when only small amounts of Ag are available, as is often the case in work on invertebrate peptides.

1. Preparation of Antigen

An aqueous extract of lyophilized eyestalks (2.5 g) from *Uca pugilator* was boiled for 3 min and centrifuged at 10,000*g* for 30 min at 4°C. The supernatant was lyophilized, yielding 0.38 g of dry material. This material was resuspended in 50% methanol and further purified by passage through a Sep-Pak cartridge (Waters Corp., MA) with 50% methanol as the eluant. After lyophilization of the eluate, 48 mg of a powder was obtained. The material was subjected to SDS polyacrylamide electrophoresis (PAGE), and the zone containing BPDH activity was located by bioassay of eluates from gel slices, Almost all (92%) of the BPDH activity was confined to a single zone corresponding to a molecular size of 1,800 daltons and containing 500 μg of protein. It formed a single band in SDS-PAGE and was shown to be free of erythrophore-concentrating or -dispersing activities (partially purified BPDH).

2. Immunization

The immunogen mixture consisted of 34 μg of the partially purified BPDH, 2 ml of 0.1% methylated BSA in water, and 2 ml Freund's complete adjuvant (Benoit et al., 1982). Two female New Zealand white rabbits were given 80 intradermal injections each according to Vaitukaitis (1981). No booster injections were given. Bleedings were begun 1 week prior to immunization and continued weekly thereafter.

The development of anti-BPDH ABs was tested by the ability of the sera to suppress the biological activity of aqueous SG extracts of *Uca pugilator*, preincubated with the serum prior to the bioassay. Sera from both rabbits developed ABs, although of a different titer. The serum of one of the rabbits (BJ) was particularly effective, 6 weeks after immunization, in blocking the chromatophore response when preincubated with SG extracts at dilutions of 1:100–1:1,000. The serum also formed an immunoprecipitate with the protein zone corresponding to the BPDH activity. Preimmune serum neither blocked the chromatophore response nor precipitated any fractions from the eyestalk extract.

3. ELISA

Partially purified BPDH (6 pg to 13.3 ng) was adsorbed to polystyrene wells in microtiter plates (Immunolon I, Dynatech Labs., Alexandria, VA). Wells were then blocked with 5% ovalbumin for 1 h at 37°C. Aliquots of serial dilutions of antiserum in PBS (0.01 M phosphate buffer, pH 7.2, 0.85% NaCl) were added to the wells, and incubation was carried out for 24 h at 4°C.

Bound rabbit IgG was detected with peroxidase-conjugated goat anti-rabbit AB (Kirkegaard and Perry Labs, Gaithersburg, MD), diluted 1:800 in 5% ovalbumin in PBS. After incubation for 2 h at 22°C, the peroxidase-immune complex was visualized with 2,2′-azinobis[3-ethylbenzothiazoline-6-sulfonate] (ABTS) (Kirkegaard and Perry Labs) and H_2O_2 and quantitatively evaluated at 420 nm. For determination of BPDH in biological samples, the antiserum was used at a dilution of 1:8,000 in 5% ovalbumin in PBS.

2. Application of the ELISA

The ELISA was used to measure immunoreactive material in hemolymph samples (50 μl) of *Uca pugilator* that were actively dispersing their black chromatophores in adaptation to a black background. At a stage of strong dispersion, the hemolymph was found to contain 8.8 ± 0.9 ng immunoreactive material in 50 μl. This would amount to a hemolymph concentration of approximately $8-10^{-8}$ M. In the hemolymph of control crabs, kept on a white background, no significant amounts of immunoreactive material were found. The isolated optic ganglia in vitro released, when challenged with high K^+, approximately 4.5 ± 0.4 ng per eyestalk.

The concentration found in the hemolymph appears rather high, considering the fact that an injection of 0.05 pmole per crab of synthetic *Uca* BPDH already produces a significant chromatophore response and that 0.3 pmole per crab produces a very strong response (Rao et al., 1985). The amounts would translate into effective hemolymph concentrations of only 5×10^{-11} to 3×10^{-10} M. Obviously, one shortcoming of the described procedure is that the standard substance was not of proven purity. Therefore, any quantification is uncertain. It can be expected that simple replacement of the partially purified BPDH by synthetic hormone would yield correct values.

Recently, an antiserum raised with synthetic *Uca* BPDH as Ag was obtained and described (Mangerich et al., 1987; Dircksen et al., 1987). It has thus far been used for immunocytochemistry only, but the development of an RIA is in progress.

Acknowledgments. The author thanks Dr. S.W. Applebaum for providing a preprint of the work on the AKH I RIA. Own work described here was supported by Deutsche Forschungsgemeinschaft grant Ke 206/7.

Reference

Abramowitz, A.A., Hisaw, F.L., and Papandrea, D.N. (1944). The occurrence of a diabetogenic factor in the eyestalk of crustaceans. *Biol. Bull.* **86**:1–5.

Beenakkers, A.T.M. (1983). Regulation of lipid metabolism. In: *Invertebrate Endocrinology* Vol, I. *Endocrinology of Insects*. R.G.H. Downer and H. Laufer (Eds.). Alan R. Liss, New York, pp. 441–450.

Benoit, R., Bohlen, P., Ling, N., et al. (1982). Presence of somatostatin-28-(1-12) in hypothalamus and pancreas. *Proc. Natl. Acad. Sci. USA* **79**:917–921.

Benson, J.A., Sullivan, P.T., Watson, W.N., III, and Augustine, G.J. Jr. (1981). The neuropeptide proctolin acts directly on *Limulus* cardiac muscle to increase the amplitude of contraction. *Brain Res.* **213**:449–454.

Borst, D.W., and O'Connor, J.D. (1972). Arthropod molting hormone: Radioimmune assay. *Science* **178**:418–419.

Bishop, C.A., O'Shea, M., and Miller, R.J. (1981). Neuropeptide proctolin: Immunocytochemical detection and neuronal localization in insect central nervous system. *Proc. Natl. Acad. Sci. USA* **78**:5899:5902.

Dircksen, H., Zahnow, C., Gaus, G., Keller, R., Rao, K.R., and Riehm, J.P. (1987). The ultrastructure of nerve endings containing pigment dispersing hormone (PDH) in crustacean sinus glands identified by an antiserum against a synthetic PDH. Cell Tissue Res. **250**:377–387.

Fernlund, P. (1976). Structure of a light adapting hormone from the shrimp, *Pandalus borealis*. *Biochim. Biophys. Acta* **439**:17–25.

Frechet, P., Roth, J., and Neville, D.M. (1971). Monoiodoinsulin: Demonstration of its biological activity and binding to fat cells and liver membranes. *Biochem. Biophys. Res. Commun.* **43**:400–405.

Holets, V.R., Hökfelt, J., Ude, J., Eckert, M., and Hansen, S. (1984). Coexistence of proctolin with TRH and 5-HT in the rat CNS. *Soc, Neurosci. Abstr.* **10**(2):692.

Keller, R., and Kegel, G, (1984). Studies on crustacean neuropeptides by use of high performance liquid chromatography. In: *Biosynthesis, Metabolism and Mode of Action of Invertebrate Hormones*. J.A. Hoffmann and M. Porchet (Eds.). Springer-Verlag, Heidelberg, pp. 145–154.

Keller, R., and Sedlmeier, D. (1987). A metabolic hormone in crustaceans: The hyperglycemic neuropeptides. In: *Invertebrate Endocrinology* Vol. II. R.G.H. Downer and H. Laufer (Eds.). Alan R. Liss, New York (in press)

Keller, R., Jaros, P.P., and Kegel, G. (1985). Crustacean hyperglycemic neuropeptides. *Am. Zool.* **25**:207–221.

Kingan, T., and Titmus, M. (1983). Radioimmunologic detection of proctolin in arthropods. *Comp. Biochem. Physiol.* **74C**:75–78.

Kishida, Y., Olsen, B.R., Berg, R.A., and Prockop, D.J. (1975). Two improved methods for preparing ferritin-protein conjugates for electron microscopy. *J. Cell Biol.* **64**:331–339.

Kleinholz, L.H., and Keller, R. (1973). Comparative studies in crustacean neurosecretory hyperglycemic hormones. I. The initial survey. *Gen. Comp. Endocrinol.* **21**:554–564.

Kleinholz, L.H., and Keller, R. (1979). Endocrine regulation in crustacea. In: *Hormones and Evolution*, Vol. I. E.J.W. Barrington (Ed.). Academic Press, New York, pp. 159–214.

Kleinholz, L.H., Rao, K.R., Riehm, J.P., Tarr, G.E., Johnson, L., and Norton, S. (1986). Isolation and sequence analysis of a pigment dispersing hormone from eyestalks of the crab, *Cancer magister*. *Biol. Bull.* **170**:135–143.

Kopec, S. (1922). Studies on the necessity of the brain for the inception of insect metamorphosis. *Biol. Bull. Woods Hole* **42**:322–342.

Mangerich, S., Keller, R., Dircksen, H., Rao, K.R., and Riehm, J.P. (1987). Localization of pigment-dispersing hormone (PDH) and coexistence with FMRFamide immunoreactivity in the eyestalks of two decapod crustaceans. Cell Tissue Res. **250**:365–375.

Martin, R., and Voigt, K.H. (1987). The neurosecretory system of the *Octopus* vena cava: A neurophemal organ. *Experientia* **43**:537–543.

Moshitzky, P., Yamashiro, D., Stuve, L., Ramachandran, J., and Applebaum, S.W. (1987). Determination of locust AKH-I by radioimmunoassay and the identification of an AKH-like factor in the locust brain. *Insect Biochem.* **17**:765–769.

Nagasawa, H., Kataoka, H., Isogai, A., et al. (1986). Amino acid sequence of prothoracicotropic hormone of the silkworm *Bombyx mori*. *Proc. Natl. Acad. Sci. USA* **83**:5840–5843.

Quackenbush, L.S., and Fingerman, M. (1985). Enzyme-linked immunosorbent assay of black pigment dispersing hormone from the fiddler crab, *Uca pugilator*. *Gen. Comp. Endocrinol.* **57**:438–445.

Rao, K.R. (1985). Pigmentary effectors. In: *The Biology of Crustacea*, Vol. 9. D.E. Bliss and L.H. Mantel (Eds.). Academic Press, New York, pp. 395–462.

Rao, K.R., Riehm, J.R., Zahnow, C.A., et al. (1985). Characterization of a pigment-dispersing hormone in eyestalks of the fiddler crab *Uca pugilator*. *Proc. natl. Acad. Sci USA* **82**:5319–5323.

Schooneveld, H., and Tesser, G.I. (1983). Adipokinetic hormone and AKH-like peptide demonstrated in the corpora cardiaca and nervous system of *Locusta migratoria* by immunocytochemistry. *Cell Tissue Res.* **230**:67–76.

Schooneveld, H., and Veenstra, J.A. (1985). Insect neuroendocrine cells and neurons containing various adipokinetic hormone (AKH)-immunoreactive substances. In: *Neurosecretion and the Biology of Neuropeptides*. H. Kobayashi et al., (Eds.). Japan Science Society press, Tokyo, pp. 425–434.

Schwarz, T.L., Lee, G.M.M., Siwicki, K., Standaert, D.G., and Kravitz, E.A. (1984). Proctolin in the lobster: The distribution, release, and chemical characterization of a likely neurohormone. *J. Neurosci.* **4**:1300–1311.

Siegert, K., Morgan, P., and Mordue, W. (1985). Primary structures of locust adipokinetic hormones II. *Biol. Chem, Hoppe-Seyler* **366**:723–727.

Siwicki, K.K., and Bishop, C.A. (1986). Mapping of proctolinlike immunoreactivity in the nervous systems of lobster and crayfish. *J. Comp. Neurol.* **243**:435–453.

Skowsky, W.R., and Fisher, D.A. (1972). The use of thyroglobulin to induce antigenicity to small molecules. *J. Lab. Clin, Med.* **80**:134–144.

Stangier, J., Dircksen, H., and Keller, R. (1986). Identification and immunocytochemical localization of proctolin in pericardial organs of the shore crab, *Carcinus maenas*. *Peptides* **7**:67–72.

Starratt, A.N., and Brown, B.E., (1975). Structure of the pentapeptide proctolin, a proposed neurotransmitter in insects. *Life Sci.* **17**:1253–1256.

Stone, J.V., Mordue, W., Batley, K.E., and Morris, H.R. (1976). Structure of locust adipokinetic hormone, a neurohormone that regulates lipid utilisation during flight. *Nature* **263**:207–211.

Vaitukaitis, J.L., (1981). Production of antisera with small doses of immunogen: Multiple intradermal injections. In: *Methods in Enzymology* Vol. 73. J.J. Langone and H. Van Vunakis (Eds.). Academic Press, New York, pp. 46–52.

Webster, S.G., and Keller, R. (1986). Purification, characterization and amino acid composition of the putative moult-inhibiting hormone (MIH) of *Carcinus maenas* (Crustacea, Decapoda). *J. Comp. Physiol.* **B 156**:617–624.

Index

T

MIX
Papier aus verantwortungsvollen Quellen
Paper from responsible sources
FSC® C105338

If you have any concerns about our products,
you can contact us on
ProductSafety@springernature.com

In case Publisher is established outside the EU,
the EU authorized representative is:
Springer Nature Customer Service Center GmbH
Europaplatz 3, 69115 Heidelberg, Germany

Printed by Libri Plureos GmbH
in Hamburg, Germany